MICROWAVE TUBES

The Artech House Microwave Library

Introduction to Microwaves by Fred E. Gardiol
Microwaves Made Simple: Principles and Applications by W. Stephen Cheung and Frederic H. Levien
Microwave Tubes by A. S. Gilmour, Jr.
Electric Filters by Martin Hasler and Jacques Neirynck
Nonlinear Circuits by Martin Hasler and Jacques Neirynck
Microwave Technology by Erich Pehl
Receiving Systems Design by Stephen J. Erst
Microwave Mixers by Stephen A. Maas
Feedback Maximization by B.J. Lurie
Applications of GaAs MESFETs by Robert Soares, et al.
GaAs Processing Techniques by Ralph E. Williams
GaAs FET Principles and Technology by James V. DiLorenzo and Deen D. Khandelwal
Dielectric Resonators, Darko Kajfez and Pierre Guillon, eds.
Modern Spectrum Analyzer Theory and Applications by Morris Engelson
Design Tables for Discrete Time Normalized Lowpass Filters by Arild Lacroix and Karl-Heinz Witte
Microwave Materials and Fabrication Techniques by Thomas S. Laverghetta
Handbook of Microwave Testing by Thomas S. Laverghetta
Microwave Measurements and Techniques by Thomas S. Laverghetta
Principles of Electromagnetic Compatibility by Bernhard E. Keiser
Linear Active Circuits: Design and Analysis by William Rynone, Jr.
The Design of Impedance-Matching Networks for Radio-Frequency and Microwave Amplifiers by Pieter L.D. Abrie
Microwave Filters, Impedance Matching Networks, and Coupling Structures by G.L. Matthaei, Leo Young, and E.M.T. Jones
Analysis, Design, and Applications of Fin Lines by Bharathi Bhat and Shiban Koul
Microwave Engineer's Handbook, 2 vol., Theodore Saad, ed.
Handbook of Microwave Integrated Circuits by R.K. Hoffmann
Microwave Integrated Circuits, Jeffrey Frey and Kul Bhasin, eds.
Computer-Aided Design of Microwave Circuits by K.C. Gupta, Ramesh Garg, and Rakesh Chadha
Microstrip Lines and Slotlines by K.C. Gupta, R. Garg, and I.J. Bahl
Advanced Mathematics for Practicing Engineers by Kurt Arbenz and Alfred Wohlhauser
Microstrip Antennas by I.J. Bahl and P. Bhartia
Antenna Design Using Personal Computers by David M. Pozar
Microwave Circuit Design Using Programmable Calculators by J. Lamar Allen and Max Medley, Jr.
Stripline Circuit Design by Harlan Howe, Jr.
Microwave Transmission Line Filters by J.A.G. Malherbe
Electrical Characteristics of Transmission Lines by W. Hilberg
Microwave Diode Control Devices by Robert V. Garver
Tables for Active Filter Design by Mario Biey and Amedeo Premoli
Active Filter Design by Arthur B. Williams
Laser Applications by W.V. Smith
Ferrite Control Components, 2 vol., Lawrence Whicker, ed.
Microwave Remote Sensing, 3 vol., by F.T. Ulaby, R.K. Moore, and A.K. Fung

MICROWAVE TUBES

A. S. Gilmour, Jr.

Artech House

ARTECH HOUSE, INC.
685 Canton St.
Norwood, MA 02062

International Standard Book Number: 0-89006-181-5
Library of Congress Catalog Card Number: 85-73445

10 9 8 7 6 5 4 3 2

CONTENTS

To
Paula and Duane

Preface

This book is based on a five-day short course that I developed in 1981 and 1982 for presentation at the Naval Ship Weapons Systems Engineering Station (NSWSES) in Port Hueneme, CA. The short course has subsequently been presented over a dozen times at Air Force and Navy installations. In addition, with emphasis on analysis, the course is a regularly scheduled graduate course at the University of Buffalo.

There have been over 300 participants in the course with backgrounds ranging from 3rd-class Navy Electronic Technicians through microwave tube experts to individuals at the chief scientist level. The short course has evolved with help from the participants so that mathematical analyses are avoided and qualitative discussions of physical principles are emphasized.

The order of presentation of the material has been arranged to try to answer questions before they are asked. As an example, a discussion of microwaves and their properties was moved to the beginning as the result of the question by a young Navy Second Lieutenant with a business degree who sheepishly approached me at the end of the first day of one of the short courses and asked, "What's a microwave?"

Much of the content is drawn from over 30 years of experience with microwave tube research, development, applications and trouble-shooting as well as activity in related fields, such as vacuum technology, high-power switching, high-power modulators and, more recently, pulse-power and space-defense applications. In addition, material is drawn from a variety of sources ranging from publications of inventions and developments through applications manuals and textbooks to several excellent review articles. In particular, I am indebted to Dr. George Farney, Dr. Armand Staprans, and Dr. John Skowron for the material provided in their many manuals and publications.

I must admit to having had a significant level of reluctance in undertaking the development of the short course and then in converting the course notes into this book because of the large amount of effort required in a field in which "everything had already been done." The efforts were finally undertaken largely as the result of the encouragement of Mr. Thomas Bekker at

NSWSES, who helped convince me of the need. Also, the very favorable reviews of the resulting short course by virtually all of the participants helped convince me of the value of the material.

I am indebted to many individuals who have provided support in one form or another for my microwave tube activities. Included are Dr. G. Conrad Dalman, who was my advisor at Cornell, Dr. Carl Miller, formerly of Wright Patterson Air Force Base, Mr. Joseph Primerano, formerly of the Naval Surface Weapons Center, and Messrs Frank Welker, Hunter Chilton, and Robert Gray of the Rome Air Development Center. Finally, I would like to thank my wife for her patience during the long nights and weekends required for the preparation of this book and Mrs. Mary Hill for typing the entire manuscript.

A. S. Gilmour, Jr.
Amherst, NY
February 1986

Chapter 1
Introduction

Microwave tubes are vacuum devices used for the generation and amplification of radio frequencies in the microwave frequency range. Microwave frequencies are generally considered to be from somewhat less than 10^9 Hz to over 10^{11} Hz as is indicated in Figs. 1.1. and 1.2 [1.1, 1.2, 1.3]. The millimeter wave frequency region is part of the microwave range and is generally considered to extend from about 30 GHz to 300 GHz.

The microwave spectrum is divided into several frequency bands. The two most commonly used band designations are shown in Fig. 1.2. The old band designation was established during World War II and is still widely used. The new designation, established to provide a more logical system, has not been widely accepted, apparently, because of the reluctance to learn the new system. The old band system is most commonly used in the microwave tube industry and is the one used in this book.

Before the advent of solid-state devices, microwave tubes were the only active devices available for use throughout the entire microwave range. Now, as indicated in Fig. 1.3, microwave tubes predominate in the high-power, high-frequency applications with solid-state devices gradually moving up in power and frequency.

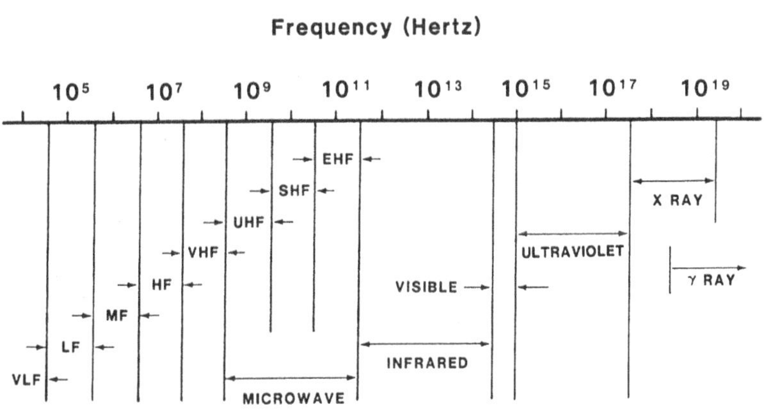

Figure 1.1 The Electromagnetic Spectrum

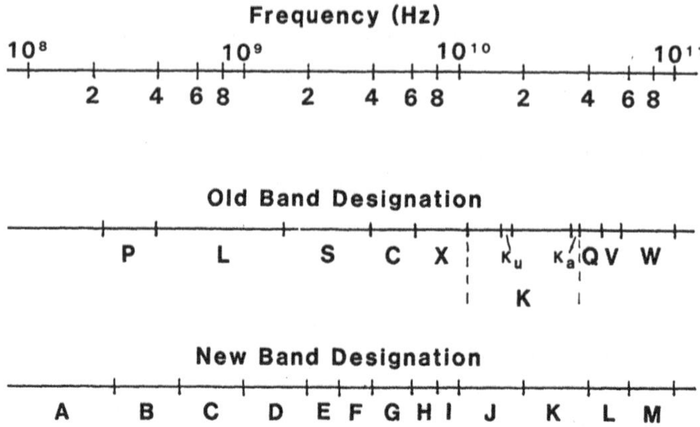

Figure 1.2 Microwave Band Designations

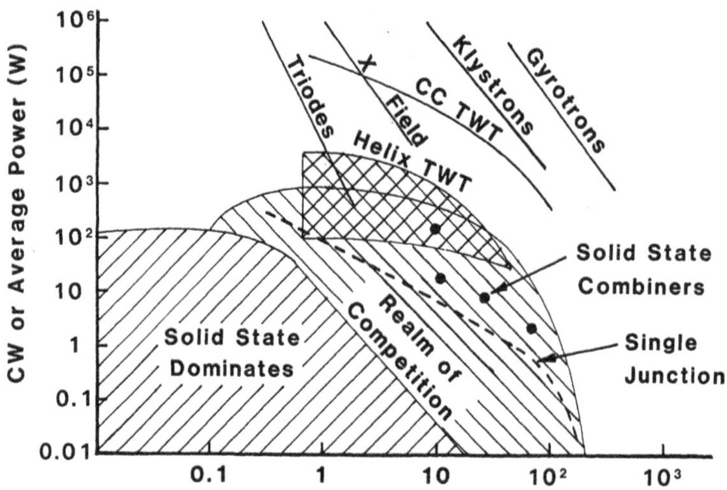

Figure 1.3 Ranges of Application of Microwave Tubes and of Solid-State Devices

(Courtesy Mr. Frank Welker, Rome Air Development Center)

A factor of great importance in many applications and not evident from Fig. 1.3 is the large bandwidth for some microwave tubes. This can be over two octaves for some helix-type traveling wave tubes (TWTs). Thus, even though there is an overlap in capabilities on a power *versus* frequency chart, the tubes may be selected to obtain the required system capability.

The predominant applications for microwave tubes are:

- Radar
- Communications
- Electronic countermeasures
- Microwave cooking

Other uses include particle accelerators, plasma heating, materials processing, and power transmission. Future developments in any of these areas could result in significant growth possibilities for the microwave tube industry. As one example, the full deployment of some proposed space defense systems would require tens of thousands of devices each with an average power capability of at least a megawatt. It appears that the cost of these tubes would at least equal the present total US sales of microwave tubes.

Microwave tubes were invented prior to World War II [1.4, 1.5, 1.6]. Two types, the magnetron and the reflex klystron, were the subjects of intense development efforts during the war for use in radar. During and after the war, numerous other microwave tube types and configurations were invented. Most did not find application for any of several reasons (low efficiency, low gain, complexity, *et cetera).*

Today, in addition to gridded tubes (triodes and tetrodes), which continue to be the most powerful sources available at low microwave frequencies, three types of microwave tubes predominate. These are:

- Klystrons
- Traveling wave tubes
- Magnetrons and crossed field amplifiers

For high power generation and amplification in the millimeter wave range, new devices such as the gyrotron are being developed.

The basic operating principles of all microwave tubes are the same and may be explained with the aid of Fig. 1.4. An electron beam is passed through the tube near a circuit, along which an electromagnetic wave is propagating in the microwave frequency range. The circuit is constructed so that the velocity of the wave is nearly the same as the velocity of the beam. The electron beam and the wave interact to produce amplification.

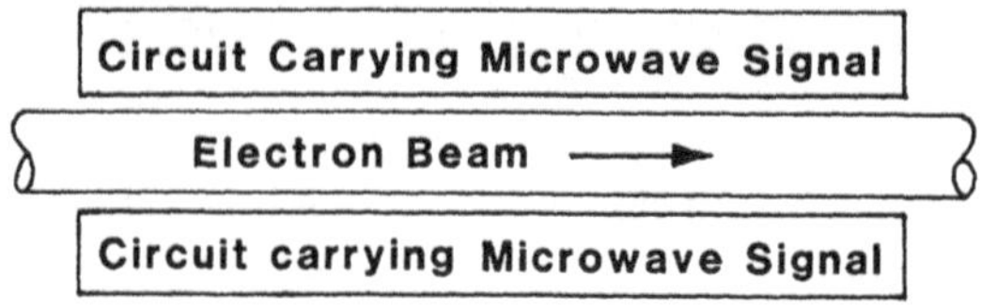

Figure 1.4 Simplistic View of Microwave Tube

Hidden in this simplistic statement are some clues to the extremely complex nature of a microwave tube. For example, the fact that an electron beam is used makes it necessary to evacuate the interior of a microwave tube to a pressure that may be over ten orders of magnitude below atmospheric pressure. To achieve and maintain this pressure over the life of a tube, it is necessary to use special materials and fabrication techniques.

As another example, because of the high electron density required in the electron beam, the source of the electrons and the beam in most microwave tubes is a very sophisticated structure. As we look further into the operation of a microwave tube, it is found that nearly every aspect of construction and operation is extremely complex.

In selecting the subject matter for this book (and the short course on which the book is based), a very serious attempt has been made to include at least basic discussions of all the technologies that are essential to a qualitative understanding of microwave tube operation. The order of presentation of the material has been selected so that the reader, in working through the book from the beginning, can develop, in a logical sequence, a basic working knowledge of microwave tubes.

Thus, the book starts with a discussion of microwaves and those characteristics of microwaves that affect the design and fabrication of a tube. Next, several essential aspects of construction, as indicated in Fig. 1.5, are examined. Included are high and ultra-high vacuum techniques and associated materials and fabrication techniques. Then, techniques for producing magnetic focusing fields are examined in preparation for a discussion of electron beams. Then, in the final step leading to descriptions of specific microwave tubes, the electron sources (cathodes) and the guns and beams required in those tubes are discussed. In discussing tube types, devices known as linear-beam tubes are covered first. Included are gridded tubes, klystrons, traveling wave tubes, and hybrid tubes. The electron-beam collectors used to enhance the efficiencies of these tubes are described.

Following linear-beam tubes, magnetrons, and crossed-field amplifiers are covered. The last devices covered are gyrotrons and other relatively novel devices such as ubitrons and peniotrons.

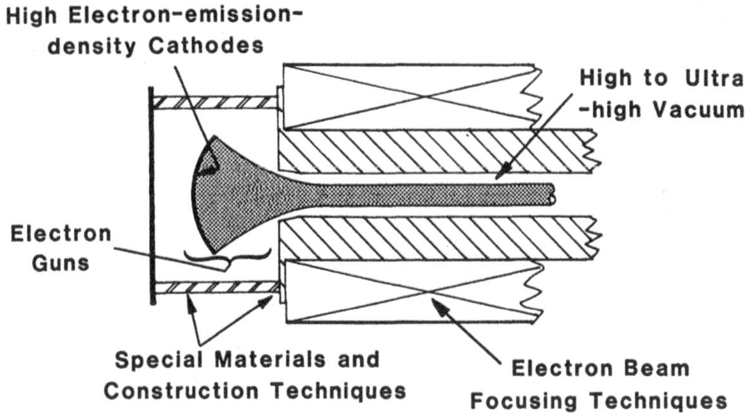

Figure 1.5 Major Aspects of Construction of a Microwave Tube

The last major chapter in the book is a limited discussion of breakdown phenomena that very often lead to the failure of microwave tubes. Protective measures that can be used to help prevent failure are described.

In concluding an introduction to microwave tubes, two additional items require comment. These are:

- The microwave tube industry
- Microwave-tube education

There are fewer than a half dozen major suppliers of microwave tubes in the US. Another half dozen companies supply relatively small numbers of special-purpose tubes. The total market is a few hundred million dollars with over 80 percent of sales being for defense. Over 50 percent of microwave tube sales is for traveling wave tubes [1.7].

Concerning microwave tube engineering, it is estimated that there are only about 300 microwave tube engineers in the US [1.8]. Many are retiring, approaching retirement age, or leaving the industry for other reasons.

The only formal microwave tube education program is the Air Force's AFTER program. AFTER stands for *Air Force Thermionic Engineering and Research.* The basic elements of the program are:

- Appointments with annual stipend of $15,000 plus full tuition.
- Two-year program leading to Degree of Electrical Engineer.
- Jointly sponsored by US Air Force and the following seven electronics firms in cooperation with the University of Utah. An appointment to participate includes employment at one of the sponsoring firms:

Hughes Aircraft Company
Litton Industries
Northrup Corporation
Raytheon Company
Teledyne MEC
Varian Associates
Watkins-Johnson Company

As shown in Fig. 1.6, the program is guided by an advisory board consisting of participants from the Air Force, industry, and the university. The objectives of the program are to produce thermionic (microwave tube) research and researchers.

The AFTER program was initiated at Stanford University in 1977 and was moved to the University of Utah in 1981. As shown in Fig. 1.7, about 30 students produced by the program remain in the microwave tube field. In the future, although the nature of the program may change, it will be important for the government to continue its support of the program to help ensure a continuing supply of microwave tube engineers.

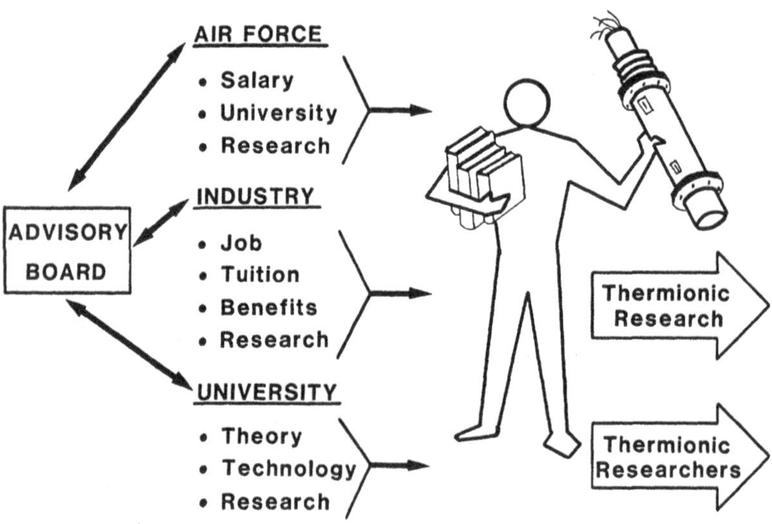

Figure 1.6 AFTER Program

(Courtesy R. Hunter Chilton, Rome Air Development Center)

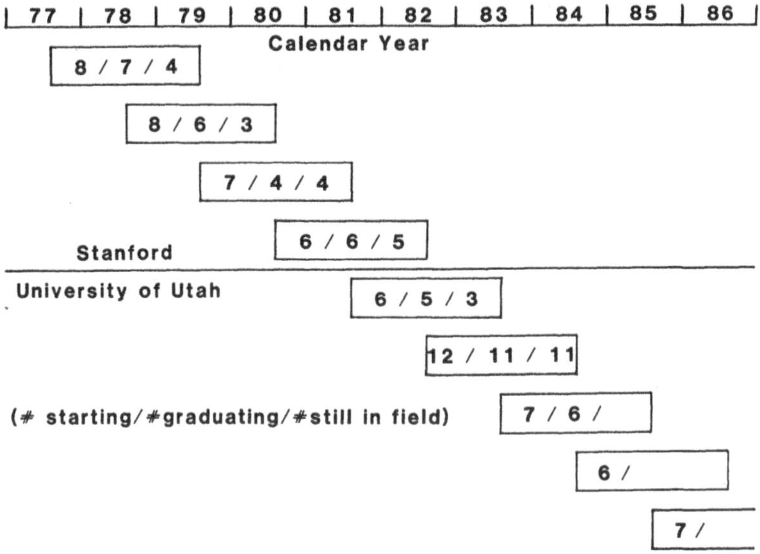

Figure 1.7 Numbers of AFTER Students

(Courtesy R. Hunter Chilton, Rome Air Development Center)

REFERENCES

[1.1] L. C. Shen and J. A. Kong, *Applied Electromagnetism.* Monterey, CA: Brooks/Cole Engineering Division, 1983, p. 34.

[1.2] *Reference Data for Radio Engineers,* New York: Howard W. Sams and Co., 1975.

[1.3] See, for example, *AFSC Design Handbook,* 1–4, Electromagnetic Compatibility, Fourth Edition, 2 March 1984.

[1.4] H. A. H. Boot and J. T. Randall, Historical notes on the cavity magnetron," *IEEE Trans. Electron Devices,* vol. ED-23, July 1976, pp. 724–729.

[1.5] G. Ross Kilgore, "Recollections of pre-World War II magnetrons and their applications," *IEEE Trans. Electron Devices,* vol. ED-31, Nov. 1984, pp. 1593–1605.

[1.6] R. H. Varian and S. F. Varian, "A high frequency oscillator and amplifier," *Journal of Applied Physics,* vol. 10, May 1939, pp. 321–327.

[1.7] G. Caryotakis, "Industry capacity," Microwave Power Tube Conference, Monterey, CA, April 26–28, 1982.

[1.8] Extrapolation based on survey of microwave tube industry by R. H. Chilton, Rome Air Development Center, at initiation of AFTER Program and continuing discussions with personnel in the microwave tube industry.

Chapter 2
Microwaves

Electromagnetic waves propagate through free space at the speed of light, c, which is 3×10^8 m/s (about 1 foot per nanosecond). The relation between wavelength, λ, of a signal and its frequency, f, is

$$\lambda = \frac{c}{f} \tag{2.1}$$

As a result, the wavelengths of electromagnetic waves in the microwave range are as shown in Fig. 2.1. Compared with the radio broadcast band, where $\lambda \cong 300$ m and with which early wavelength comparisons were made, these wavelengths are short, hence the term microwaves.

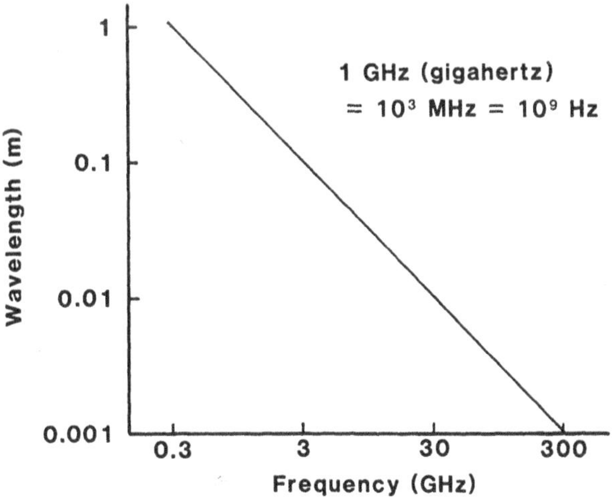

Figure 2.1 Wavelengths of Signals in the Microwave Range (1 GHz (gigahertz) = 10^3 MHz = 10^9 Hz)

2.1 PROPAGATION CHARACTERISTICS [2.1–2.5]

In nearly all of the microwave tubes discussed in this book, the microwave signals are periodic with waveforms that are essentially the same for large numbers of successive cycles. While the waveforms may not be sinusoidal, they can be easily broken down into sinusoidal components having fundamental and harmonic frequencies. As an example, the fundamental, third, fifth, and seventh harmonic frequencies of the infinite set that make up a square wave are shown in Fig. 2.2

The effects of, and on, the various harmonics in a microwave circuit can be determined for each harmonic, and then the harmonics can be added back together to determine the overall effect on the waveform. As a result, at this point in explaining those aspects of microwave signals of importance in microwave tubes, it is only necessary to consider sinusoidal signals.

The propagation of an electromagnetic wave can be expressed very simply by

$$E = E_0 \sin(\omega t \pm \gamma z) \tag{2.2}$$

where E_0 is the peak value of the electric field, ω is the angular frequency, γ is called the propagation constant* and, of course, t and z are time and distance.

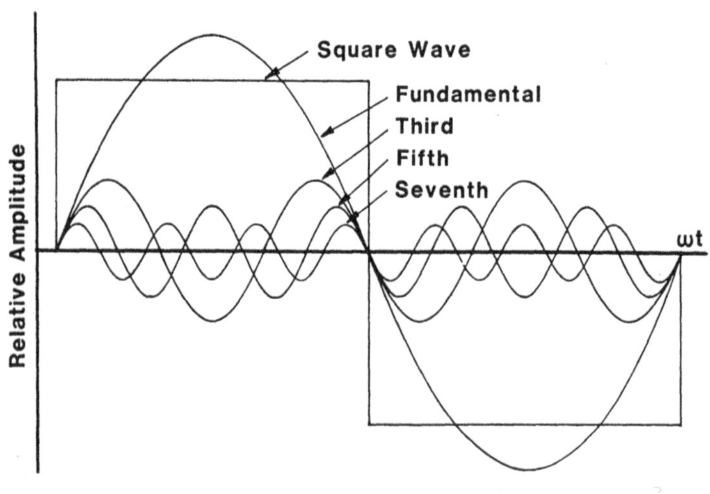

Figure 2.2 Fundamental and Low Harmonic Frequency Components of a Square Wave

* A word of caution is in order at this point. While the symbol γ is widely used for the propagation constant, it is often chosen to be an imaginary number for convenience in mathematic analyses. The imaginary γ must be multiplied by $-j$ $(\sqrt{-1})$ to equate to the γ used here.

To show how the above equation represents the propagation of a wave, a constant value of the electric field, E, may be chosen. For the field to remain constant at some value E_1, the angle of the sine wave, $\omega t \pm \gamma z$, must be constant at some value ϕ_0, so

$$\phi_0 = \omega t \pm \gamma z \tag{2.3}$$

If the minus sign is chosen first, then

$$z = \frac{\omega t - \phi_0}{\gamma} \tag{2.4}$$

Thus, as t increases, z must increase, and so a wave moving in the $+z$ direction is represented as shown in Fig. 2.3. Similarly, when the plus sign is chosen so that $\phi_0 = \omega t + \gamma z$, a wave moving in the $-z$ direction is described.

The velocity of the wave, u, can be determined from (2.4) taking the derivative with respect to time, which is

$$u = \frac{dz}{dt} = \frac{\omega}{\gamma} \tag{2.5}$$

Therefore, the propagation constant, γ, is

$$\gamma = \frac{\omega}{u} \text{ radians/meter} \tag{2.6}$$

and is a measure of the phase characteristic of a wave moving through a medium. For free space, $\gamma = \omega / c$.

In practice, the propagation constant will also usually contain information that describes the rate of attenuation or growth of a wave as it moves through a medium. When this is the case

$$\gamma = \beta + j\alpha \tag{2.7}$$

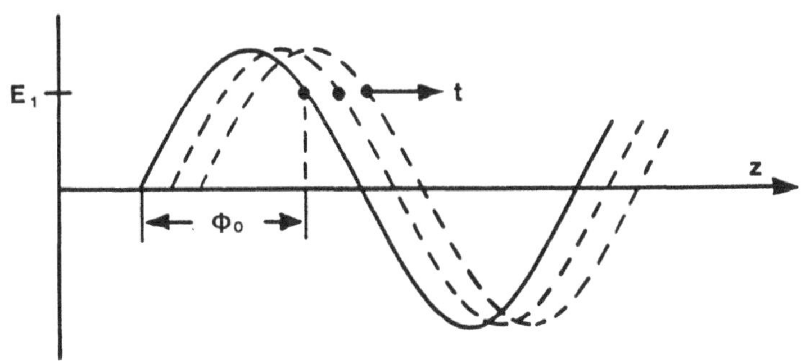

Figure 2.3 Wave Propagating in $+z$ Direction

where, β is now the phase characteristic, ω/u, and α gives the rate of attenuation or growth. Thus, for an attenuating electric wave traveling in the $+z$ direction,

$$E = E_o e^{-\alpha z} \sin(\omega t - \gamma z) \tag{2.8}$$

and appears as shown in Fig. 2.4.

An electromagnetic wave has both electric and magnetic components. The variations of amplitude and phase with position can be very complex. A particularly simple case is the plane electromagnetic wave in a lossless medium. At one instant in time, the magnitudes of the electric and magnetic fields are constant in any plane perpendicular to the direction of travel of the wave. The relation of the electric field to the magnetic field for a plane wave is indicated in Fig. 2.5. Both the electric and magnetic fields are sine waves with the same propagation constant γ. The direction of the magnetic field is at a right angle to the direction of the electric field.

When an electromagnetic wave, such as that shown in Fig. 2.5, impinges on a sheet of lossy material, voltages and currents are induced in the material. If the resistance of the material is such that all of the power in the wave is absorbed and none is reflected, then there is a perfect impedance match.

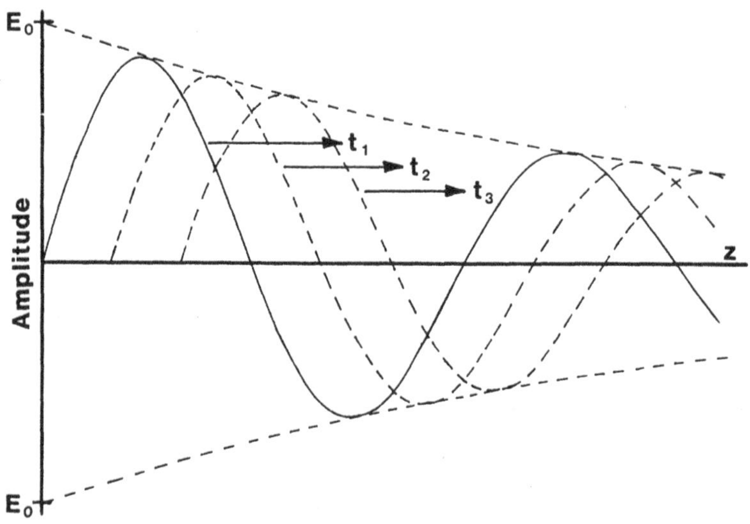

Figure 2.4 Attenuating Electric Wave Traveling in the $+z$ Direction

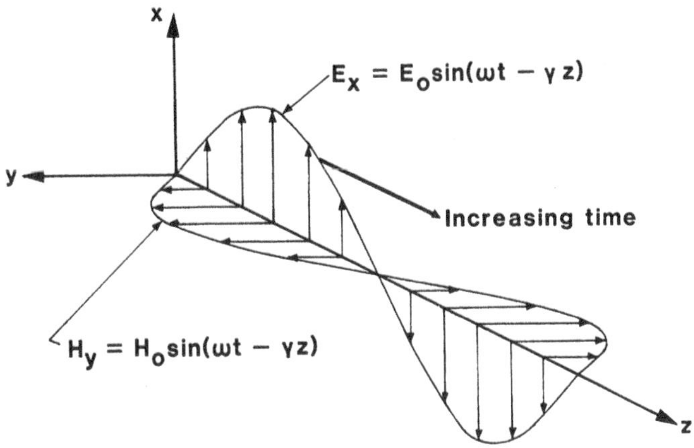

Figure 2.5 Electric and Magnetic Fields in a Plane Electromagnetic Wave

For a wave propagating in free space, the amplitude of the electric and magnetic fields are such that the resistance of the material must be 377 Ω to prevent reflections. The impedance of free space is 377 Ω and is related to the magnetic permeability, μ_0, and the dielectic constant, ϵ_0, by the relation

$$Z_0 = \sqrt{\frac{\mu_0}{\epsilon_0}} = 377\ \Omega \tag{2.9}$$

The value of μ_0 is $4\pi \times 10^{-7}$ Henrys/m and ϵ_0 is $1/36\pi \times 10^{-9}$ Farads/m. The quantity Z_0 is often called the characteristic impedance.

At this point, it should be noted that the speed of light in free space is also determined from μ_0 and ϵ_0 by the relation

$$c = \sqrt{\frac{1}{\mu_0 \epsilon_0}} = 3\times 10^8 \text{ m/s} \tag{2.10}$$

For values of μ of ϵ other than those for free space, the impedance and speed change accordingly. For example, for aluminum oxide, a material commonly used in microwave components, the dielectric constant is about nine times that of free space. As a result, the characteristic impedance and speed of light are one-third of the values for free space.

2.2 SKIN EFFECT

When an electromagnetic wave propagating through air impinges on a good conductor, most of the energy is reflected. This is because of the impedance mismatch between the surface of the conductor, which acts essentially as a short circuit, and the air. The impinging wave causes current to flow along the surface of the conductor. The current density is a maximum at the surface and decreases exponentially with depth into the conductor. The depth at which the current density has dropped to 1/e of the value at the surface is called the skin depth. The skin depth is usually denoted by δ and is given by

$$\delta = \frac{1}{2\pi} \sqrt{\frac{\rho}{f \mu_r}} \text{ cm} \tag{2.11}$$

where f is the frequency of the electromagnetic wave in gigahertz, ρ is the resistivity in ohm-cm, and μ_r the relative permeability (relative to the free-space value) of the conductor.

At microwave frequencies, the skin depth is very small as shown in Fig. 2.6. Notice, for example, in the case of copper for which ρ is 1.72×10^{-6} Ω-cm and μ_r is unity, the skin depth is only 2.1×10^{-4} cm at 1 GHz. This is approximately one-tenth of one-thousandth of an inch (0.1 mil). For magnetic materials, skin depth may be much smaller. For example, in Fig. 2.6, skin depth for a material with a μ_r of 5000 is shown. The resistivity used for this material was ten times the value for copper and typical of the resistivity for a magnetic material. Caution must be exercised in attempting to determine the skin depth of magnetic materials. In the first place, permeability is strongly dependent on magnetic field level. Thus, not only does the permeability at the surface of the conductor vary with the intensity of the impinging field, it also varies with distance into the metal as the field decays. Secondly, the hysteresis effects and the hysteresis loss are not included in (2.11).

A small skin depth is beneficial when a conductor is being used to shield a region from electromagnetic waves. However, for conductors used in transmission lines for electromagnetic waves, the small skin depth leads to high transmission-line losses. The losses are the same as if the total current were of uniform density to a depth δ.

The resistance of a conductor resulting from the flow of current near the surface is given by

$$R_s = \frac{\rho}{\delta} = 2\pi \sqrt{\rho f \mu_r} \tag{2.12}$$

This is surface resistance and has units of ohms per unit width per unit length, or simply ohms per square. Thus, resistance increases with frequency and resistivity. The problem of loss resulting from a small skin depth is compounded by the fact that conductor surfaces are often not as smooth as they

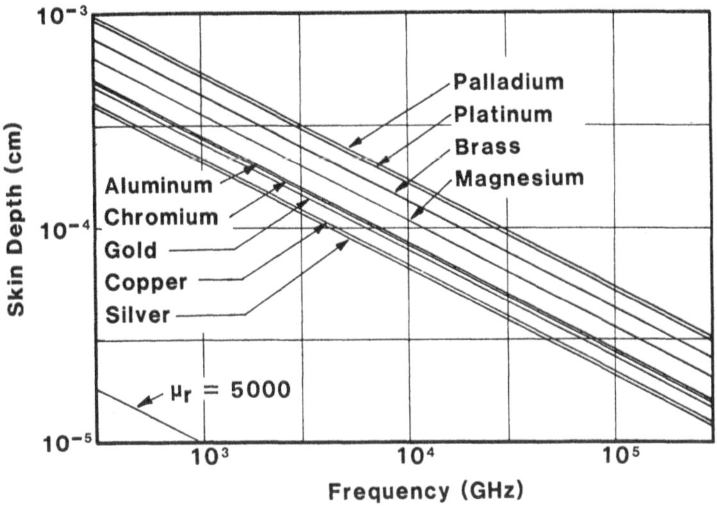

Figure 2.6 Skin Depth for Several Commonly Used Metals

should be. Irregularities with dimensions comparable to the skin depth are often present because of the machining technique used, corrosion, *et cetera*. Because the current must flow over these irregularities, the current path is lengthened and so the loss is increased.

2.3 COAXIAL LINES AND WAVEGUIDES

Coaxial lines and waveguides are used for carrying microwave power to and from microwave tubes. The waveguides used for this purpose are metallic pipes, which are usually made from copper and have either rectangular or circular cross sections. A coaxial line is a circular waveguide with a center conductor.

The need for coaxial lines and waveguides arises from deficiencies in the open-wire conductors used at lower frequencies. At dc and very low frequencies, the configurations of the wires used in putting a circuit together have little or no effect on circuit performance. As frequency is increased, the following three effects limit the use of open-wire systems:

1. The electric and magnetic fields surrounding the wires (illustrated in Fig. 2.7 for two wires spaced a short distance apart) couple signals from one set of wires to another. This leads to "crosstalk" and power loss to nearby conductors. This problem may be alleviated at low and

medium frequencies by twisting together pairs of wires carrying a common signal to confine the fields, and then by providing separation or metallic shields between wire pairs with different signals.

2. As frequency is increased, the wires begin to act like antennas and may radiate signals. This problem is largely alleviated by using metallic shields around the wires.
3. When frequency is high enough and wavelength short enough so that the lengths of the conductors are a reasonable fraction of a wavelength, inductive and capacitive effects become important. Each wire has inductance and there is capacitance between wires. As a result, a wire has impedance, and this impedance is a function of the size of a wire and its location relative to other conductors. If variations in impedance occur, then a signal propagating along a wire may be partially reflected at each change in impedance.

With care, the above effects can be eliminated through the use of coaxial lines or waveguides.

2.3.1 Coaxial Lines

Coaxial lines normally have circular cross sections with a center conductor that is concentric with the outer conductor. Depending on the frequency of

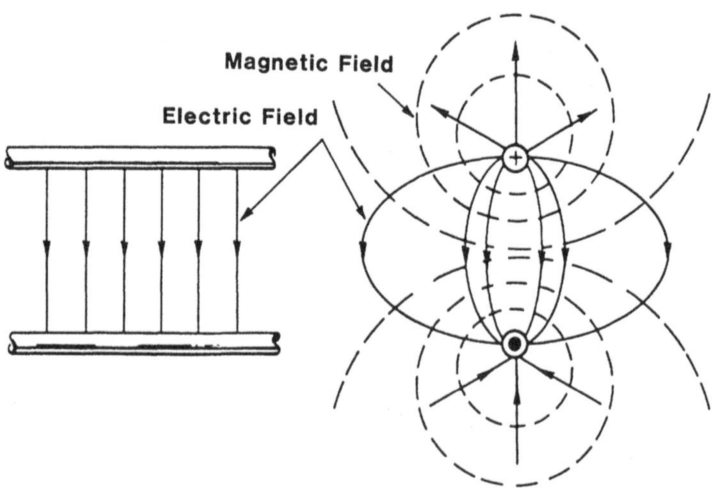

Figure 2.7 Electric and Magnetic Fields Surrounding Current-Carrying Wires

the signal in the line, the fields may have any of an infinite set of configurations. The various configurations are called modes and are described in the following section.

At low and relatively low frequencies, only one mode, called the principle mode can exist. The field pattern for that mode is shown in Fig. 2.8. The electric field lines extend radially between the center and outer conductor, and the magnetic field lines are circular and concentric with the conductors. This mode is called the transverse electric and magnetic (TEM) mode because the electric and magnetic fields are both transverse (perpendicular) to the direction of propagation of power (coaxial with the axis of the transmission line) along the line. For this mode, propagation is at the speed of light.

The fields propagate in the region between the center conductor and the outer conductor and decay exponentially within the surfaces of the conductors, as was described in the previous section concerning the skin effect.

A coaxial line has inductance and capacitance and a characteristic impedance given by

$$Z_0 = \sqrt{\frac{L}{C}} \tag{2.13}$$

where L is the inductance per unit length and C is the capacitance per unit length. For an air-insulated line, L and C, and thus Z_0, depend only on the ratio of the radius of the outer conductor to the radius of the inner conductor. The relationship of Z_0 to conductor radii is shown in Fig. 2.9

The insulating material of the line affects Z_0 because L is proportional to the permeability and C is proportional to the dielectric constant of the material. Normally, insulating materials are used which have a relative permeability of one and a dielectric constant that is the same as free space

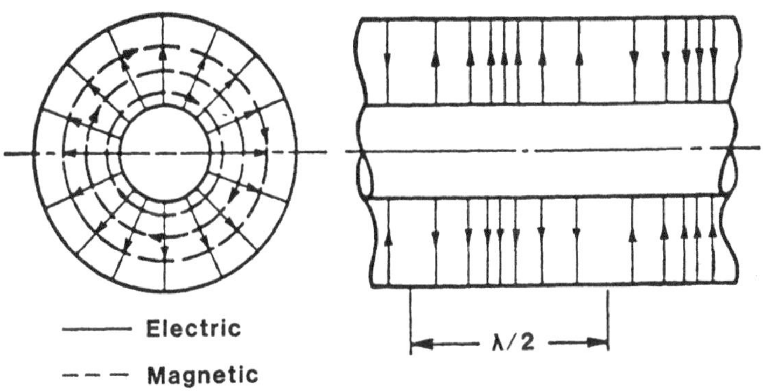

Figure 2.8 Field Distributions for TEM Mode in Coaxial Line

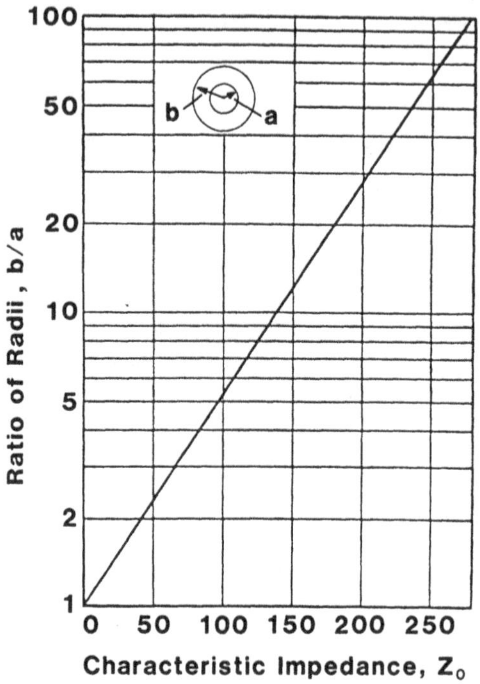

Figure 2.9 Characteristic Impedance of an Air-Insulated Coaxial Line

(such as for air or other gas-insulated lines), or is somewhat larger (up to a factor of nine for aluminum oxide) than that for free space.

As long as the dimensions and materials of a coaxial line are such that the characteristic impedance is constant, then a signal will propagate along the line without being reflected. The signal will decay slowly as it propagates because of the resistive losses in the surfaces of the conductors and because of dielectric losses in the insulating material. Signal variations caused by impedance changes and losses are discussed in Section 2.5.

The optimum impedance for a coaxial line is somewhat dependent on the application for the line. At least three factors are of importance: attenuation, breakdown voltage, and power capability. Shown in Fig. 2.10 are variations of these quantities with respect to optimum values (the optimum value is indicated by a relative value of 1.0) for various ratios of radii (or values of impedance from Fig. 2.9) in an air-filled line.

The minimum value of attenuation occurs at a characteristic impedance of 77 ohms. At smaller impedances, current increases and so I^2R losses increase. At larger impedances, the inner conductor radius becomes so small that resistance and I^2R losses increase. The maximum breakdown voltage occurs at a characteristic impedance of 60 ohms, The reduced conductor spacing associated with smaller impedances reduces breakdown voltage. At higher impedances, the reduced radius of the inner conductor results in an increased field at the inner conductor surface and to a reduced breakdown voltage. Finally, the maximum power that can be transferred by a coaxial line with no reflections depends on V^2/Z_0, where V is the maximum voltage. As Fig. 2.10 shows, the breakdown voltage has a fairly broad maximum. As a result, as Z_0 is decreased, at first, V^2 falls more slowly than Z_0, so V^2/Z_0 increases. A maximum is reached by a characteristic impedance of 30 ohms, below which V^2/Z_0 falls very rapidly.

In practice, it is necessary to restrict maximum voltages and powers to values well below the theoretical limits. Voltage is normally limited by more than an order of magnitude and power by more than two orders of magnitude

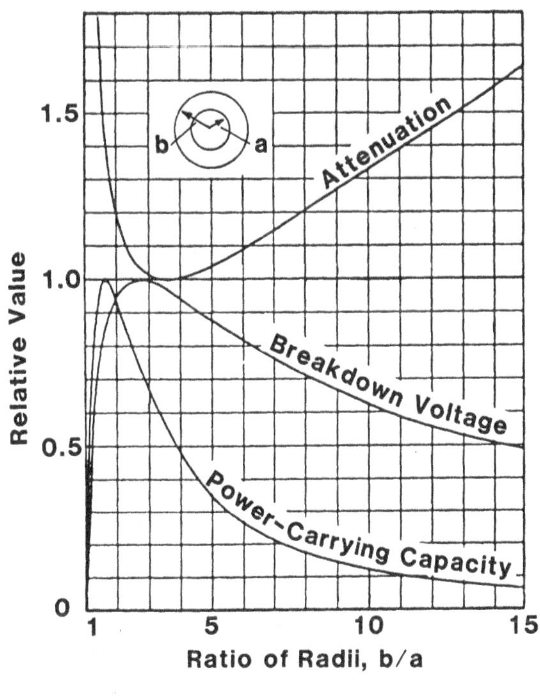

Figure 2.10 Characteristics of Air-Filled Coaxial Lines

below the theoretical limits. Many factors, such as reflections, heating, sharp corners on conductors, and center-conductor supports, enter into the determination of practical limits.
corners on conductors, and center-conductor supports, enter into the determination of practical limits.

Probably the most common characteristic impedance for practical coaxial lines is 50 ohms, which represents a reasonable compromise among the values of 77, 60, and 30 ohms given by Fig. 2.9 and 2.10 for attenuation, breakdown field, and power handling capability, respectively.

2.3.2 Waveguides

As stated previously, a waveguide is a metallic pipe usually made from copper and having either a rectangular or circular cross section. Electromagnetic waves can be thought of as being guided through a waveguide by reflections from the waveguide walls. As an example, consider the situation shown in Fig. 2.11. Two plane waves of equal intensity are assumed to be propagating to the right. The direction of propagation of one is at an angle of $+\alpha$ with respect to the z axis, and the other is at an angle of $-\alpha$. The solid lines represent positions of wave crests (maximum value of E field out of paper) and the dashed lines represent positions of wave troughs (maximum value of E field into paper.)

Whenever wave crests and troughs (solid and dashed lines) intersect, the electric field is zero and only a magnetic field exists. In fact, all along a plane through zero-field points and parallel to the y axis (out of paper), the electric

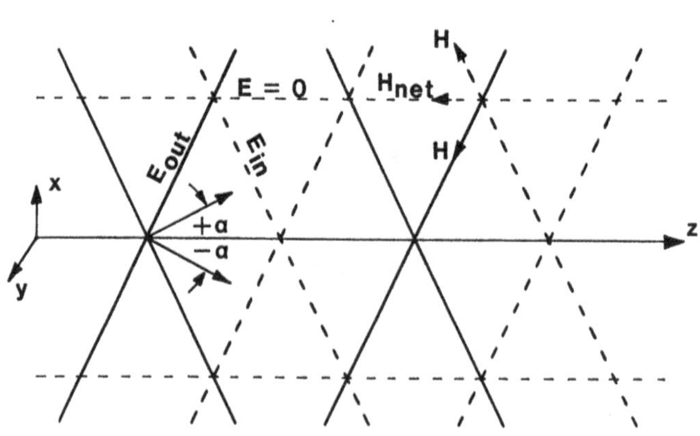

Figure 2.11 Plane Waves of Equal Intensity Propagating at Angles of $\pm\alpha$ with Respect to the z Direction

field is zero. Not only is the E field zero for the instant of time implied in Fig. 2.11, but also at other times as long as the waves propagate at the angles shown. Thus, conductors can be inserted along these planes without disturbing the E fields as long as the currents in the conductors correspond to those that would be induced by the magnetic fields, which, it should be noted, are parallel to the conductors.

The magnetic field lines are also parallel to any x-y plane and the electric field lines are perpendicular to these planes. As a result, conductors containing currents and charges corresponding to those that would be induced by the fields can be placed in the x-y planes.

Finally, two conductors in the x-y plane can be joined to the conductors in the y-z plane to form a rectangular pipe (waveguide) containing portions of the two waves. These two waves can propagate as though they were in free space, except that, in reality, they are continually reflected back and forth from one side of the guide to the other. In fact, there is only one wave because each wave is a reflection of the other.

The amplitudc of the resulting electric field at an instant in time is indicated by the vertical deflections of the grid pattern above and below the center plane of the waveguide in Fig. 2.12. The distribution of both the electric and magnetic field components of this wave are shown in Fig. 2.13. The field pattern shown in Figs. 2.12 and 2.13 is the field pattern that normally occurs in rectangular waveguide, and is referred to as the dominant mode or the TE_{10} mode. The TE designation refers to the fact that the electric fields are transverse to the axis of the waveguide. The magnetic fields have components in the direction of propagation along the guide, and the wave is sometimes referred to as an H wave. The indices on the mode designation refer first to the number of half-wave variations of the electric field in the wide dimension of the guide and second to the number of half-wave variations in the narrow dimension.

The velocity with which the signal propagates along the waveguide, u_g, in the TE_{10} mode can easily be determined with the aid of Fig. 2.14. Each component of the wave travels at the speed of light, c, at an angle α with respect to the axis of the guide. As a result, the velocity of the wave along the axis is smaller than c and, in fact, is

$$u_g = c \cos \alpha \tag{2.14}$$

This velocity is called the group velocity and is the velocity with which the energy propagates along a circuit.

For a given size of waveguide, as the signal frequency is decreased and wavelength is increased, the angle that the wave components make with respect to the axis of the guide increases as shown in Fig. 2.15.

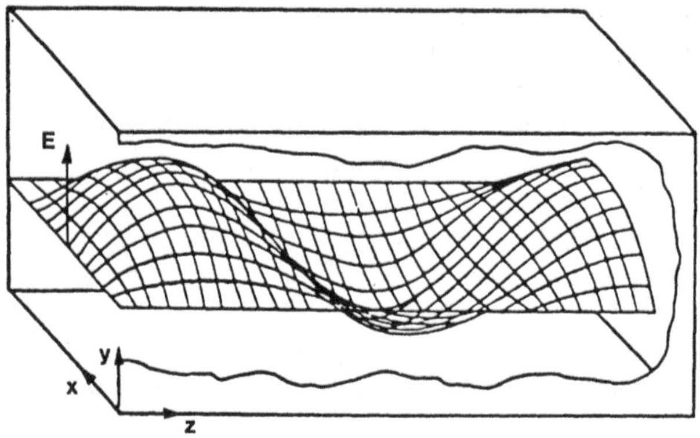

Figure 2.12 Amplitude of Electric Field as a Function of Position in Waveguide

(From L.J. Chu and W.L. Barrow, *Proc. IRE*, December 1938.)

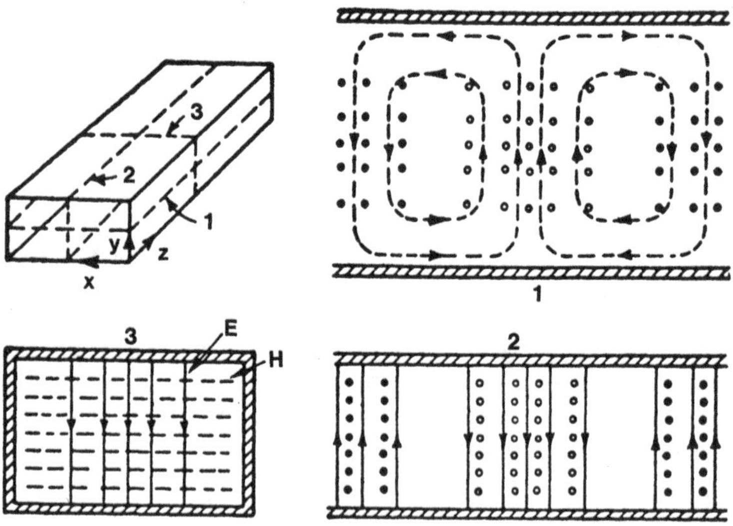

Figure 2.13 Electric and Magnetic Field Components in Waveguide for TE_{10} Mode

(From *Fields and Waves in Communication Electronics* by Simon Ramo, John R. Whinnery and Theodore Van Duzer. Copyright 1984 by John Wiley and Sons. Reproduced with permission.)

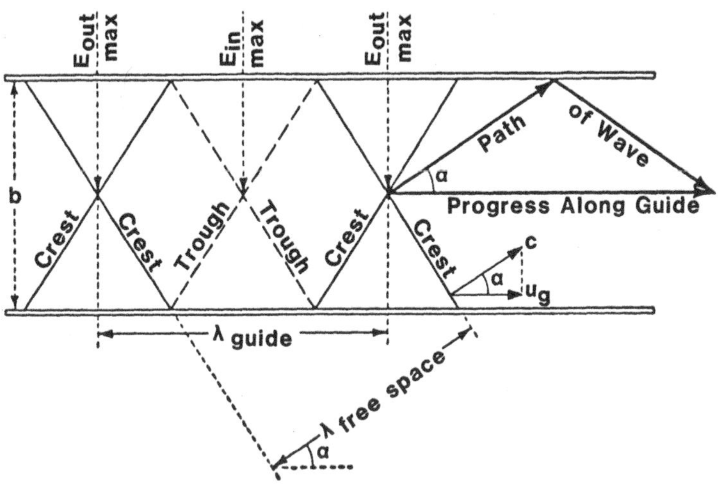

Figure 2.14 Velocities and Wavelengths of TE_{10} Mode

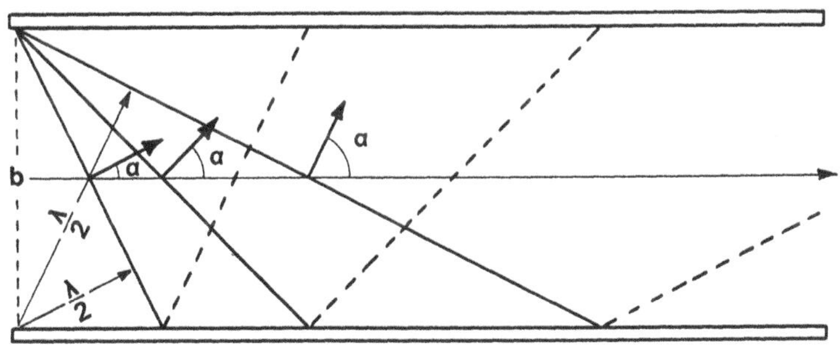

Figure 2.15 Approach of Cut-Off Condition for TE_{10} Mode

As the half-wavelength, $\lambda/2$, of the wave components approaches the width of the guide, *b*, the angle of component propagation approaches 90°. Finally, when $\lambda/2 = b$, the wave components are reflected back and forth across the guide and there is no propagation along the guide. The frequency at which this condition occurs is called the cut-off frequency for the waveguide. Signals at this frequency and below cannot propagate through the waveguide.

The cut-off frequency of rectangular waveguide can be decreased by changing the cross section to the single-ridge or double-ridge configuration shown in Fig. 2.16. The change in cut-off frequency can be understood qualitatively by examining an equivalent-circuit model. As shown in Fig. 2.15, cut-off corresponds to a resonance condition in which the wave propagates back and forth across the guide. If the equivalent inductance and capacitance of the waveguide can be determined, then cut-off would correspond to the resonance of the inductance and capacitance.

As indicated in Fig. 2.16, regions A of the single-ridge and double-ridge waveguides are primarily inductive, and region B is primarily capacitive. As a result, the waveguides can be modeled as parallel LC circuits as indicated in Fig. 2.16. Cut-off occurs at resonance of the circuit, for which the resonant frequency, f_c, is

$$f_c = \frac{1}{2\pi \sqrt{C_B L_A/2}} \tag{2.15}$$

If the dimensions of the ridges are changed to increase C_B (the ridges could be made wider or the spacing across region B could be made narrower), in principle, the cut-off frequency can be made as low as desired.

Single-ridge and double-ridge waveguides are commonly used with microwave tubes in applications requiring high power and broad bandwidths.

Circular waveguides are used in conjunction with microwave tubes for many reasons. For example, the ceramic disks used as windows through which power is propagated out of a tube are frequently circular and are brazed into a section of circular waveguide. As another example, in some fast-wave tubes such as the gyrotron, parts of the microwave circuit and the output waveguide are circular waveguides.

The dominant mode in circular waveguide is the TE_{11} mode illustrated in Fig. 2.17. The field configuration corresponds to the dominant TE_{10} mode in rectangular waveguide. The indices for modes in circular waveguide indicate the following: the first index denotes the total number of full-period variations of the field along a circular path; the second index is one more than the total number of sign reversals along a radial path.

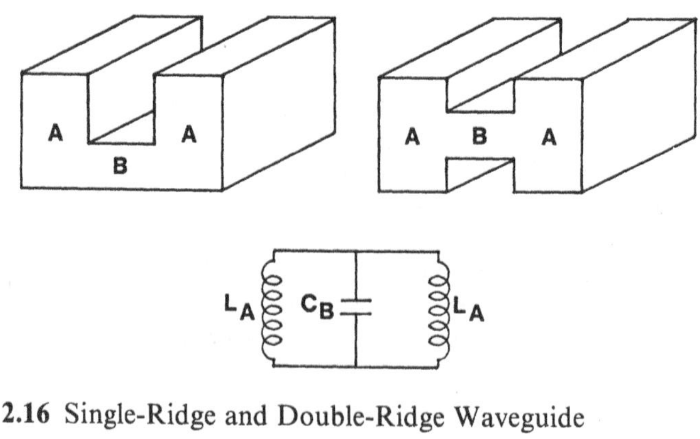

Figure 2.16 Single-Ridge and Double-Ridge Waveguide

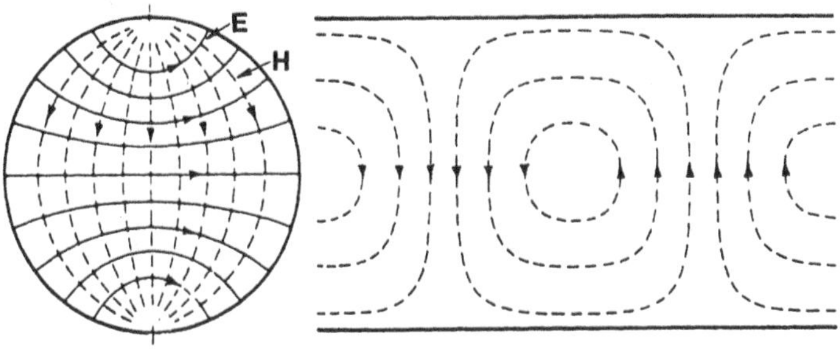

Figure 2.17 Field Configurations for Dominant TE_{11} Mode in Circular Waveguide

2.4 MODES

Up to this point, only dominant modes have been discussed. These are the lowest frequency modes of operation for coaxial lines and waveguides and are the modes usually used in their operation. For the coaxial line, this is the TEM mode; for the rectangular waveguide, it is the TE_{10} mode; and, for the circular waveguide, it is the TE_{11} mode.

At higher frequencies, other modes can exist. For example, the electric field wave crests and troughs for two modes at a given frequency are shown in Fig. 2.18. One of these is the dominant TE_{10} mode that has already been discussed. The other is also a transverse electric mode, but has two half-wave variations of the electric field in the wide dimension of the waveguide. The cut-off frequency for this mode occurs when the free-space wavelength is the same as the width of the guide.

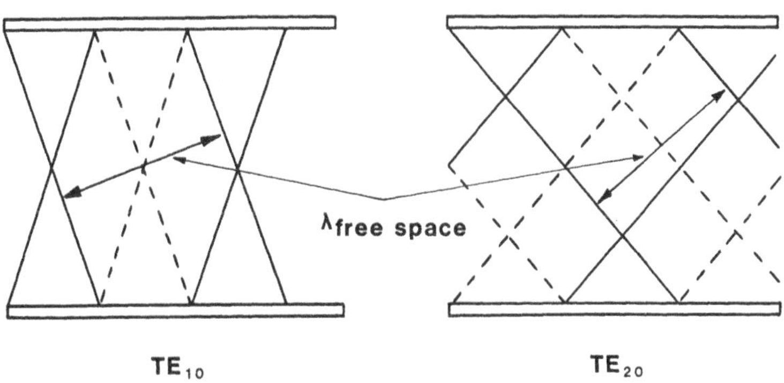

Figure 2.18 Electric Field Wave Crests and Troughs in Two Modes at a Given Frequency in Rectangular Waveguide

Depending on frequency, many other modes can exist. Some of these are shown in Fig. 2.19.

Similarly, in circular waveguide, depending on frequency, many modes can exist. Some of these are shown in Fig. 2.20. Other than the dominant TE_{11} mode, the TE_{01} mode is of particular interest. The attenuation for this mode decreases indefinitely as frequency increases. As a result, this mode is of importance in millimeter devices where large waveguide is needed for operation at high power levels and low attenuation is desired.

In coaxial lines, when the frequency becomes high enough so that the circumference halfway between the conductors is approximately a wavelength, the first higher-order mode can propagate. Thus, even though it is desirable to use lines that are as large in diameter as possible to minimize losses, size is limited by the first higher-order mode, for which losses are above those for the dominant mode.

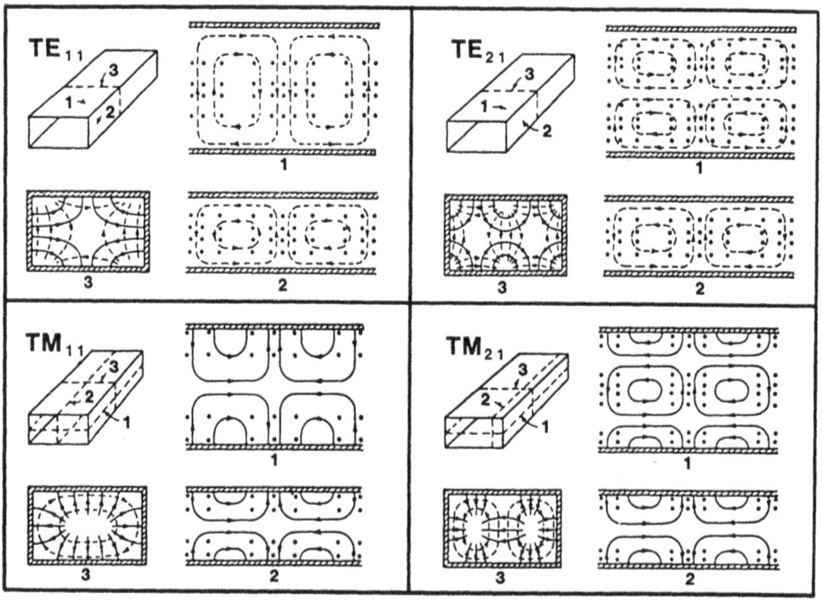

Figure 2.19 Modes in Rectangular Waveguide

(From *Fields and Waves in Communication Electronics* by Simon Ramo, John R. Whinnery and Theodore Van Duzer.Copyright 1984 by John Wiley and Sons. Reproduced with permission.)

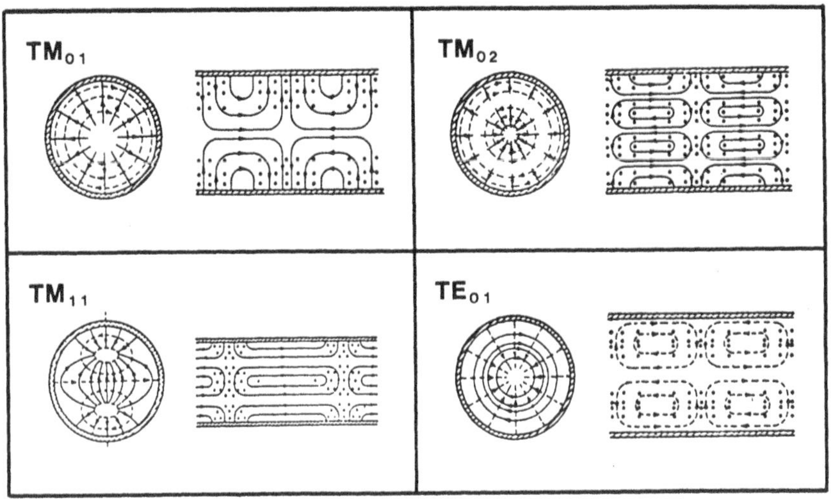

Figure 2.20 Modes in Circular Waveguide

(From *Fields and Waves in Communication Electronics* by Simon Ramo, John R. Whinnery and Theodore Van Duzer.Copyright 1984 by John Wiley and Sons. Reproduced with permission.)

2.5 TRANSMISSION-LINE EFFECTS

A signal propagates along a lossless constant-impedance transmission line without attenuation or reflection. When the transmission line is terminated in a resistive load with a resistance equal to the characteristic impedance, Z_0, of the line, the signal is completely absorbed so that no reflections occur. Thus, all of the power delivered by a signal generator connected to the line is delivered to the load. This situation is called the matched-line condition.

If the transmission line is terminated in an impedance that is not equal to Z_0, then reflection of the signal occurs and some of the incident power returns to the generator, where re-reflection can occur if the generator is not matched to the line.

As a result of the reflections, signals propagate in both directions on the transmission line and combine to produce a standing wave pattern. This is illustrated in Fig. 2.21 for the case when the amplitude of the reflected signal is equal to the amplitude of the incident signal. The incident and reflected waves have equal amplitudes when the load on the line is a short circuit, an open circuit, or a lossless reactance. No power is absorbed by these loads and so it must be reflected. In the case of the reactive load, energy may be stored for a part of each cycle and then returned to the line so that the reflected wave is not in phase with the incident wave.

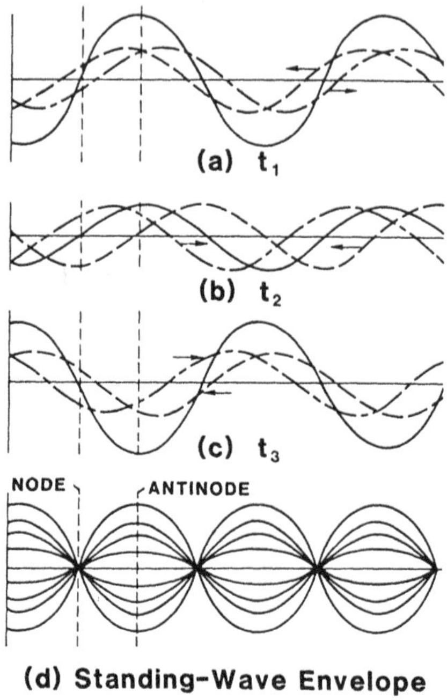

Figure 2.21 Formation of Standing Wave Pattern for Equal-Amplitude Incident and Reflected Waves

(From *Microwave Theory and Applications* by Stephen F. Adam. Copyright 1969 by Prentice-Hall. Reproduced with permission.)

In Fig. 2.21, the incident and reflected waves indicated by dashed curves are shown at three different times. The resulting wave is the sum of the incident and reflected waves, as indicated by the solid curve. It should be noted that only the amplitude, and not the position of the resulting wave, varies with time. Thus, the resulting wave is called a standing wave. Standing wave patterns at several instants of time are superimposed in Fig. 2.21(d). The points where the signals cancel to produce a minimum field are called nodes. The maximum-field points are antinodes. Normally, when standing waves are illustrated, only the maximum values of voltage or current are shown.

The standing wave pattern in Fig. 2.21 represents an extreme case where the signal level at the node is zero, and at the antinode it is twice the amplitude of the incident or reflected wave. At the other extreme, when the load is matched to the line so that no signal is reflected, there is no standing wave and the signal level along the line is constant. In practice, a situation somewhere between these two extremes always exists because perfect open and short circuits are never encountered, and therefore a perfect match of a load to a line never occurs. Also, loads with impedances from values closes to zero to values close to infinity can occur for any one of a number of reasons.

The impedance of a termination on a transmission line appears to have a wide range of values depending on the position along the transmission line where the effect of the termination is determined. As an example, the standing waves of voltage and current on a lossless line that is short-circuited are shown in Fig. 2.22. Because the load impedance is zero, no voltage can be developed across it, and the incident and reflected waves must cancel at the load while the corresponding current waves add.

At a distance $\lambda/4$ from the short circuit, the situation is reversed and identical to that occurring at the end of an open-circuited line. That is, no current can flow through the load, so the current waves must cancel and

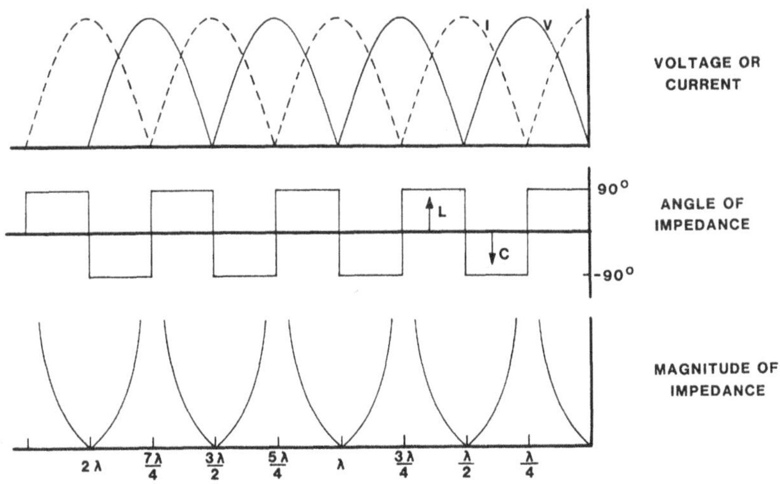

Figure 2.22 Voltage and Current Standing Waves and Impedance Along a Short-Circuited Transmission Line

the voltage waves add. A person presented only with voltage and current measurements at this point could very easily assume that the line was open-circuited and would have no way of knowing that there was a short circuit at a distance $\lambda/4$ from the point of the measurements. Identical conditions exist at distance $3\lambda/4$, $5\lambda/4$, *et cetera* from the load. On the other hand, measurements made at $\lambda/2$, λ, *et cetera*, produce voltages and currents identical to those at the short circuit and so the load impedance at these points appears to be a short circuit.

At points along the line other than at multiples of $\lambda/4$, the load impedance appears to be reactive. For example, between the load and the position $\lambda/4$, the current lags the voltage by 90° and so the load appears to be inductive. The magnitude of the inductive impedance at any point is the ratio of voltage to current and varies from zero at the short circuit to infinity at a distance $\lambda/4$ away. In the region between $\lambda/4$ and $\lambda/2$, the current leads the voltage by 90° and so the load appears to be capacitive. Again, the magnitude of the capacitive reactance is the ratio of voltage to current.

From the above example, it is clear that in order to determine the impedance of a load on a transmission line, voltage and current measurements must be accompanied by a knowledge of distance to the load. Also of extreme importance is the fact that a mismatched load can take on a wide range of impedance values at various positions along a transmission line.

The maximum voltage, V_{max}, at the antinode of a standing wave is the sum of the peak voltages for the incident wave, V_i, and for the reflected wave, V_r. The minimum votage, V_{min}, at the node of a standing wave is the difference between V_i and V_r. The voltage standing wave ratio (VSWR) is the ratio of V_{max} to V_{min}. The VSWR is usually denoted by σ, and is, therefore,

$$\sigma = \text{VSWR} = \frac{V_{max}}{V_{min}} = \frac{V_i + V_r}{V_i - V_r} \tag{2.16}$$

The VSWR for the standing wave in Fig. 2.21 is ∞. For a perfectly matched load, the VSWR is unity.

The magnitude of the reflected voltage depends on the mismatch in impedance between the load and the transmission line. The reflection coefficient is the ratio of the reflected voltage to the incident voltage. The reflection coefficient is usually a complex number because the load usually contains reactance. The magnitude of the reflection coefficient is usually denoted by ρ and is simply the ratio of V_r to V_i, that is,

$$\rho = \frac{V_r}{V_i} \tag{2.17}$$

The reflection coefficient that produced the standing wave in Fig. 2.21 had a value of one. For the perfectly matched case, ρ is zero.

Now the relation between σ and ρ is readily found by combining (2.17) and (2.16), and is

$$\sigma = \frac{1+\rho}{1-\rho} \quad \text{or } \rho = \frac{\sigma-1}{\sigma+1} \tag{2.18}$$

The impedance, Z, of a termination usually contains reactance as well as resistance, and therefore has a complex value given by

$$Z = R + jX \tag{2.19}$$

As has been indicated, the effect of this impedance can vary over a wide range depending on the position along the transmission line where its effect is observed.

The complex reflection coefficient, Γ (for which ρ in (2.17) is the magnitude), is the ratio of reflected voltage to incident voltage with phase angles taken into consideration. The ratio of reflected to incident voltage is, in turn, directly determined by the load impedance and Z_0. The value of Γ is given by

$$\Gamma = \frac{Z-Z_0}{Z+Z_0} = \frac{\dfrac{Z}{Z_0}-1}{\dfrac{Z}{Z_0}+1} \tag{2.20}$$

The value of Z can be the impedance at the load or the effective value of load impedance at some other point on the line. Thus, Γ takes on different values for a given load, depending on where it is determined on a transmission line. In fact, the magnitude, ρ, of Γ remains constant, but it is the phase angle that changes. The usefullness of Γ will become apparent in the following discussion of impedances and their determination.

While the effective value of load impedance at any point along a transmission line can be calculated, the calculation is cumbersome at best and the mathematics provide very little guidance on what has been done or what should be done next. As a result, impedance charts and, in particular, the Smith chart, have been developed as aids to the solution of transmission-line problems.

2.5.1 The Smith Chart

In rectangular coordinates, impedance is plotted as shown in Fig. 2.23. The resistive part is plotted along the horizontal axis and reactance is plotted vertically. Inductive reactances are positive and capacitive reactances are negative. A rectangular impedance chart is difficult to use for transmission-line calculations because very large values of impedance are not on the chart and standing wave ratios cannot be conveniently determined.

In the Smith chart, which is universally used for transmission line calculations, the rectangular impedance chart is transformed to circular coordinates. This transformation is indicated in Fig. 2.24. First of all, impedances are normalized with respect to the characteristic impedance of the transmission line. That is,

$$\frac{Z}{Z_0} = \frac{R}{Z_0} + j\frac{X}{Z_0} \tag{2.21}$$

On the rectangular chart, the lines of constant resistance are vertical lines with the line at 1 representing a resistance equal to characteristic impedance. As shown in Fig. 2.24, lines of constant resistance become circles of constant resistance on the Smith chart. Resistance is zero for the outer circle. Smaller and smaller circles represent larger and larger resistances, and finally converge to infinity at a point on the right-hand side of the chart. The $R/Z_0 = 1$ circle passes through the center of the chart.

On the rectangular chart, lines of constant reactance are horizontal lines. These are transformed to the fan-shaped pattern on the Smith chart, where each curved line is a portion of a circle. The zero-reactance line is horizontal through the center of the chart and, in fact, the center point of the chart represents the matched load condition because there $R = Z_0$ and X is zero.

In the completed Smith chart shown in Fig. 2.25, the lines of constant resistance and constant reactance are overlayed. The usefulness of the Smith chart becomes apparent only through practice in using it. Some of its properties can, perhaps, be best illustrated with specific examples.

In Fig. 2.22, the standing wave pattern for a short-circuited transmission line was shown, and subsequently the variation of apparent load impedance along the line was discussed. The impedance varied from zero at the load, through a wide range of inductive values to infinity at a position $\lambda/4$ from the load. Then, in moving from $\lambda/4$ to $\lambda/2$, the impedance varied through capacitance values and then back to a value of zero at $\lambda/2$.

On the Smith chart, the short circuit condition is on the left-hand side where the resistive and reactive components are zero. Movement away from the load (toward the generator) is indicated by the arrows at the outer circumference of the chart and is in the clockwise direction. A movement of $\lambda/4$ is halfway around the chart. At this point, the resistive and reactive components of impedance are infinity and this is in agreement with the actual situation described in Fig. 2.22. In the region between the short circuit and $\lambda/4$, the Smith chart shows the impedance to be inductive, again in agreement with Fig. 2.22.

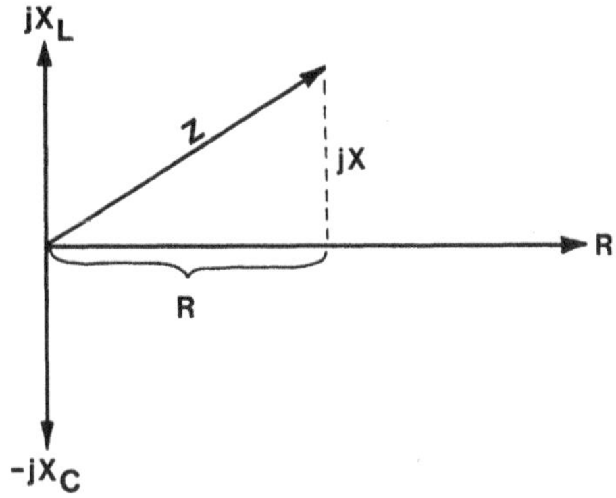

Figure 2.23 Impedance Plot on Rectangular Coordinates

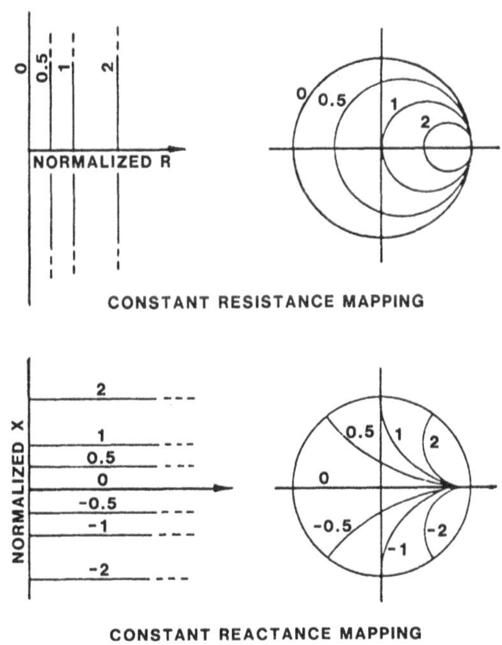

Figure 2.24 Development of Smith Chart

In moving further around the Smith chart, from $\lambda/4$ to $\lambda/2$, the impedance on the Smith chart is capacitive as it should be. Finally, at $\lambda/2$, the impedance is again a short circuit. Subsequent cycles around the Smith chart toward the generator indicate repeats of the impedance pattern, just as was the case in Fig. 2.22.

As another example of the properties of the Smith chart, consider a termination of a 50-ohm transmission line that is purely resistive and has a resistance of 100 ohms. The normalized resistance is 2.0 and is to the right of the center of the Smith chart. The apparent load impedance at various distances along a lossless transmission line falls on the circle drawn about the center of the chart and through the $R/Z_0 = 2.0$ point. This circle represents a constant value of VSWR for this particular load.

It should be noted here that for a purely resistive load with $R/Z_0 \geq 1$, from (2.20), we have

$$\rho = \Gamma = \frac{R/Z_0 - 1}{R/Z_0 + 1} \tag{2.22}$$

and so, from (2.18), the VSWR is

$$\sigma = \frac{R}{Z_0} \tag{2.23}$$

Thus, for $R/Z_0 = 2.0$, the VSWR is 2.0 and the circle about the center of the chart in Fig. 2.25 is a circle of constant VSWR of 2.0. (VSWR does not change along a lossless, constant-impedance transmission line.)

In moving around the Smith chart away from the load (toward the generator) on the $\sigma = 2$ circle, the apparent resistive component of the load impedance decreases, and the capacitance reactance component at first increases (to a normalized value of about 0.75), and then decreases again. At a distance of $\lambda/4$ from the termination, the apparent normalized impedance is purely resistive with a value of 0.5. At this point, the 100-ohm load on the end of the line appears to be a 25-ohm load. In continuing away from the load, from $\lambda/4$ to $\lambda/2$, the resistive component increases again and is accompanied by an inductive reactance component that peaks at a normalized value of about 0.75.

In moving from the load to a position $\lambda/2$ from the load, the apparent normalized resistive component of load impedance varies from 2.0 through 1.0 to 0.5, and then through 1.0 and back to 2.0. If the reactive component could be cancelled at one of the points where $R/Z_0 = 1.0$, then, in continuing from that point further away from the load, the normalized load impedance would appear to be 1.0 and purely resistive. Thus, the load would have been matched to the transmission line.

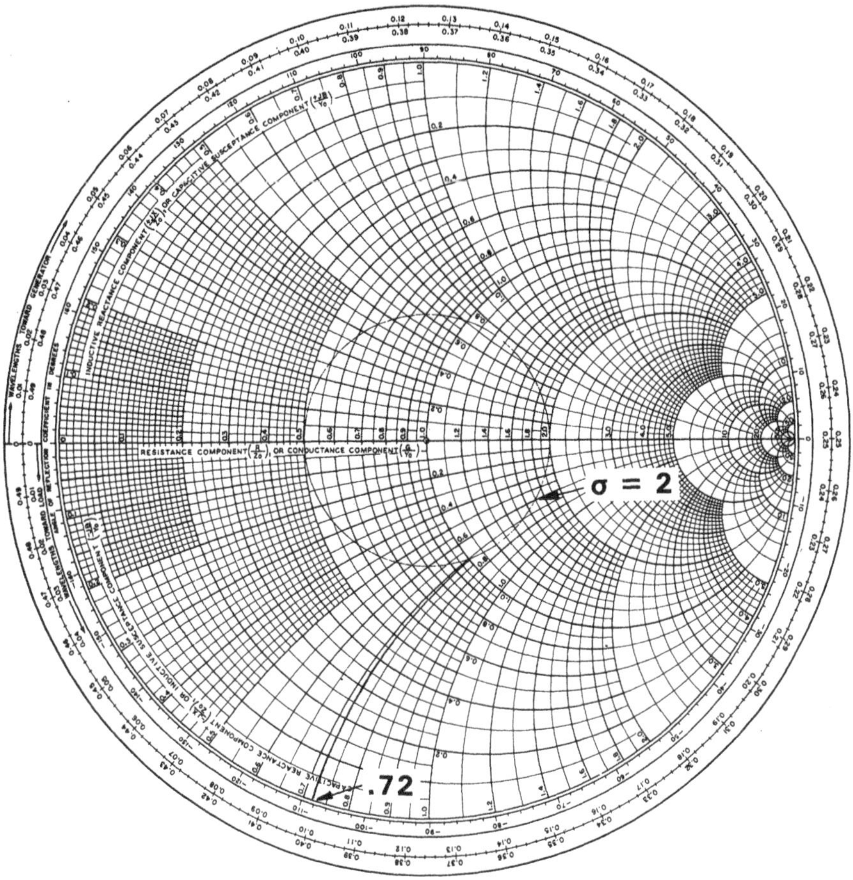

Figure 2.25 Smith Chart

2.5.2 Impedance Matching

There are numerous ways in which reactance can be inserted into a transmission line to change the impedance. One of the most common techniques utilizes the reactive impedance produced by a short circuit (as was illustrated in Fig. 2.22 and in the first example of Smith chart properties). For example, the normalized inductive reactance at a distance of about .099 λ from a short is 0.72. This inductive reactance could be used to cancel the capacitive reactance (with a normalized value of 0.72) at which the VSWR = 2 circle for the $R/Z_0 = 2.0$ termination crosses the $R/Z_0 = 1.0$ resistance circle (at about .097 λ from the termination). An example of the configuration of a shorted coaxial line used to match a 100-ohm load to a 50-ohm transmission line is shown in Fig. 2.26. This configuration is frequently referred to as a single-stub tuner.

A problem that is immediately obvious with the impedance-matching technique illustrated in Fig. 2.26 is that the transmission-line lengths used are functions of wavelength. Unfortunately, most impedance-matching techniques are frequency-sensitive and this makes impedance matches over broad bandwidths difficult to achieve.

If the length and position of the shorted line in Fig. 2.26 are variable, then, after a change in frequency, adjustments can be made to return to the impedance-match condition. In a similar way, the slide-screw tuner shown in Fig. 2.27 can be used for impedance matching at various frequencies. The adjustable screw appears as a variable capacitance, the position of which can be moved along the transmission line.

Double- and triple-stub or screw tuners are sometimes used to eliminate the necessity of changing the position of the tuner along the transmission line. With these tuners, one of the tuning elements is used in conjunction with the spacing between tuning elements, so that another element is in the correct position to cancel the remaining mismatch.

Adjustable stub- and screw-type tuners are used in waveguides as well as in coaxial lines. In addition, it is common to place diaphragms like those shown in Fig. 2.28 in waveguides. As was noted in Fig. 2.16, the sides of the waveguide are sometimes modeled as inductances. The inductive doors

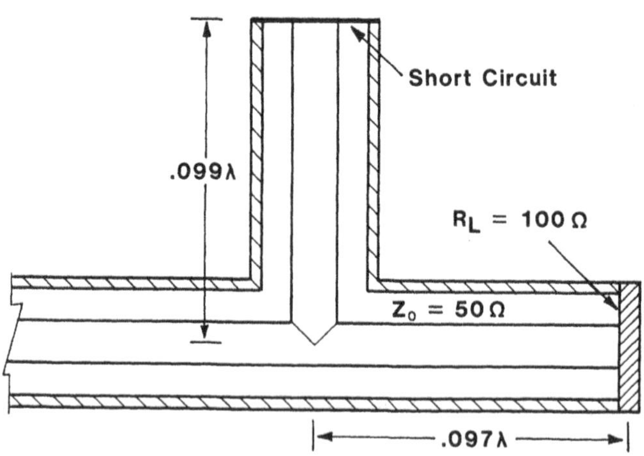

Figure 2.26 Use of Shorted Coaxial Line to Match 100Ω Load to 50Ω Transmission Line

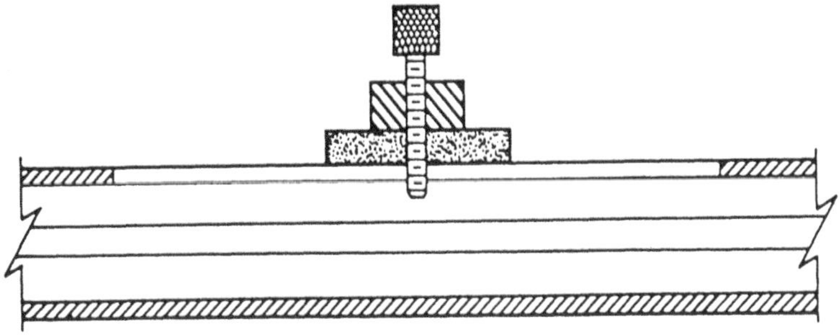

Figure 2.27 Cross Section of Coaxial Slide Screw Tuner

(From *Microwave Theory and Applications* by Stephen F. Adam. Copyright 1969 by Prentice-Hall. Reproduced with permission.)

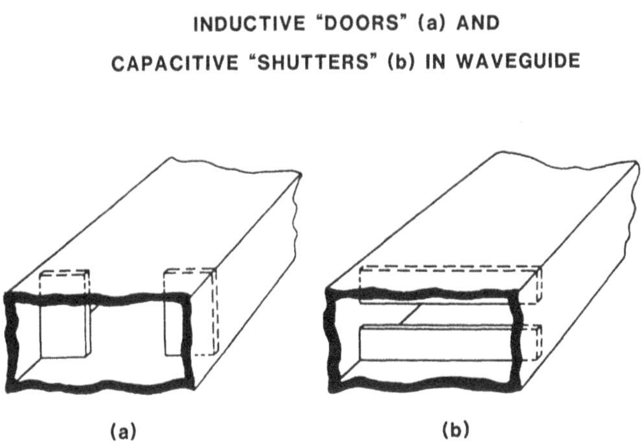

Figure 2.28 Commonly Used Reactances in Waveguides

have the effect of reducing these inductances. The capacitive shutter reduces the height of the waveguide, and thus increases capacitance.

2.6 RESONANT CAVITIES

Resonant cavities are commonly used in microwave tubes, especially klystrons and magnetrons. The circuit models frequently used for resonant cavities are parallel LC circuits as indicated in Fig. 2.29. In fact, at many points in this book, cavities are modeled as parallel resonant LC circuits in order to facilitate discussions or analyses. As shown in Fig. 2.29 the resonant frequency is inversely proportional to the square root of inductance and capacitance. While these quantities are relatively easy to measure in a lumped-element circuit, they are very difficult to obtain for a resonant cavity. The Q (quality factor) of a cavity has the same definition as the Q of a resonant circuit and is basically the ratio of energy stored to the energy lost by dissipation. The Q of a circuit is also related to the bandwidth, Δf, through

$$\frac{\Delta f}{f_0} = \frac{1}{Q} \tag{2.24}$$

where f_0 is the center frequency of oscillation. Bandwidth and Q are both quantities of considerable importance in microwave tube cavities.

While the operation of resonant cavities can be approached from a circuit point of view, it can just as well be based on the knowledge of waves on transmission lines discussed previously in this chapter. For example, as shown in Fig. 2.22, at each half-wavelength from a short circuit on a transmission line, the voltage and current correspond to short-circuit values. That is, the voltage is zero and the current is finite. At any one of these points, a short circuit could be placed on the transmission line without affecting the voltage and current patterns (as long as the line is lossless).

A short circuit at a position $\lambda/2$ from the short at the end of the line, for example, would have the effects of trapping energy between the short and the end of the line and of reflecting energy from the generator. There would be no change of any part of the standing-wave pattern. The trapped energy between the short at $\lambda/2$ and the short at the end of the line could be thought of as energy that is bouncing back and forth. In reality, the equal-amplitude waves propagating in opposite directions are components of a single standing wave in the rectangular box-shaped resonant cavity formed by the two short circuits on the rectangular waveguide.

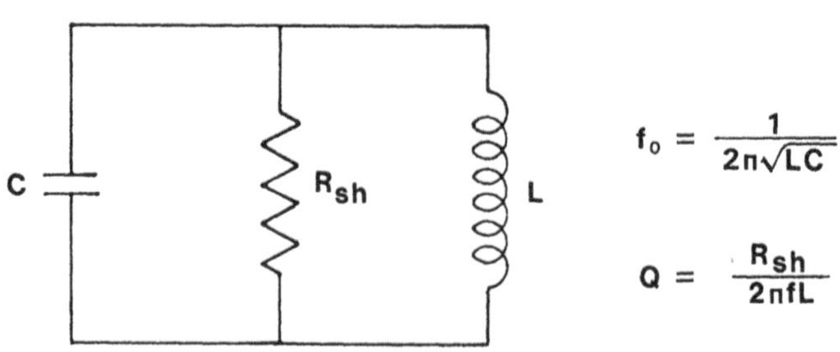

Figure 2.29 Circuit Model for Resonant Cavity

It should be noted at this point that a short circuit could be placed at any number of half-wavelengths from the short at the end of the line in Fig. 2.22, and the effect would be the same as for a short at $\lambda/2$.

Similarly, for a short at any given position, the frequency could be varied so that any number of half-wavelengths could fit between the short and the end of the line. Thus, the resonant cavity formed by the two shorts on the transmission line can have any number of resonant frequencies (modes of oscillation). The mode designation for a cavity is established by using the designation for a waveguide and adding an index that gives the number of waveguide half-wave variations of the field along the length of the cavity. Thus, for rectangular waveguide operating in the TE_{10} mode, if short circuits are separated by one-half waveguide wavelength, the resulting cavity mode is TE_{101}. For the corresponding situation is circular waveguide, the mode is TE_{111}.

Resonant cavities can have a wide range of shapes. Most of the cavities used in the fabrication of klystrons and TWTs are circular, while those used in magnetrons may have any of several configurations. Specific cavity configurations are discussed in detail along with the microwave tubes in which they are used.

REFERENCES

For additional reading, the following are recommended.

[2.1] Theodore Moreno, *Microwave Transmission Design Data,* New York: Dover Publications, 1958.

[2.2] Simon Ramo, John R. Whinnery, and Theodore Van Duzer, *Fields and Waves in Communication Electronics,* New York: John Wiley and Sons, 1984.

[2.3] Hugh Hildreth Skilling, *Fundamentals of Electric Waves,* New York: John Wiley and Sons, 1948.

[2.4] M. Marcuvitz, *Waveguide Handbook,* New York: McGraw-Hill, 1951.

[2.5] Stephen F. Adam, *Microwave Theory and Applications,* Englewood Cliffs, NJ: Prentice-Hall, 1969.

Chapter 3

Vacuum Techniques

Before discussing microwave tubes in detail, it is important to examine the harsh environment that must exist within the tubes if they are to operate properly. A high or ultra-high vacuum must be maintained throughout the life of the tube. As a result, special techniques, processes, and materials must be used.

The emphasis in this chapter is on vacuum techniques. Subjects to be discussed are:

- Units of measurement
- Ranges of operation
- Pumping techniques
- Vacuum measuring techniques
- Microwave tube materials
- Fabrication techniques

3.1 UNITS OF MEASUREMENT

Several years ago, the most commonly used unit of pressure was the millimeter of mercury (mmHg). Atmospheric pressure is 760 mmHg.

Now, the accepted units of pressure are the Torr and the Newton/m^2 (Pascal). The relation between these units is

1 Torr = 1mmHg = 1.33×10^2 Pascal

Another unit of pressure sometimes used at relatively low vacuum (high pressure) levels is the micron:

1 micron = 10^{-6} meters of Hg = 10^{-3} mmHg $\approx 10^{-3}$ Torr

Still another unit of pressure sometimes used in European literature is the mbar (millibar). Formally, 1.33 mbar is equal to 1 Torr.

In microwave tube work, it is rarely and perhaps never necessary to know the pressure to an accuracy better than a factor of two or so. As a result, it is safe to remember that

1 Torr = 1 mmHg $\cong$ 1 mbar

3.2 RANGES OF OPERATION

In evacuating a microwave tube, the pressure changes from atmospheric to a value as low as 10^{-8} to 10^{-9} Torr. Thus, the pressure is reduced by over eleven orders of magnitude.

It is common to divide this very wide pressure variation into the following four ranges:

760 Torr to ~ 1 Torr — Rough vacuum
1 Torr to ~ 10^{-3} Torr — Medium vacuum
10^{-3} Torr to ~ 10^{-7} Torr — High vacuum
< 10^{-7} Torr — Ultra-high vacuum

In most electron tubes, the pressure is at the low end of the high vacuum range or the high end of the ultra-high vacuum range, i.e., from 10^{-7} to 10^{-9} Torr. In some cases, for example, some high power klystrons that have been in operation for more than 50,000 hours, the pressure is even lower.

Listed in Figs. 3.1 and 3.2 are the pressure ranges for several other vacuum applications.

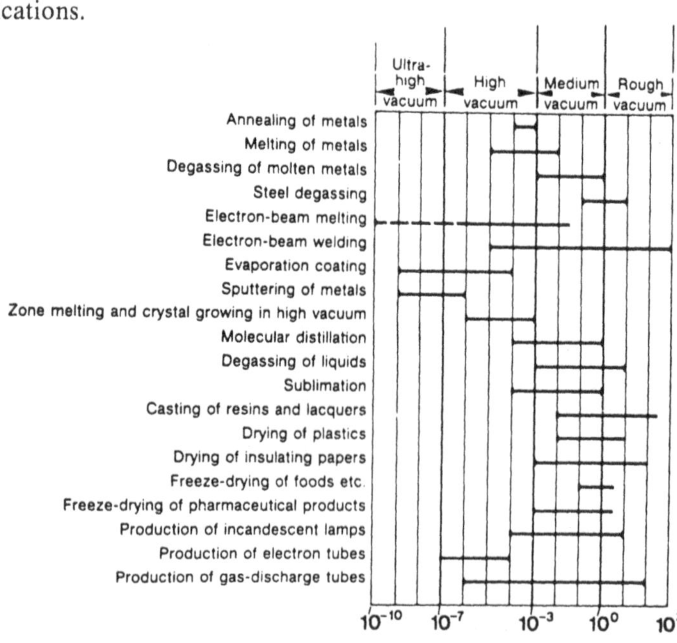

Figure 3.1 Pressure Ranges for Various Vacuum Applications

(Courtesy of Leybold Heraeus, Inc.)

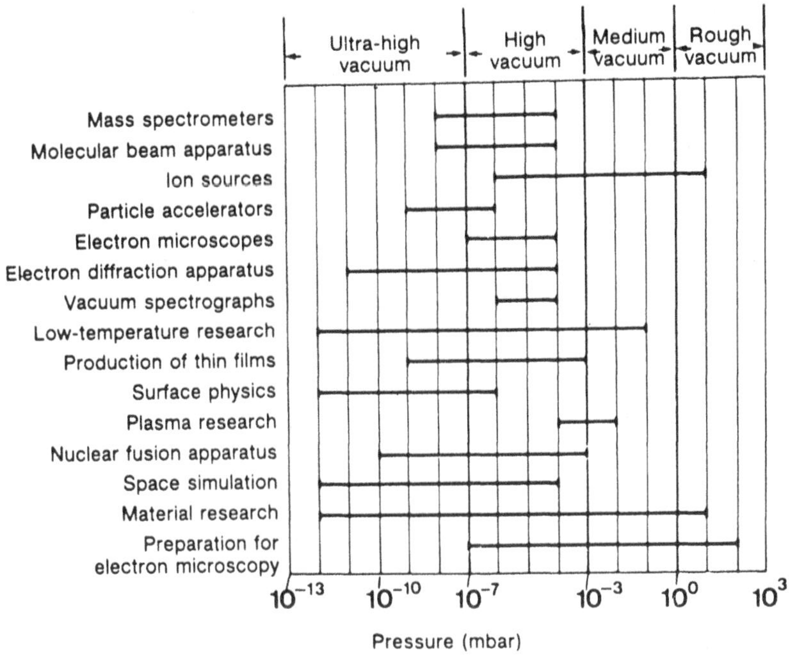

Figure 3.2 Pressure Ranges for Various Vacuum Applications

(Courtesy of Leybold Heraeus, Inc.)

3.2.1 Gas Density and Mean Free Path Length [3.1]

At atmospheric pressure, the gas density is about 2.7×10^{19} molecules/cm^3. As shown in Fig. 3.3, density drops linearly with decreasing pressure and is about 3×10^8 molecules/cm^3 at 10^{-8} Torr. While this density appears to be high, 3×10^8 molecules will form a monolayer on an area of only 10^{-6} cm^2.

Mean free path length is the mean distance traveled by a molecule between collisions. Figure 3.4 shows the mean free path length for air from the rough- to the ultra-high-vacuum range.

Notice that at 10^{-8} Torr, the mean free path length is $\sim 10^4$ meters. Thus, the average molecule could pass through a one-meter vacuum chamber 10,000 times without colliding with another molecule.

At 1 Torr, on the other hand, the mean free path length is only $\sim 10^{-4}$ meters. Thus, molecules collide with each other far more frequently than they collide with the walls of a vacuum system.

A relatively high pressures, where molecules collide with each other far more frequently than with vacuum chamber walls, gas flow is referred to as *viscous flow*.

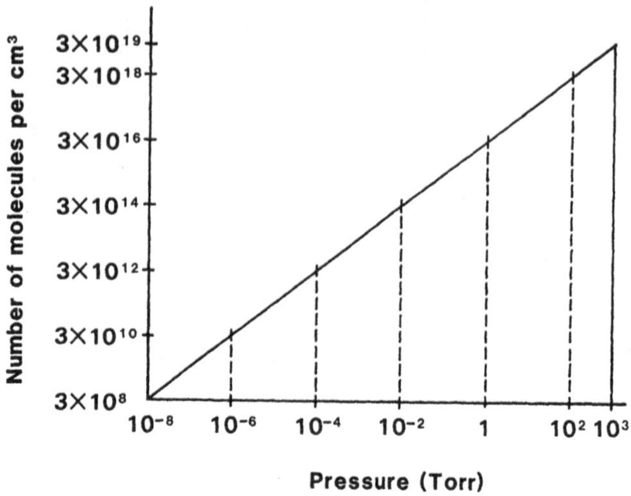

Figure 3.3 Gas Density as a Function of Pressure

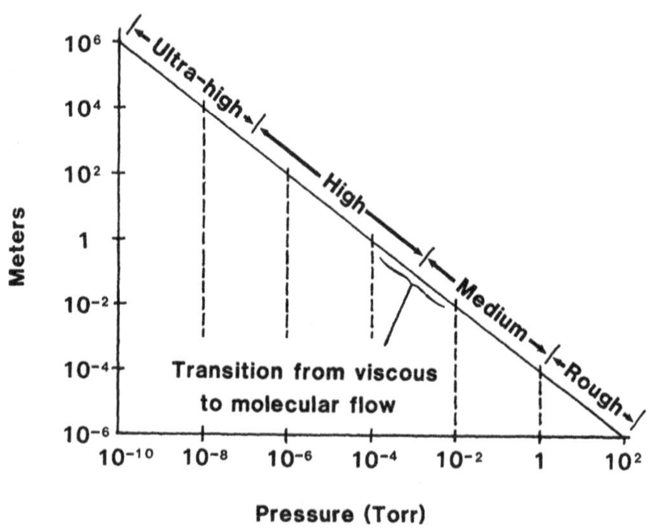

Figure 3.4 Mean Free Path Length as a Function of Pressure

At low pressures, where molecules collide with vacuum system walls more frequently than with each other, gas flow is referred to as *molecular flow*. In the viscous-flow and molecular-flow regions, the analysis of gas flow is relatively straightforward. In the pressure range from ~ 10^{-2} Torr to 10^{-4} Torr, a transition from viscous to molecular flow occurs, and gas flow behavior is difficult to analyze.

The effects of viscous and molecular flow in vacuum-system design are illustrated in Fig. 3.5. In the viscous-flow region, molecules gradually migrate in the direction of the pump because of the change in pressure. In the case of molecular flow, molecules collide only with the walls of the chamber and only occasionally are directed toward the pump orifice. As a result, a large pump orifice is required if a high pump speed is to be achieved.

For example, if 1-inch (2.5 cm) tubulation were used to connect the vacuum system to the pump, the speed would be limited to a few liters per second. If a 6-inch (15 cm) diameter orifice were used, the speed could be well over 1000 liters per second.

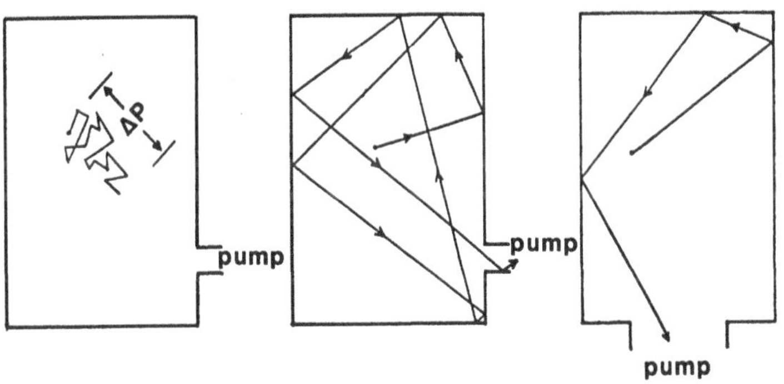

a) Viscous flow b) Molecular flow (small aperture pump) c) Molecular flow (large aperture pump)

Figure 3.5 Comparison of Viscous and Molecular Flow

3.2.2 Fundamentals of Vacuum Systems

To understand the fundamentals of vacuum systems, it is useful to examine analogous electrical circuits. A simple vacuum system and the corresponding electrical circuit are shown in Fig. 3.6.

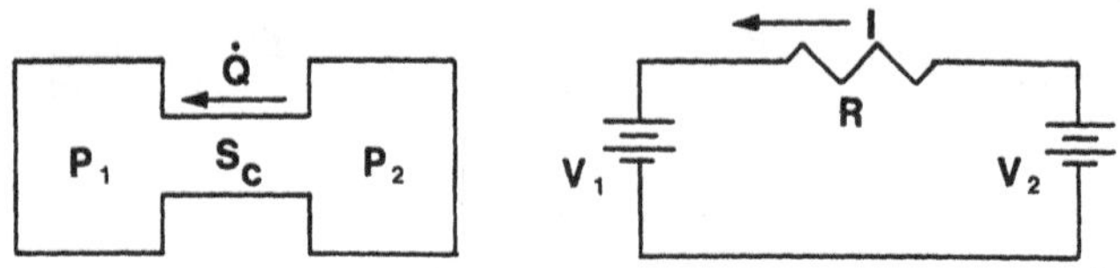

$$P_2 - P_1 = \frac{\dot{Q}}{S_c} \qquad V_2 - V_1 = IR$$

P = pressure, Torr — V = voltage, Volts

$\dot{Q}$ = gas flow rate, T l/s — I = charge flow rate, C/s

S_c = conductance, l/s — R = resistance, Ohms

Figure 3.6 Analogy Between a Simple Vacuum System and an Electrical Circuit

Note that pressure corresponds to voltage, gas flow rate (in Torr liters per second) corresponds to charge flow rate (in Coulombs per second), and conductance (in liters per second) corresponds inversely to resistance.

A vacuum pump may be modeled as shown in Fig. 3.7. The ultimate pressure of the pump, P_u, corresponds to the open circuit voltage, V_u, of a voltage source. The speed of the pump, S_p, has the same units as conductance (liters per second) and corresponds inversely to the internal resistance of the voltage source, R_p. Vacuum pumps are normally rated by their speed.

In reality, as shown in Fig. 3.8, there is always a constriction between the pump and the volume to be evacuated. The conductance of this constriction corresponds inversely to the resistance of the circuit used to connect a voltage source to load.

In the volume, V_{ev} to be evacuated, the quantity of gas, Q, and the pressure, P, are related by

$$P = \frac{Q}{V_{ev}}$$

where the units are:

P — Torr
Q — Torr-liters
V_{ev} — liters

The corresponding electrical relation is

Figure 3.7 Analogy Between Vacuum Pump and Voltage Source

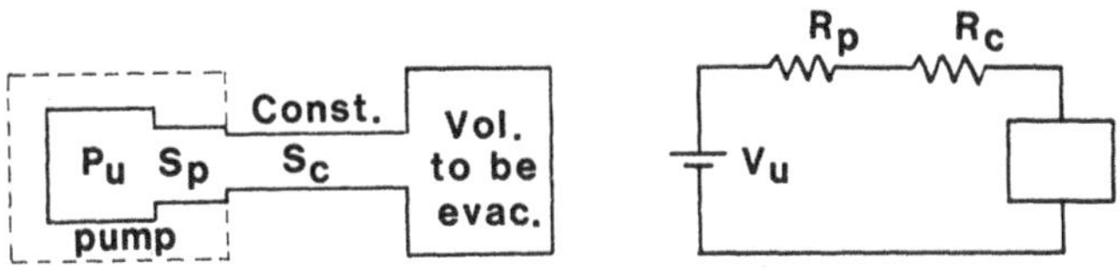

The total conductance is:

$$\frac{1}{S} = \frac{1}{S_p} + \frac{1}{S_c}$$

The total resistance is

$$R = R_p + R_c$$

Figure 3.8 Relation Between Total Conductance and Total Resistance

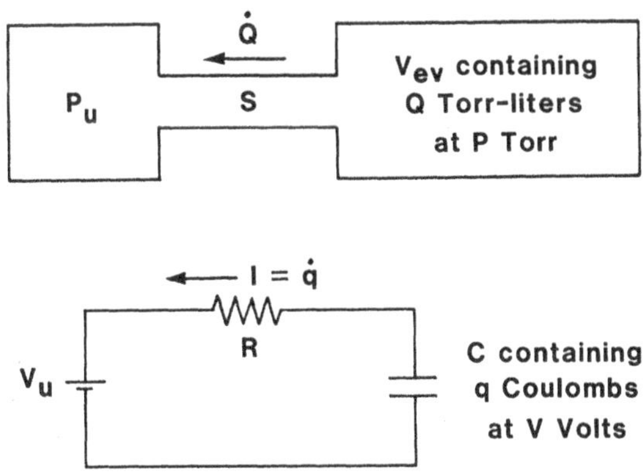

Figure 3.9 Correspondence Between Volume and Capacitance

$$V = \frac{q}{C}$$

where q is charge in Coulombs and C is capacitance in Farads. Thus, the volume to be evacuated, V_{ev}, corresponds to capacitance as shown in Fig. 3.9.

Because of the flow of charge ($\dot{q}$ or I) from the capacitor, V changes with time, i.e.,

$$V - V_u = IR = \dot{q}R$$

but $$q = CV \text{ or } \dot{q} = C\frac{dV}{dt}$$

so $$V - V_u = RC\frac{dV}{dt}$$

or $$\frac{dV}{dt} = \frac{1}{RC}(V - V_u)$$

The quantity RC is, of course, the time constant for the electrical circuit.

Correspondingly, because of the flow of gas from the volume, V_{ev}, P changes with time and is given by

$$\frac{dP}{dt} = \frac{S}{V_{ev}}(P - P_u)$$

The quantity V_{ev}/S is the time constant for the vacuum system.

At equilibrium, $dP/dt = 0$, and so the equilibrium pressure is

$$P_e = P_u$$

In reality, there are always leaks or "outgassing" in V_{ev}. resulting in gas flow, $\dot{Q}_{lo}$, into the system as indicated by Fig. 3.10.

At equilibrium $\dot{Q} = \dot{Q}_{lo}$ and the equilibrium pressure is $P_e = P_u + \dot{Q}_{lo}/S$. Frequently, at very low pressures, the equilibrium pressure is determined by the rate of gas flow into the system, $\dot{Q}_{lo}$, rather than by the ultimate pressure of the pump.

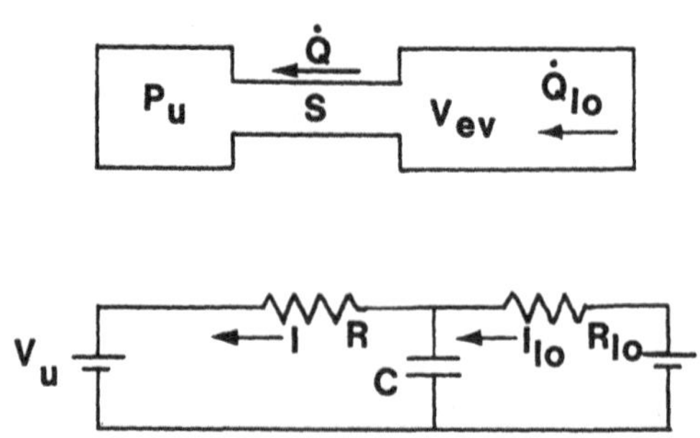

Figure 3.10 Illustration of Leakage of Gas into a System

In the processing of microwave tubes, great care is taken to minimize $\dot{Q}_{lo}$ by making sure that the envelope of the tube if free of leaks and by outgassing the internal parts of the tube. Outgassing is accomplished by baking the tube at a very high temperature (usually greater than 450C) for several hours.

3.3 PUMPING TECHNIQUES

As shown in Fig. 3.11, many pumping techniques exist. The pressure ranges at which the various pumps operate are given in Fig. 3.12.

The pumping techniques most commonly used in the fabrication of microwave tubes are:

Gas Transfer

1. Sliding-vane rotary pump — Rough and medium vacuum
2. Diffusion pump — High and ultra-high vacuum

Entrapment

1. Adsorption pump — Rough and medium vacuum
2. Getter ion pump — High and ultra-high vacuum

Turbomolecular pumps are in the gas transfer category and are being used in some applications. In addition, we will discuss the use of getters to maintain a high vacuum level once it has been established. Nonevaporable getters have recently found application in microwave tubes.

In the gas transfer pump category, the diffusion pump and the turbomolecular pump require an exhaust pressure in the medium vacuum range. The sliding-vane rotary pump is used in conjunction with the diffusion and turbomolecular pumps to maintain this medium vacuum level. The rotary pump used in this way is frequently referred to as a backing pump.

In the entrapment pump category, the ion pump requires a pressure in the low end of the medium vacuum range before it can be started. An adsorption pump is normally used to generate this medium vacuum level. Once an ion pump has been started, it can operate without assistance from other pumps.

3.3.1 Sliding Vane Rotary Pump

The essential elements of a sliding-vane rotary pump are illustrated in Fig. 3.13. A rotating cylinder containing spring-loaded vanes transfers gas from the inlet port to the output port.

A high-grade vacuum oil is used in the assembly containing the rotating cylinder and vanes. The oil serves as a lubricant and also provides a vacuum seal between moving parts.

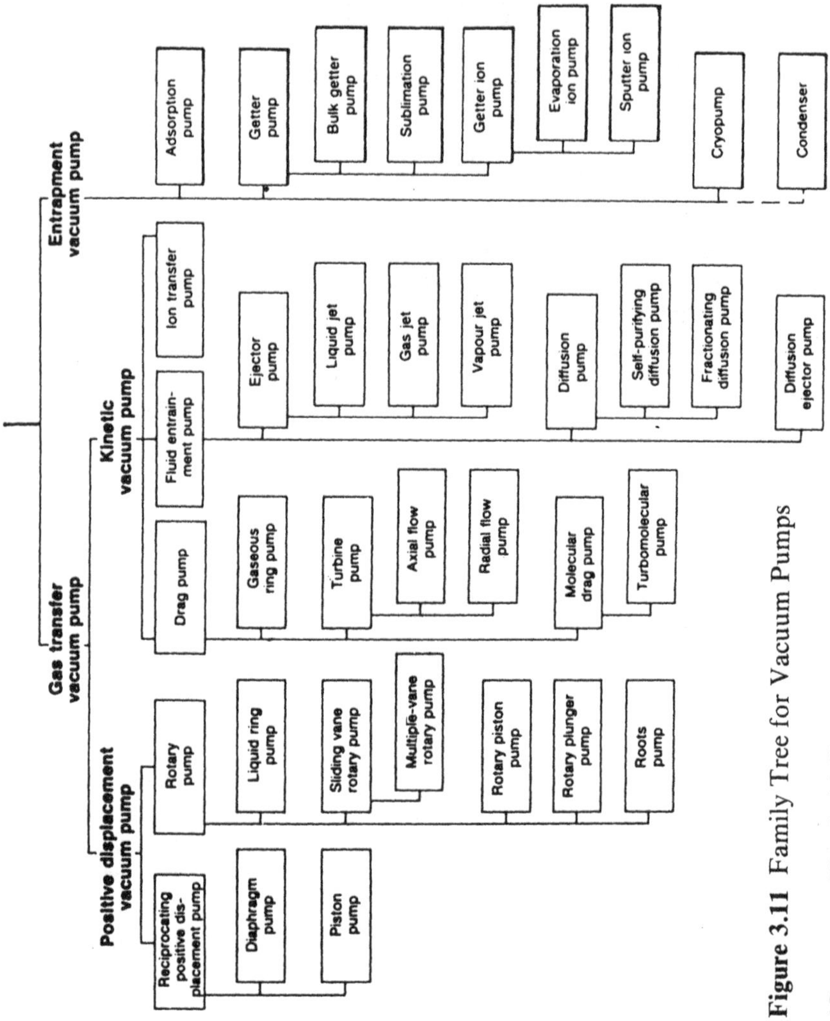

Figure 3.11 Family Tree for Vacuum Pumps

(Courtesy of Leybold Heraeus, Inc.)

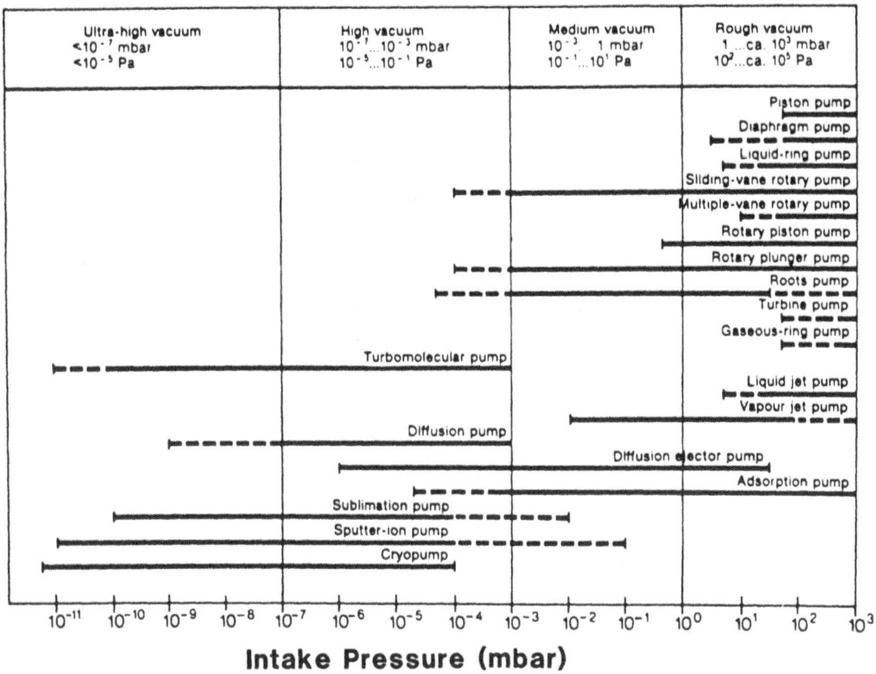

Figure 3.12 Working Pressure Ranges of Vacuum Pumps

(Courtesy of Leybold Heraeus, Inc.)

Two-stage sliding-vane pumps, as illustrated in Fig. 3.14, are sometimes used for improved ultimate pressure capability. Typical pumping speed curves for two-stage pumps are shown in Fig. 3.15.

A problem with mechanical pumps is the pumping of condensable vapors such as water. If these vapors condense during the compression cycle of pump operation, they may emulsify with the pump oil and cause deterioration of the lubricating properties. This problem may be avoided with the use of a gas ballast device that operates during the compression cycle.

Figure 3.16 illustrates the use of a gas ballast. In a_1 through a_4, gas and vapors enter the pump (a_1), are compressed (a_3) causing condensation, and are partially exhausted (a_4).

In b_1 through b_4, a gas ballast is used to inject additional air before compression (b_2). With the additional air, the discharge valve is forced open (b_3) causing gas and vapor to escape (b_4) before condensation can occur.

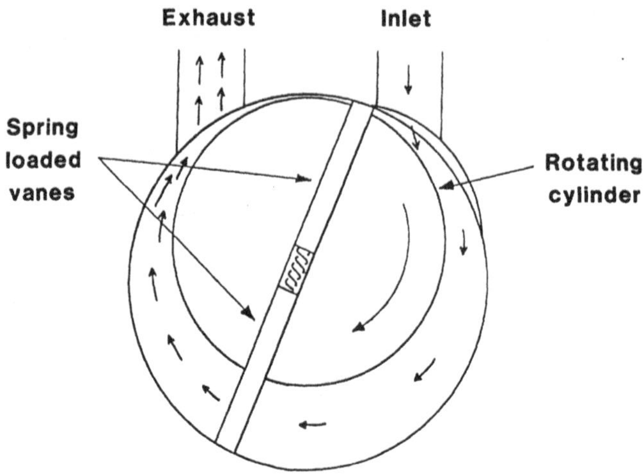

Figure 3.13 Essential Elements of a Sliding-Vane Rotary Pump

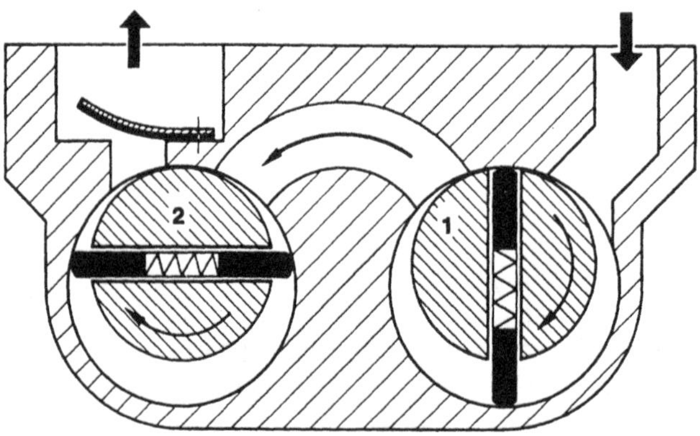

Figure 3.14 Two-Stage Sliding-Vane Rotary Pump

(Courtesy of Leybold Heraeus, Inc.)

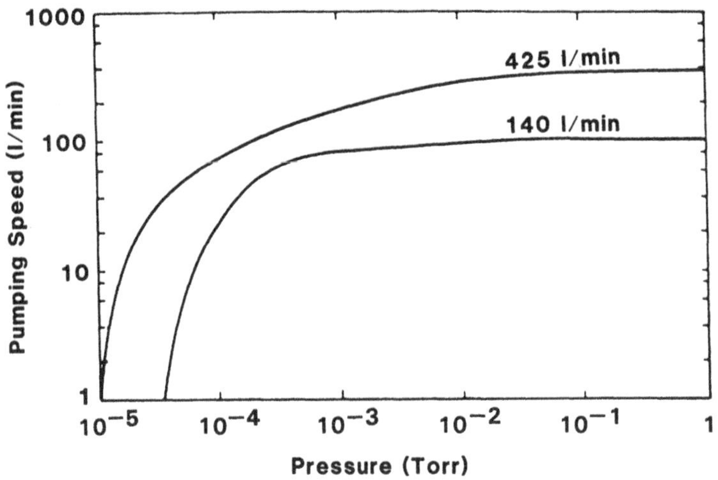

Figure 3.15 Pumping Speed Curves for Two-Stage Pumps

(From *Handbook of Thin Film Technology* by Maissel and Glang, Copyright 1970 by McGraw-Hill, Inc. Reproduced with permission.)

3.3.2 Diffusion Pump

A diffusion pump is normally used in conjunction with a rotary pump to achieve pressures in the high or ultra-high vacuum range. An oil-filled diffusion pump is shown in Fig. 3.17.

A very high quality silicon oil is vaporized by a heating element in the bottom of the pump. The oil vapors are formed into jets by a chimney structure.

Collisions of oil molecules with gas molecules cause the gas molecules to diffuse toward the exhaust port. A pressure difference (between inlet and exhaust) of approximately six orders of magnitude can be produced with a diffusion pump, which is normally water cooled. The condensed oil returns to the oil bath in the bottom of the pump.

Typical speed and throughput for a diffusion pump are shown in Fig. 3.18. Notice that speed is constant over a wide range of pressures.

At very low pressures, pressure is limited by diffusion of oil molecules back into the chamber being pumped. This is frequently referred to as "backstreaming."

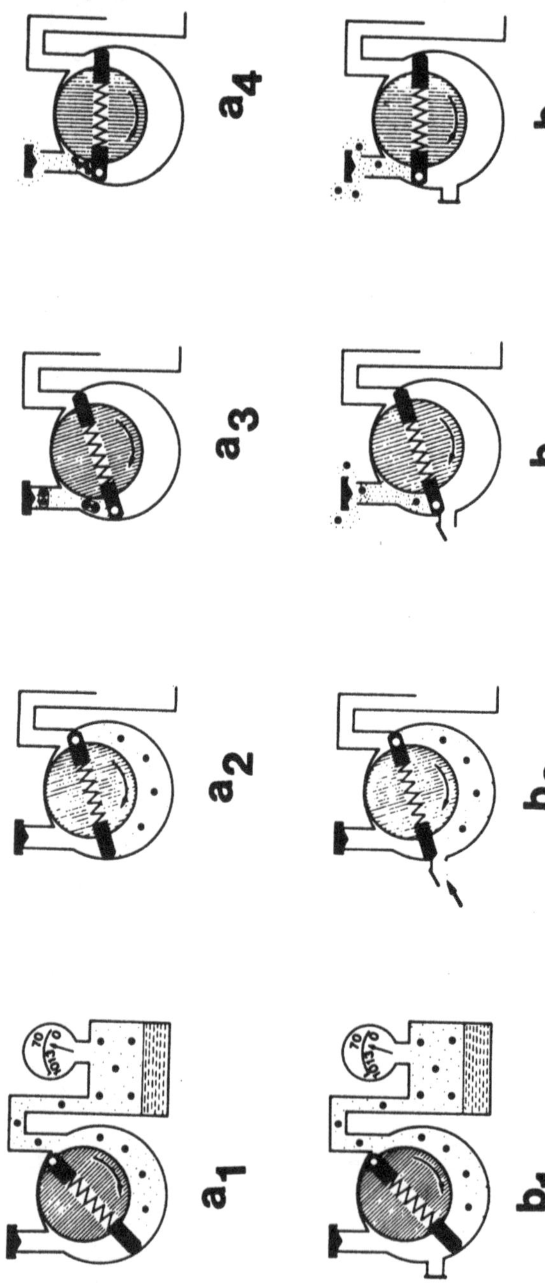

Figure 3.16 Illustration of Pumping Process in a Rotary Vane Pump without (Top) and with (Bottom) Gas Ballast

(Courtesy of Leybold Heraeus, Inc.)

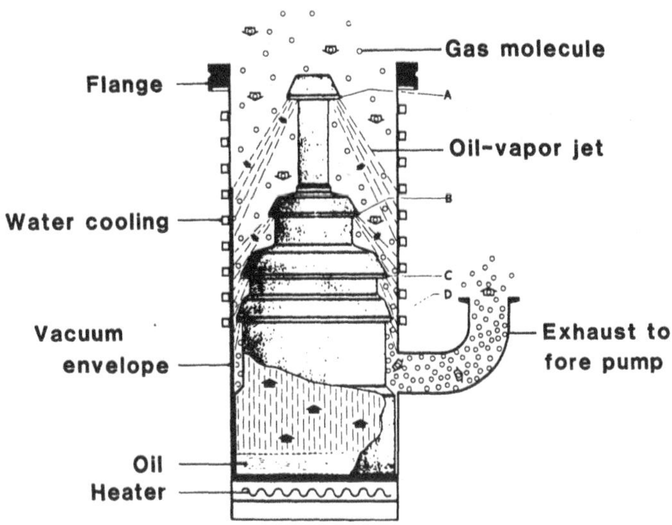

Figure 3.17 Oil-Filled Diffusion Pump

(Courtesy of Leybold Heraeus, Inc.)

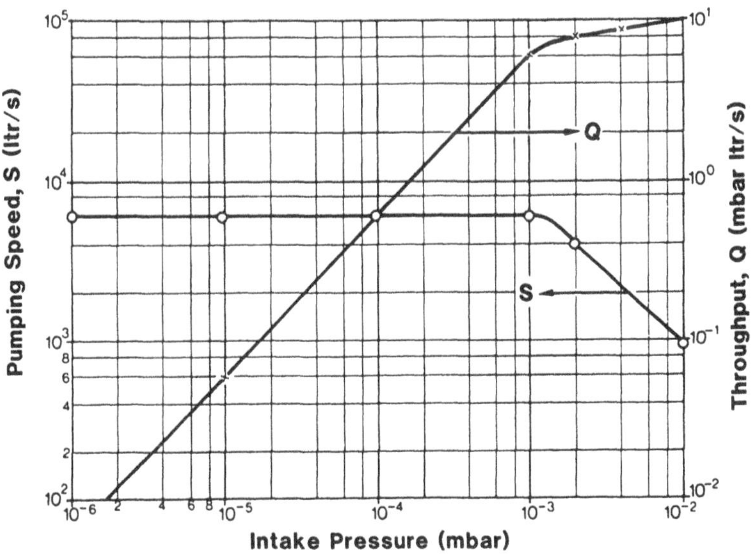

Figure 3.18 Typical Plots of Pumping Speed, S, and Throughput, Q, for a Diffusion Pump

(Courtesy of Leybold Heraeus, Inc.)

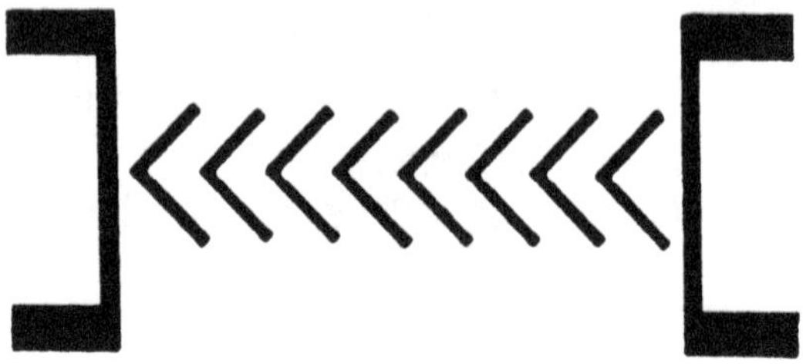

Figure 3.19 Chevron-Type Baffle

A baffle is sometimes used between the pump and the chamber to limit backstreaming. Metallic elements are arranged as indicated in Fig. 3.19 so that there is no line-of-sight path through the baffle. Baffles are normally cooled, sometimes by liquid nitrogen, to enhance the condensation of oil vapors. With a baffle it is sometimes possible to reduce the ultimate pressure by over two orders of magnitude.

3.3.3 Turbomolecular Pump

In the turbomolecular pump, a rotor, spinning at a high speed, imparts a tangential velocity to impinging molecules. As indicated in Fig. 3.20, many rotor discs are used. The gas channels for the discs are connected in series, so that, even though the pressure gradient generated by each disc is small, the pump is capable of maintaining a high pressure ratio. At 16,000 rpm, the ratio of outlet to inlet pressure is 250:1 for hydrogen and 10^6:1 for air. The recommended fore pressure is 10^{-2} Torr and the ultimate pressure is 10^{-8} to 10^{-9} Torr.

3.3.4 Adsorption Pump

An adsorption pump is capable of evacuating a chamber from atmospheric pressure to the medium vacuum range. A typical pump is shown in Fig. 3.21.

The operating principle is the physical adsorption of gases on the surfaces of molecular sieves. Zeolite, an alkali alumino-silicate, is frequently used.

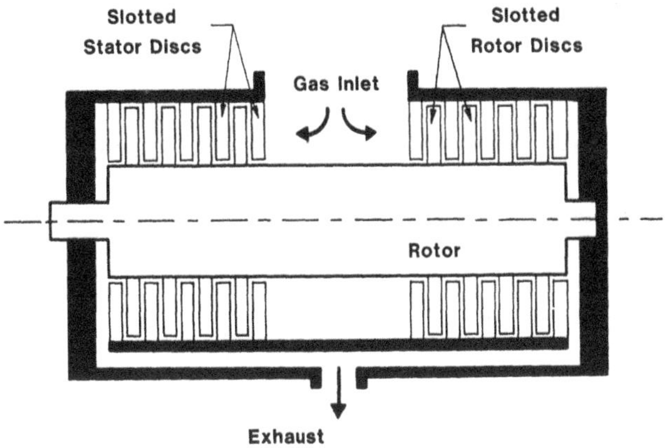

Figure 3.20 Turbomolecular Pump

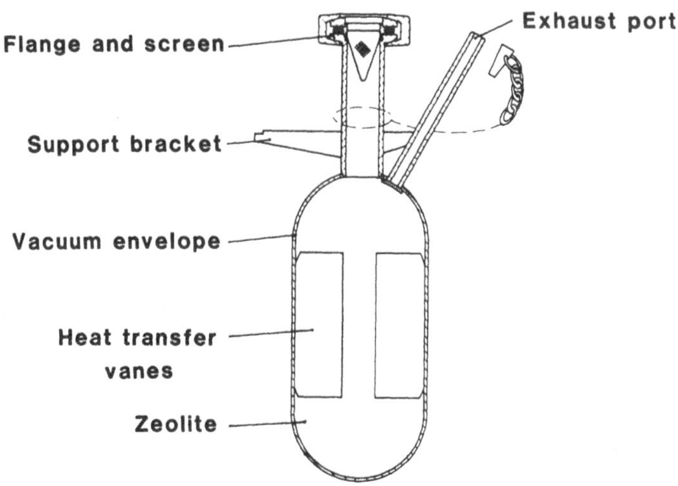

Figure 3.21 Typical Adsorption Pump

(Courtesy of Leybold Heraeus, Inc.)

The surface is about 10^3 m^2 per gram. The pore diameter of zeolite 13X (about 13 Å) is approximately the size of water vapor, oil vapor, and large gas molecules. Large molecules are, therefore, readily adsorbed. Small molecules such as hydrogen, helium, and neon are very poorly adsorbed.

The adsorption of gases is dependent on pressure and temperature as shown in Fig. 3.22. Notice that the difference between the quantity of nitrogen adsorbed and the quantity of neon or helium adsorbed is at least four orders of magnitude.

In practice, hydrogen and helium are normally present in the gases to be evacuated from a microwave tube. This is because a hydrogen atmosphere is used during brazing processes and an inert gas atmosphere containing helium is frequently used for inert gas welding.

Since hydrogen and helium are normally present, it is necessary to be cautious in using an adsorption pump. Because the pump operates mostly in the viscous flow range, hydrogen and helium are dragged along with the gases readily pumped by the zeolite. As long as the pumping process takes place fairly quickly, hydrogen and helium can be pulled into the adsorption pump, and then a valve can be closed before the hydrogen and helium diffuse back into the chamber being pumped.

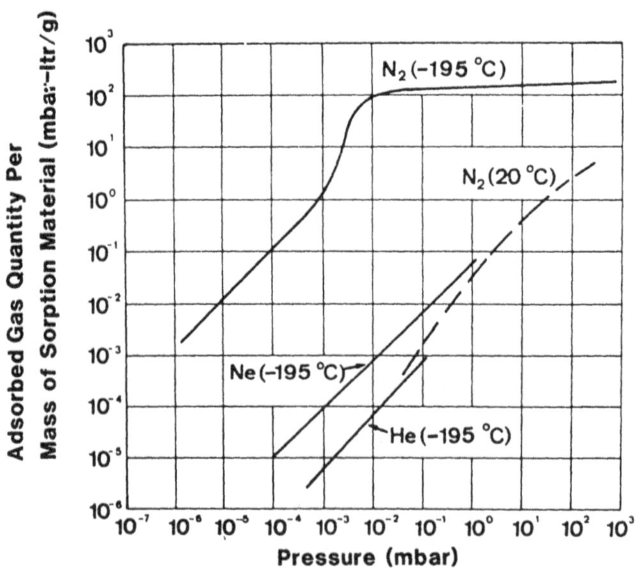

Figure 3.22 Adsorption Isotherms of Zeolite 13X for Nitrogen, Neon, and Helium

(Courtesy of Leybold Heraeus, Inc.)

In practice, adsorption pumps are immersed in liquid nitrogen to achieve large pumping capacities. To regenerate zeolite after pumping, the pump is warmed to room temperature. Gas is released from the zeolite and escapes from the pump through an exhaust port. If a great deal of water vapor or other condensibles has been pumped, it may be necessary to bake the zeolite for a few hours at 200C or more.

3.3.5 Ion Pump

Ion pumps are capable of evacuating a chamber from a pressure of approximately 10^{-2} Torr to the ultra-high vacuum range. The configuration of a diode-type ion pump is shown in Fig. 3.23 and the operation of the pump is indicated in Fig. 3.24.

A cellular anode structure is located between two planar cathodes. A magnetic field is applied normal to the cathode planes and parallel to the cell axes. Electrons spiral around the magnetic field lines and ionize gas molecules. Electrons released during ionization participate in further ionization of other gas molecules. During collisions, electrons give up energy and gradually migrate to the anode. The positive ions are accelerated to the negative electrode, which is fabricated from titanium.

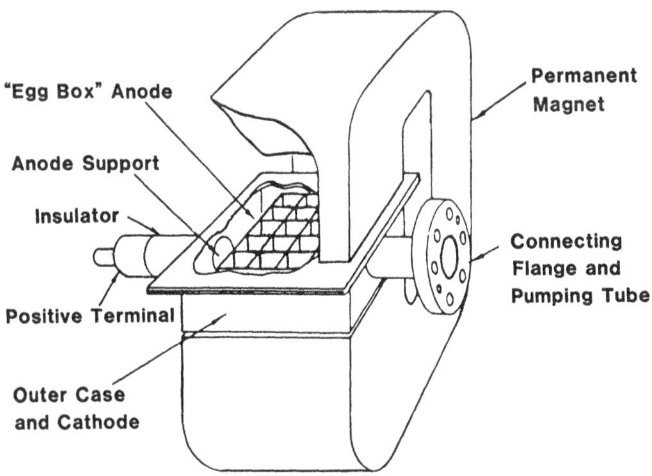

Figure 3.23 Typical Ion Pump

(From *Power Travelling-Wave Tubes* by J.F. Gittins, Published 1965 by American Elsevier Publishing Company, Inc.)

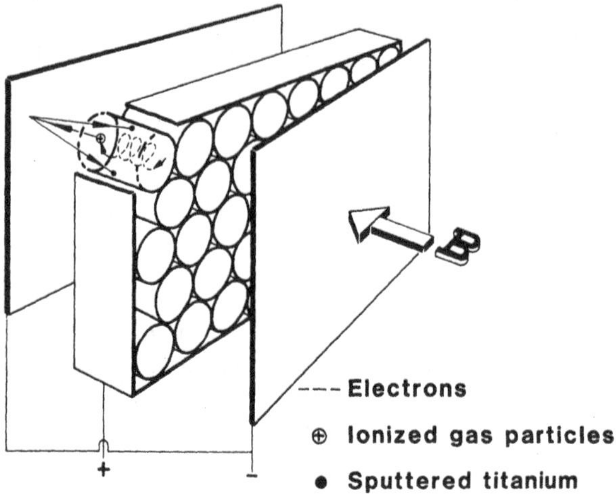

Figure 3.24 Operation of Diode-Type Ion Pump

(Courtesy of Leybold Heraeus, Inc.)

As a result of ion interaction with the titanium, at least three pumping actions occur as are illustrated in Figure 3.25:

1. Ions sputter titanium cathode material, which deposits and acts as a getter film, adsorbing reactive gases (such as nitrogen, oxygen, and hydrogen);
2. The energy of some ions is sufficient for them to be buried in the electrode surfaces where they are adsorbed by ion implantation. This process is effective for all types of ions including those of rare or inert gases;
3. Enhanced pumping of noble gases is possible with a triode ion pump (Fig. 3.26) because sputtered titanium from the cathodes buries gas atoms on the target electrode.

The pumping speed of ion pumps depends on the type of gas being pumped. Compared with the pumping speed for air, the speed for other gases is

Hydrogen	150 to 200%	Oxygen	80%
Methane	100%	Argon	30%
Other light hydrocarbons	80 to 120%	Helium	28%

The voltage of operation is typically about 3.5 kV for small pumps and about 7 kV for large pumps. Ion pumps as small as 0.2 l/s are available for

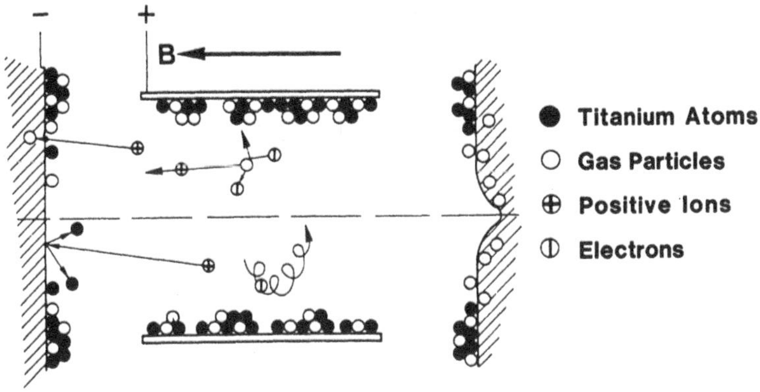

Figure 3.25 Illustration of Pumping Processes in Diode-Type Ion Pump

(Courtesy of Leybold Heraeus, Inc.)

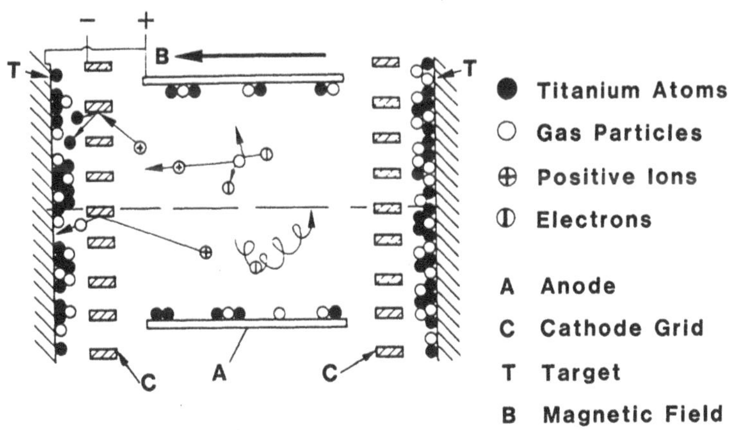

Figure 3.26 Illustration of Pumping Processes in Triode-Type Ion Pump

(Courtesy of Leybold Heraeus, Inc.)

use as appendage pumps, that is, as pumps that are an integral part of a microwave tube. Ion pumps as large as several thousand liters per second are used to pump large vacuum chambers.

As shown in Fig. 3.27, pump current is directly proportional to pressure and, after calibration, may be used as a measure of pressure.

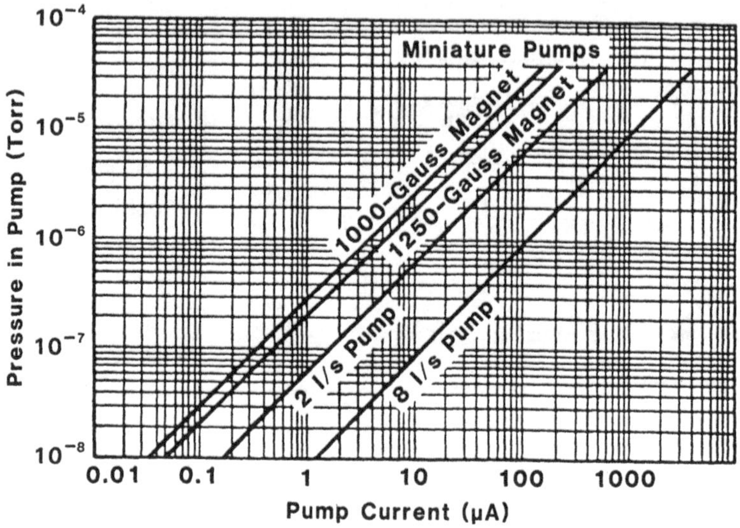

Figure 3.27 Ion Pump Pressure *versus* Current

(Courtesy of Varian Associates)

The magnetic flux density varies somewhat with the size of the pump and is 0.125 Tesla (1250 Gauss) for an 8 l/s vacion pump.

The stray magnetic field of the pump is sometimes of concern in electron beam applications. Typical plots of magnetic flux density are shown in Fig. 3.28.

3.3.6 Getters

The phenomenon of gettering has long been used in vacuum tubes. In conventional getters, a chemically active metal is evaporated in the presence of residual gas. Gases that react with the metal are said to be gettered by the evaporated film. Conventional getters have not found application in high-power microwave tubes primarily because the evaporated film may cause electrical leakage or breakdown in high-voltage structures.

During the past decade, nonevaporable getters have become available. In particular, the ST172 getters (available from SAES GETTERS/USA Inc., 1122 E. Cheyenne Mountain Blvd., Colorado Springs, CO 80906) are very attractive for use in microwave tubes.

The ST172 getters utilize a mixture of zirconium and Zr-V-Fe alloy to form a porous sintered structure. This material has excellent strength and compactness, and is advertised to be completely free from the release of particulates when subjected to shock and vibration.

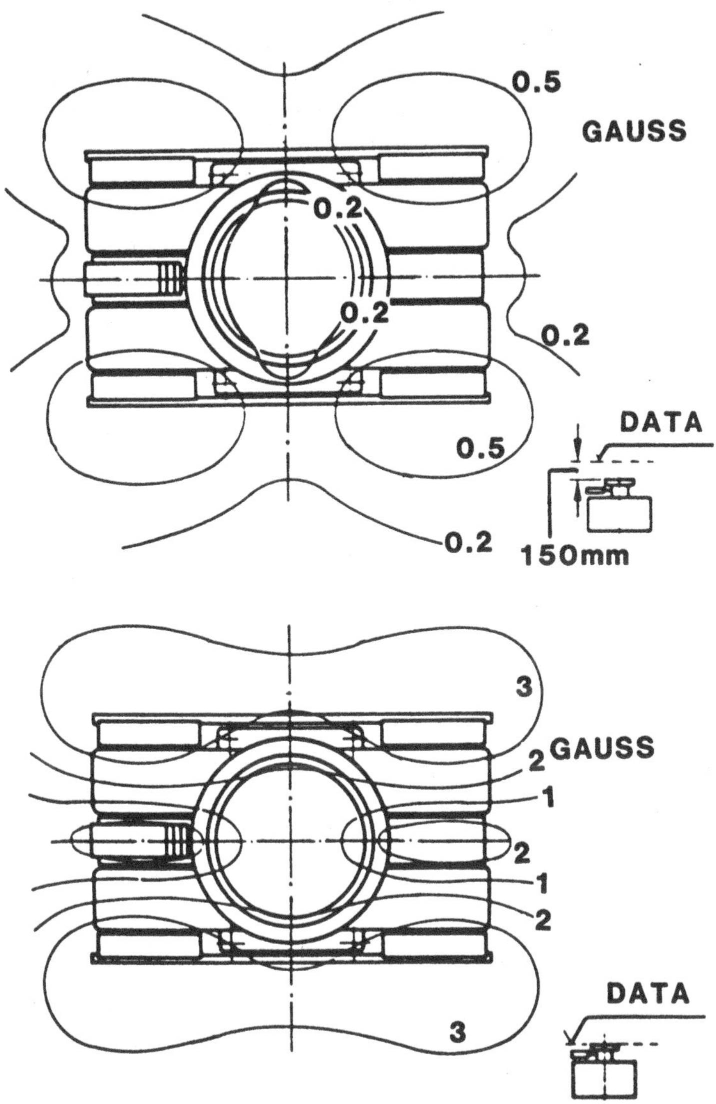

Figure 3.28 Stray Magnetic Flux of an Ion Pump in Two Planes Parallel to the Inlet Flange

(Courtesy of Leybold Heraeus, Inc.)

While optimal results are obtained after activation at 800 to 900° C, this material performs well after activation at 400 to 500° C. Thus, the ST172 getter can be activated during the normal bakeout process of a microwave tube. Figure 3.29 contains illustrations of getters with and without heating elements. The effect of activation temperature on pumping speed and quantity of gas sorbed is shown for carbon monoxide and hydrogen in Fig. 3.30.

At operating temperatures above room temperature, the sorption capacity for CO is increased significantly as shown in Fig. 3.31.

Other gases frequently of interest in microwave tube applications are CO_2, O_2, N_2, water vapor, argon, and helium. The sorption characteristics of ST172 getters for these gases are as follows:

1. CO_2 is about the same as CO;
2. O_2 is somewhat higher than CO;
3. N_2 is approximately one-half of that of CO;
4. Sorption data for H_20 are shown in Fig. 3.32. Note that the increase in capacity resulting from activation at 800° C, rather than 450° C, is a factor of about five;
5. Inert gases such as argon and helium are not sorbed by ST172 getters.

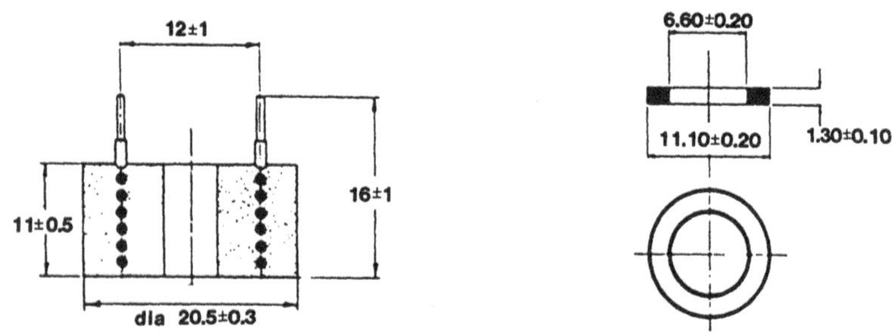

a) St 172/HI/20-10/650 C **b) St 172/OU/11-6.8/200**

Figure 3.29 SAES ST172 Getters (a) With and (b) Without Heating Elements (Other geometries are also available; all dimensions are in millimeters)

(Courtesy of SAES GETTERS/USA, Inc.)

Some of the microwave-tube experience with SAES 172 getters installed in the electron gun is as follows:

1. Appendage ion pump current, which might sometimes be reasonably high, is virtually zero;
2. Arcing in the electron gun is greatly reduced;
3. After idle periods of several days, a tube can be switched on and voltage can be applied immediately without gassing or arcing;
4. Apparently, the getters are recoverable. After activation and then exposure to air, getters have been reactivated by baking at 450 to 500° C.

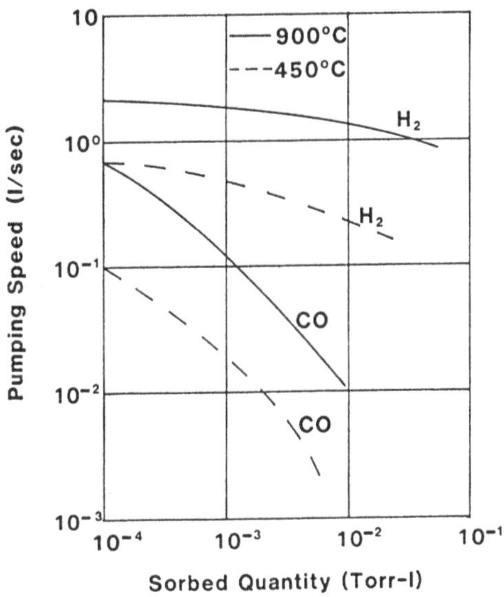

Figure 3.30 Effects of Activation Temperature for SAES ST172 Getter Material (Geometrical Surface Area of 1 cm^2, Activation Time of 10 min, and Operation at 25° C and 3×10^{-6} Torr)

(Courtesy of SAES GETTERS/USA, Inc.)

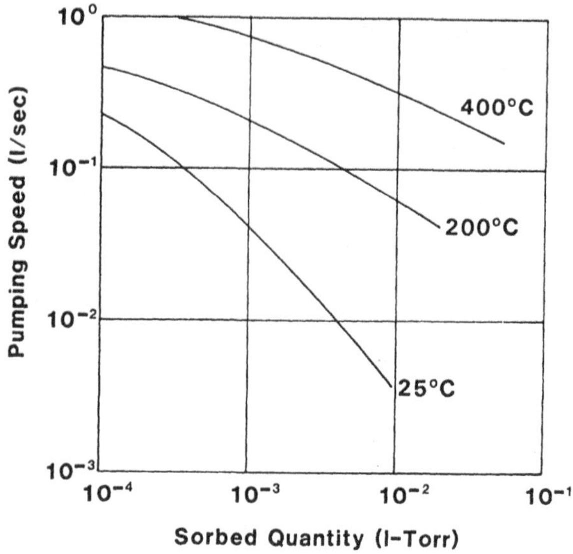

Figure 3.31 CO Sorption Characteristics at Different Temperatures for SAES ST172 Getters

(Courtesy of SAES GETTERS/USA, Inc.)

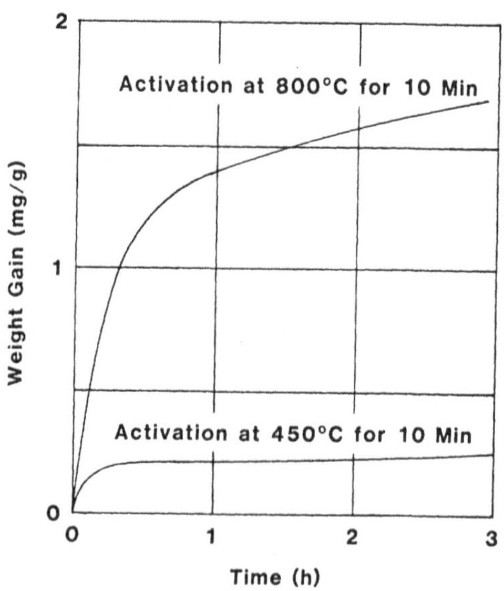

Figure 3.32 Room Temperature Sorption of Water Vapor for ST172 Getters

(Courtesy of SAES GETTERS/USA, Inc.)

3.4 VACUUM MEASURING TECHNIQUES

Many of the techniques available for measuring pressures in the rough- to ultra-high vacuum ranges are listed in Fig. 3.33 along with the pressure ranges in which they operate. Those most commonly used in the microwave tube industry are:

Thermocouple gauge	Rough and medium vacuum
Ionization gauge	High and ultra-high vacuum
Ion-pump current	High and ultra-high vacuum

The use of ion-pump current as a measure of pressure was discussed in the section on pumps. Thermocouple and ionization gauges will be described in this section.

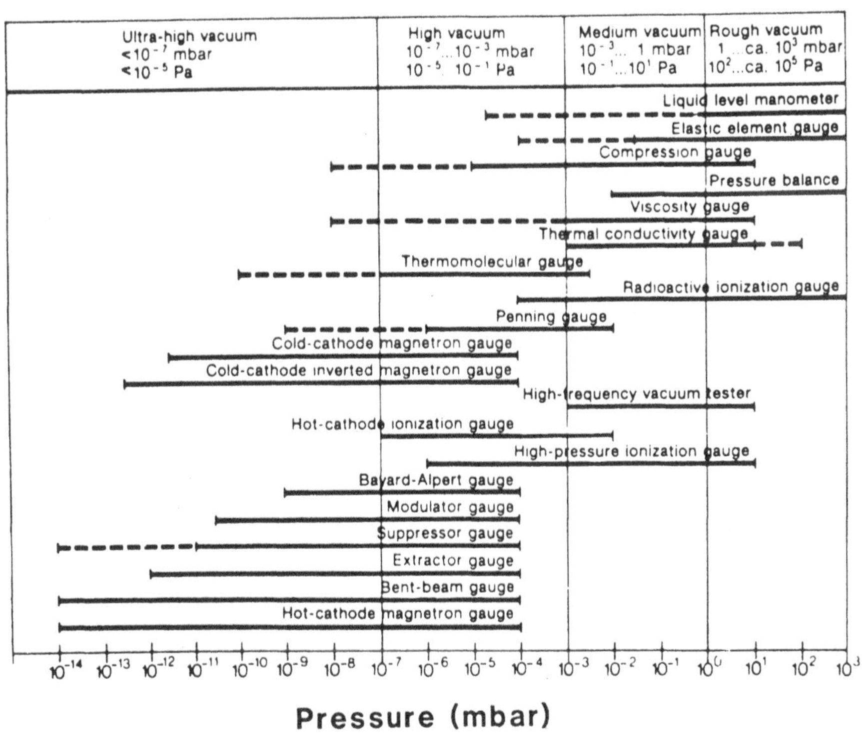

Figure 3.33 Working Pressure Ranges of Various Vacuum Gauges

(Courtesy of Leybold Heraeus, Inc.)

3.4.1 Thermocouple Gauge

The thermocouple gauge is one of several gauges that uses the thermal conductivity of gas to provide the pressure indication. As indicated in Fig. 3.34, a filament of wire heated by an electric current is mounted in an envelope attached to the vacuum system. The temperature of the wire depends on heat loss through the gas, which depends on gas pressure.

The wire temperature is monitored with a thermocouple, which is calibrated to indicate pressure. Thermocouple gauges operate over the pressure range of approximately 10 to 10^{-3} Torr.

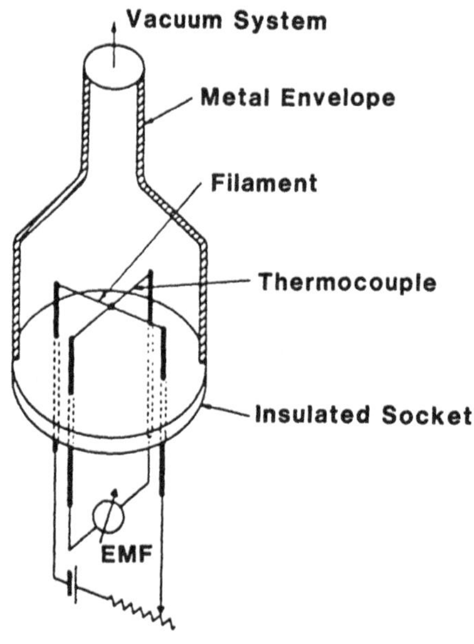

Figure 3.34 Thermocouple Gauge

(From *Handbook of Thin Film Technology* by Maissel and Glang, Copyright 1970 by McGraw-Hill, Inc. Reproduced with permission.)

3.4.2 Ionization Gauge

Ionization gauges of the type shown in Fig. 3.35 are commonly used to measure pressures in the high and ultra-high vacuum ranges.

Electrons, accelerated by a positive grid, produce ions which are collected by an ion collector. The resulting ion current is a measure of gas pressure.

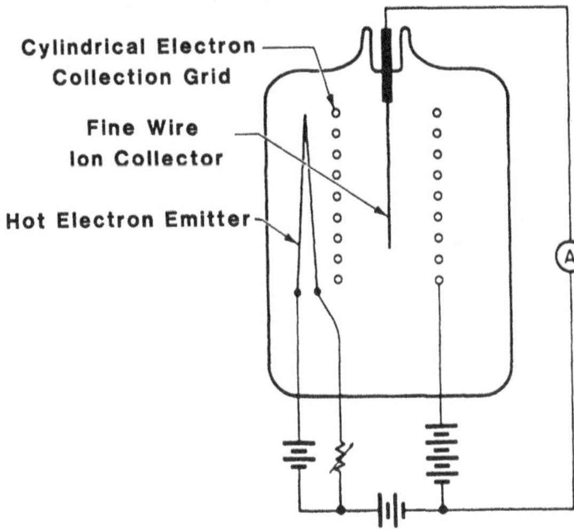

Figure 3.35 Ionization Gauge

(From *Methods of Experimental Physics*, Volume 14, "Vacuum Physics and Technology" by G. L. Weissler and R. W. Carlson, Copyright 1979 by Academic Press, Inc. Reproduced with permission.)

These gauges are sometimes referred to as inverted triode ion gauges, or Bayard-Alpert gauges. In these gauges, a very small diameter wire is used as the ion collector. This is to minimize the x-rays produced when electrons strike the positive grid. Some of these x-rays impinge on the collector wire, causing it to release electrons. The resulting flow of electron current away from the collector causes an error in pressure indication.

3.5 MICROWAVE TUBE MATERIALS

For proper performance, microwave tubes must operate in the high to ultra-high vacuum range. At best, a getter or ion pump may be used as an integral part of a tube to help maintain the vacuum level. At worst, no pump is used and the tube may be required to operate properly after several years in storage.

An extremely important factor in achieving and maintaining an acceptable vacuum level is the selection of the proper materials for the fabrication of a tube. Many of the metallic elements commonly used inside the vacuum envelope of a tube are listed in the chart in Fig. 3.36.

	ALUMINUM	BERYLLIUM	BARIUM	BORON	CALCIUM	CARBON	CHROMIUM	COPPER	GOLD	IRIDIUM	IRON	MANGANESE	MOLYBDENUM	NICKEL	OSMIUM	PALLADIUM	PLATINUM	RHENIUM	RHODIUM	SCANDIUM	SILVER	STRONTIUM	TANTALUM	THORIUM	TITANIUM	TUNGSTEN	VANADIUM	ZIRCONIUM
HEATER	■																									■		
CATHODE	■		■		■					■				■	■					■		■		■		■		
GRIDS								■	■				■													■		
ANODE								■																				
CAVITY								■																				
HELIX								■					■					■								■		
SUPPORT	■	■		■		■																						
ATTENUATOR						■																						
COLLECTOR						■		■																				
ENVELOPE							■	■			■			■														
BRAZES								■	■			■	■			■	■		■		■							■
POLEPIECE											■																	
WINDOW	■	■																										
GETTER			■																						■		■	■

Figure 3.36 Metallic Elements Commonly Used Inside the Vacuum Envelope of Microwave Tubes

(Courtesy of Mr. Frank Welker, Rome Air Development Center.)

The primary factor considered in selecting a metal for use inside the vacuum envelope is vapor pressure. The three charts in Fig. 3.37 give vapor pressure as a function of temperature for the metallic elements.

As an example of the use of these charts, let us consider zinc. At low temperatures, the vapor pressure is reasonable (10^{-9} Torr at 100° C). At a somewhat elevated temperature (400° C), the vapor pressure is only 10^{-1} Torr. This clearly shows why brass (which contains zinc) must not be used in a vacuum system that may be heated. If brass were used and baked at 400° C or higher, zinc vapor would permeate the entire system.

Non-metallic materials such as the neoprene, rubber, and teflon, shown in Fig. 3.38, are sometimes used in demountable vacuum systems. These materials severely limit the pressure and temperature at which the system can operate.

When a non-metallic material (such as an insulator, for example) is required in a system that must be baked, one of a very small number of materials is selected. If the non-metallic material is part of the vacuum envelope, then either aluminum oxide, beryllium oxide, or a glass is usually selected. These materials all have extremely lower vapor pressures. Inside the vacuum envelope, in addition to aluminum oxide or beryllium oxide, boron nitride or, in very special cases, diamond is used.

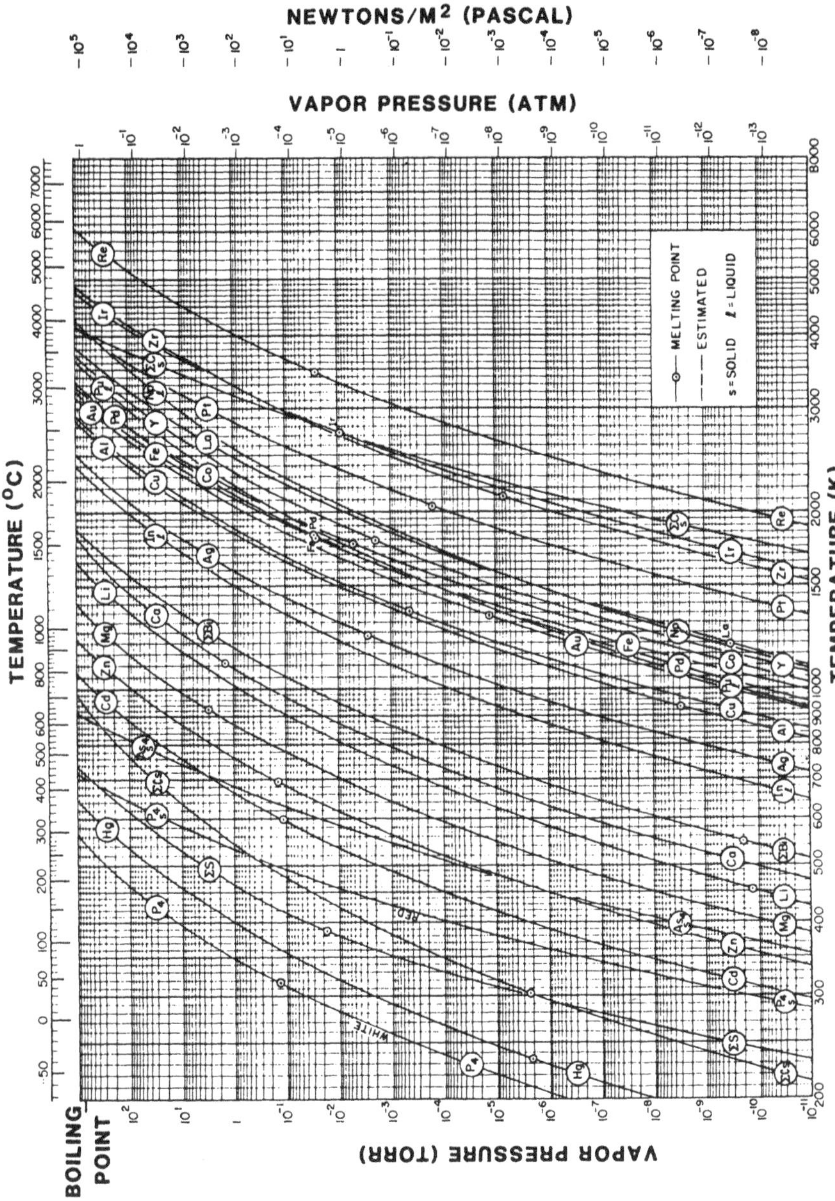

Figure 3.37 Vapor Pressures of the Elements

(From R.E. Honig and D.A. Kramer, *RCA Review*, June 1969, Copyright 1969 by Radio Corporation of America. Reproduced with permission.)

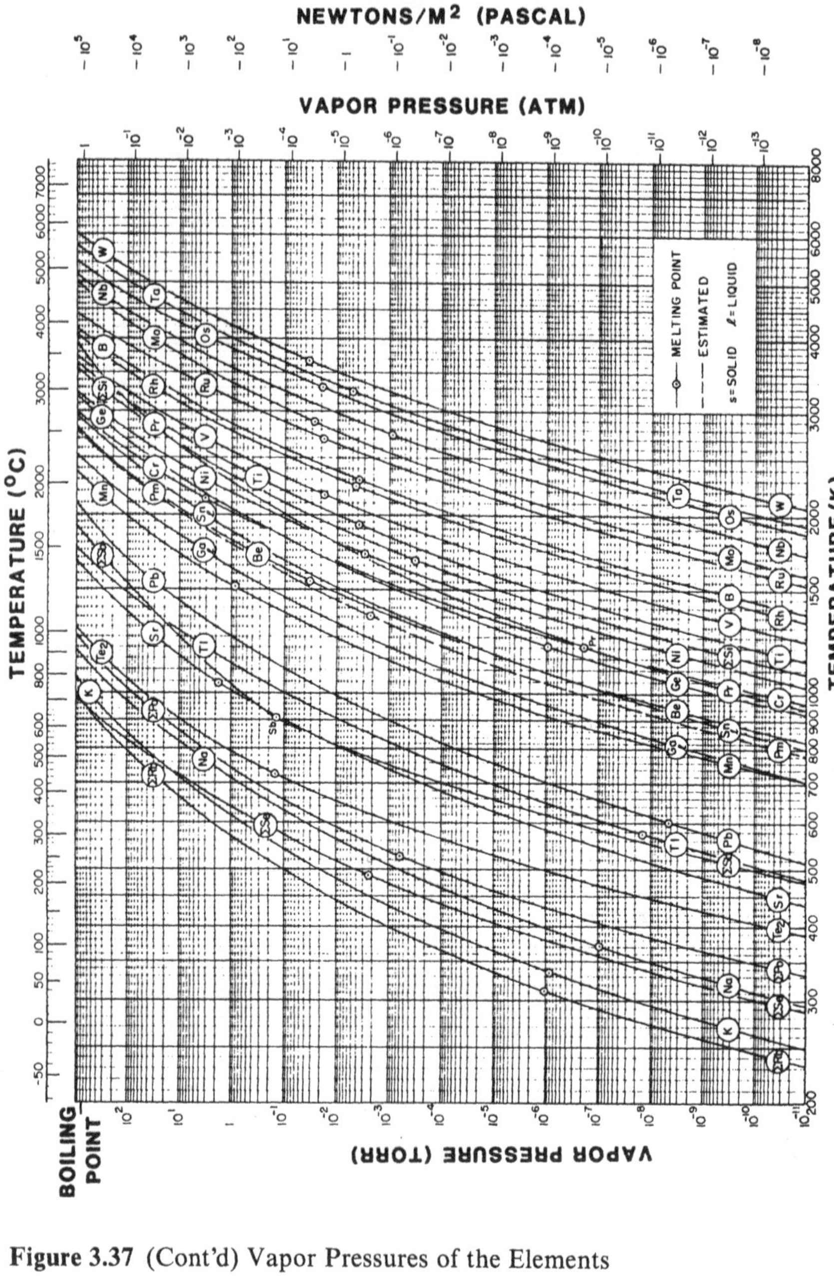

Figure 3.37 (Cont'd) Vapor Pressures of the Elements

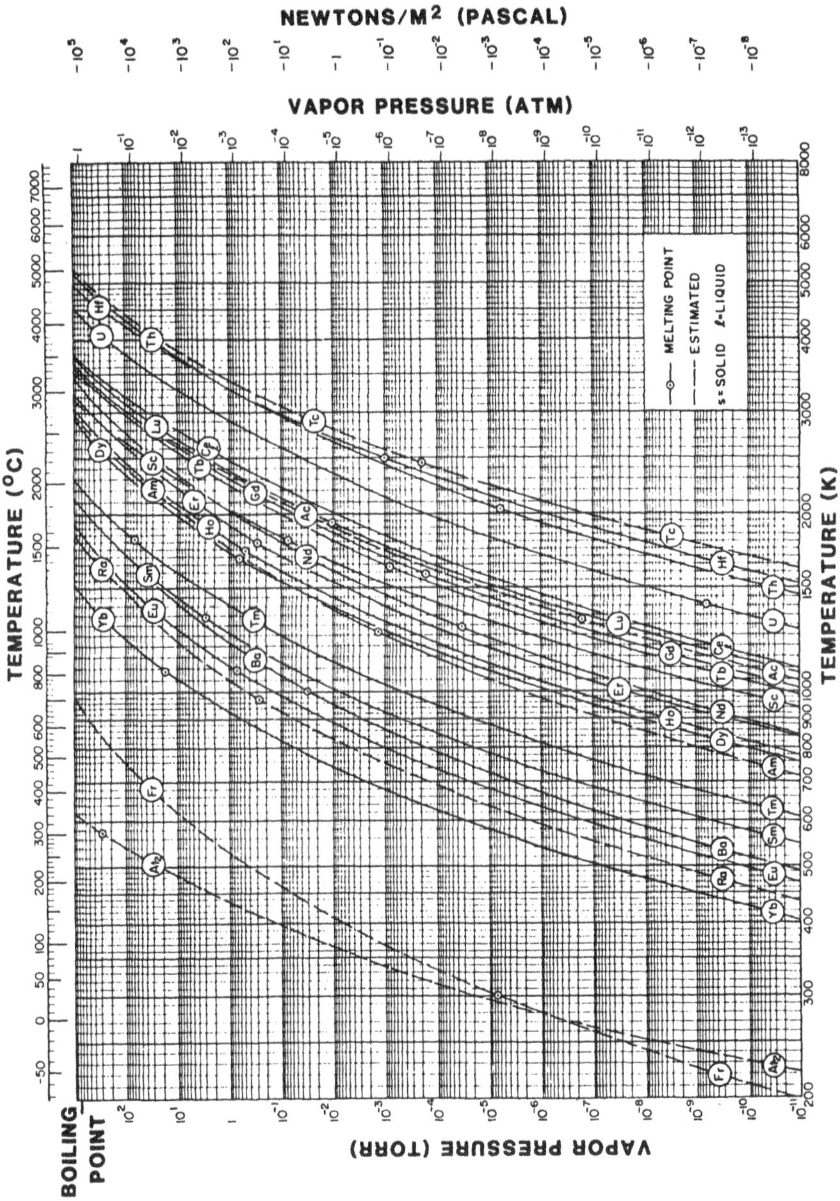

Figure 3.37 (Cont'd) Vapor Pressures of the Elements

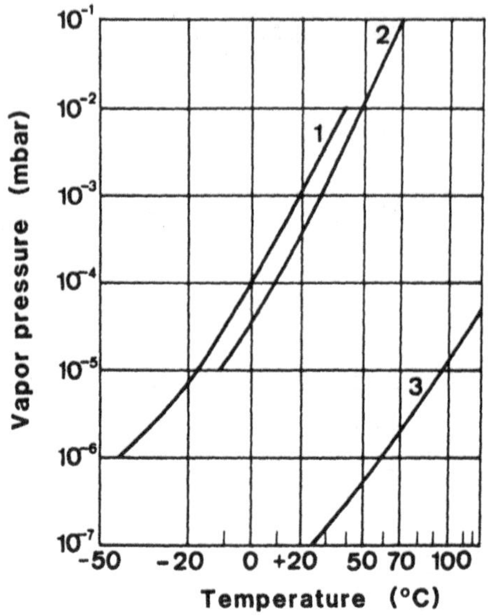

Figure 3.38 Vapor Pressure of Non-Metallic Sealing Materials: 1-Neoprene; 2-Silicone rubber; 3-Teflon

(Courtesy of Leybold Heraeus, Inc.)

Some liquids such as cleaning fluids are sometimes not completely removed before system evacuation. As shown in Fig. 3.39, these have a disasterous effect on attainable pressure.

A particularly aggravating problem with water is that as it vaporizes, it is cooled and may freeze. The vapor pressure may drop to the 10^{-2} to 10^{-3} Torr range, where it appears very much like a leak.

Other liquids may have extremely low vapor pressures. Examples are diffusion pump oils (DC 704, DC 705, and Diffelen U in Fig. 3.40). Still, there is no liquid suitable for use in the high or ultra-high vacuum environment of a microwave tube.

It is also appropriate to point out that there are no lubricants that are suitable for use in a microwave tube. Thus, in the spin-tuned magnetron, where a portion of the anode rotates at 4000 rpm, no lubricants are used. Instead, metals are carefully selected, treated, and precisely machined so that galling and excessive wear do not occur.

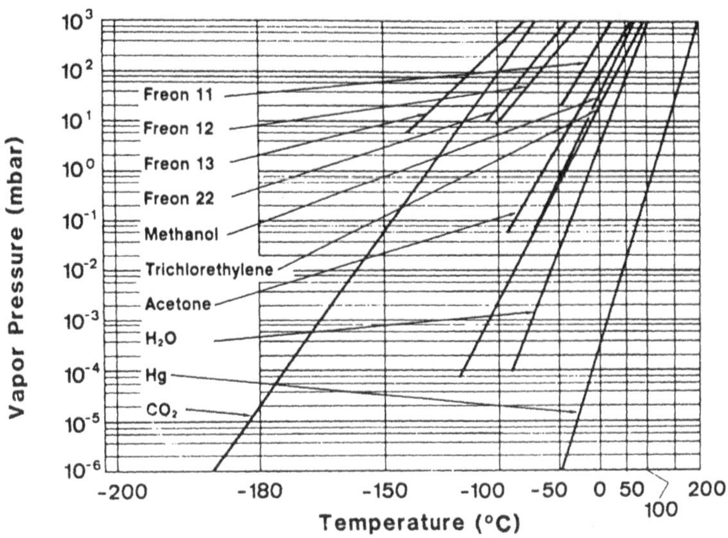

Figure 3.39 Saturation Vapor Pressure of Various Substances

(Courtesy of Leybold Heraeus, Inc.)

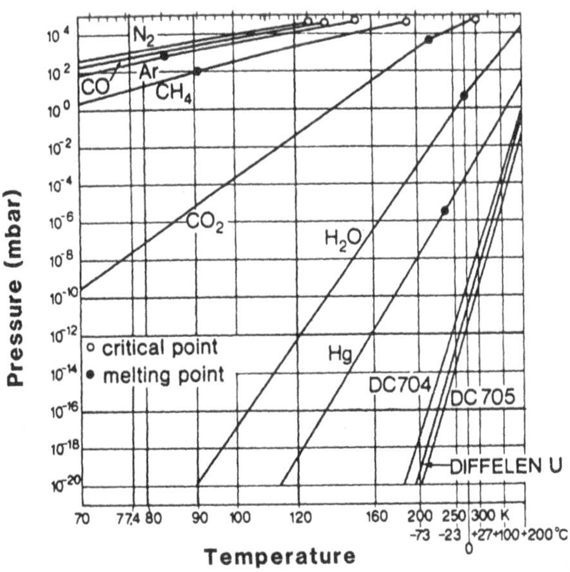

Figure 3.40 Vapor Pressure Curves of Some Gases and Pump Fluids

(Courtesy of Leybold Heraeus, Inc.)

3.6 FABRICATION TECHNIQUES

References [3.2–3.4] contain detailed information concerning many of the fabrication techniques and processes used in the microwave tube industry. It is only possible in this section to describe briefly some of the more important aspects of microwave tube fabrication.

First of all, as was pointed out in the previous section, it is essential that materials to be used are properly selected and cleaned. Then, metals may be joined by welding or brazing. Ceramics may be joined to metals by brazing. In no case is it possible to use fluxes because they may become trapped in joints and then may very slowly leak out of joints to produced contamination.

In the absence of fluxes, inert gas welding is used. A drawing of an inert gas welding torch is given in Fig. 3.41. Care must be taken in designing weld joints if adequate strength is to be obtained and gas and dirt traps are to be avoided.

Examples of the precautions that must be taken in the formation of a welded joint between a tube and a flange are shown in Fig. 3.42. Whenever possible, the weld should be made on the vacuum side of the joint. Then, weld slag or other contaminants that may be trapped between the mating surfaces of the cylinder and flange will be outside of the vacuum weld seal. Additional examples of good and bad practices in the formation of welded joints are given in Fig. 3.43.

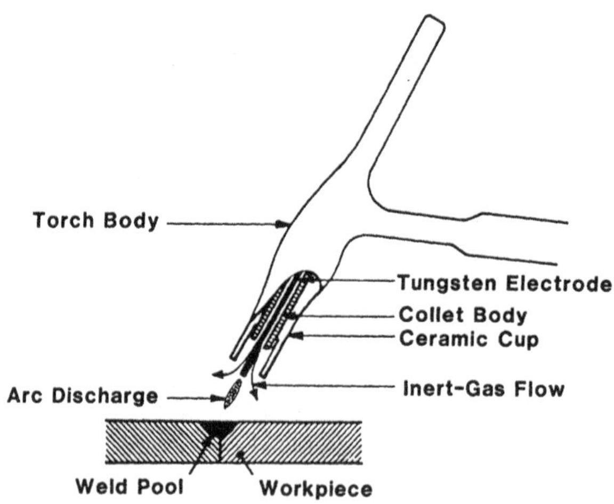

Figure 3.41 Schematic of Tungsten Inert Gas (TIG) Welding Torch

(From *Handbook of Thin Film Technology* by Maissel and Glang, Copyright 1970 by McGraw-Hill, Inc. Reproduced with permission.)

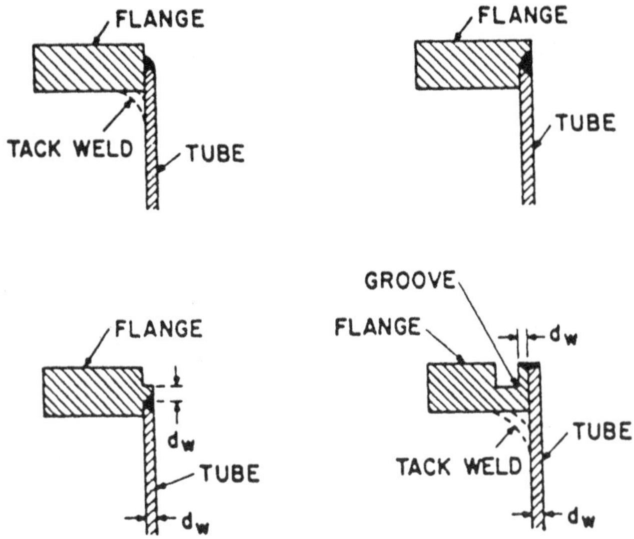

Figure 3.42 Common Tube-to-Flange Weld Designs

(From *Handbook of Thin Film Technology by Maissel and Glang*, Copyright 1970 by McGraw-Hill, Inc. Reproduced with permission.)

Similar precautions are necessary for brazed joints. Since fluxes cannot be used, brazes must be performed in vacuum or in a reducing atmosphere of hydrogen. Hydrogen is normally used rather than vacuum because the brazing process can be performed much more rapidly. In some cases, such as in the brazing of stainless steel, extremely dry hydrogen must be used to prevent oxidation. Examples of induction heating systems for brazing in hydrogen are shown in Fig. 3.44.

When the alignment of parts is critical, self-jigging may be necessary. Figure 3.45 contains some examples, both correct and incorrect, of brazing configurations.

Ceramic-to-metal seals are made by first metallizing the ceramic and then brazing the metallic part to the metallized surface.

Perhaps the most important factor to be considered in the design of ceramic-to-metal seals is the thermal expansion characteristics of the parts. Figure 3.46 shows how thermal expansion coefficients vary with temperature.

When ceramics and metals that have thermal expansion coefficients are used, it is necessary to design flexible metallic elements. Several examples of ceramic-to-metal seals are given in Figs. 3.47 and 3.48.

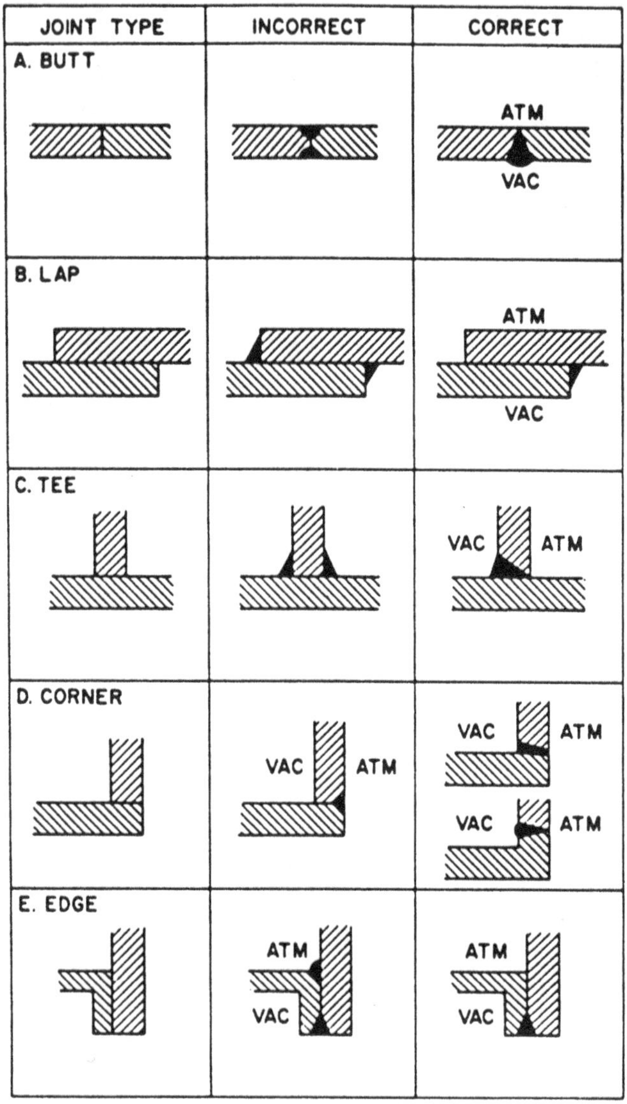

Figure 3.43 Common Types of Welded Joints

(From *Handbook of Thin Film Technology by Maissel and Glang*, Copyright 1970 by McGraw-Hill, Inc. Reproduced with permission.)

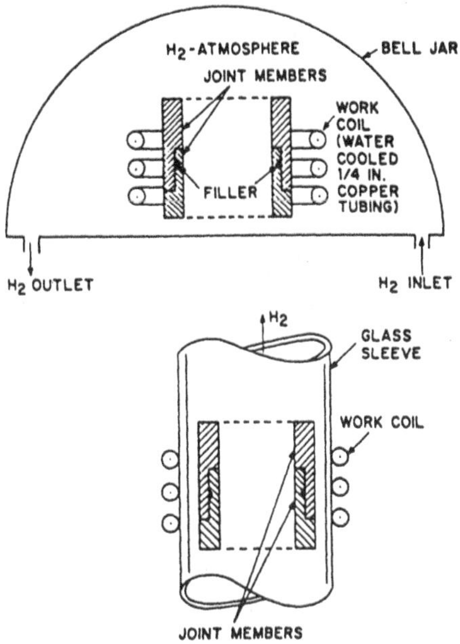

Figure 3.44 Brazing by Induction Heating in Hydrogen

(From *Handbook of Thin Film Technology by Maissel and Glang*, Copyright 1970 by McGraw-Hill, Inc. Reproduced with permission.)

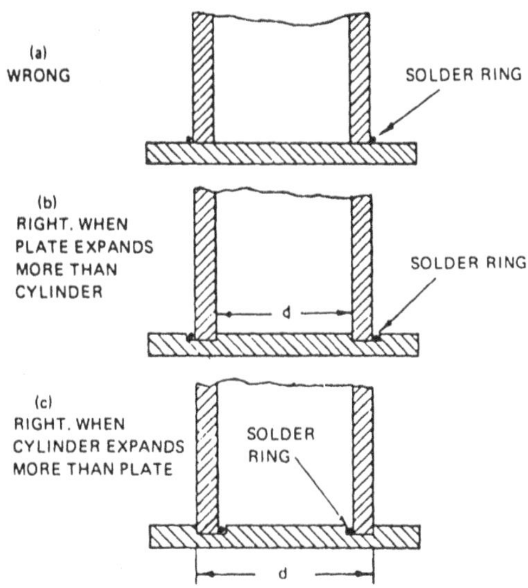

Figure 3.45 Self-Jigging Between a Cylinder and a Plate

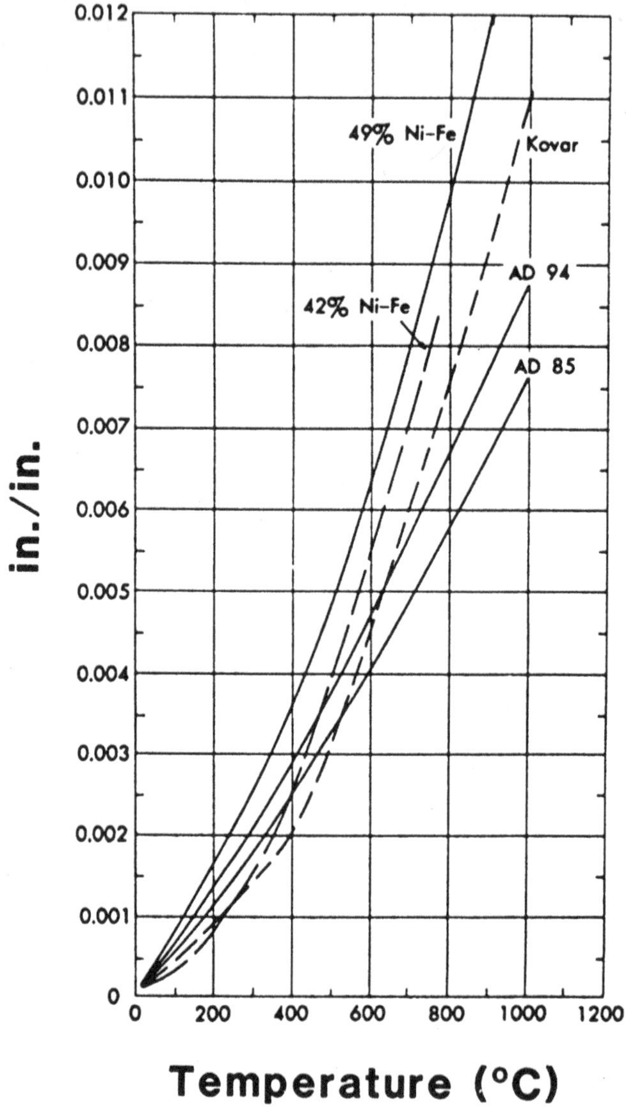

Figure 3.46 Thermal Expansion Characteristics

(From *Handbook of Electron Tube and Vacuum Techniques* by Fred Rosebury, Copyright 1965 by Addison-Wesley Publishing Company, Inc. Reproduced with permission.)

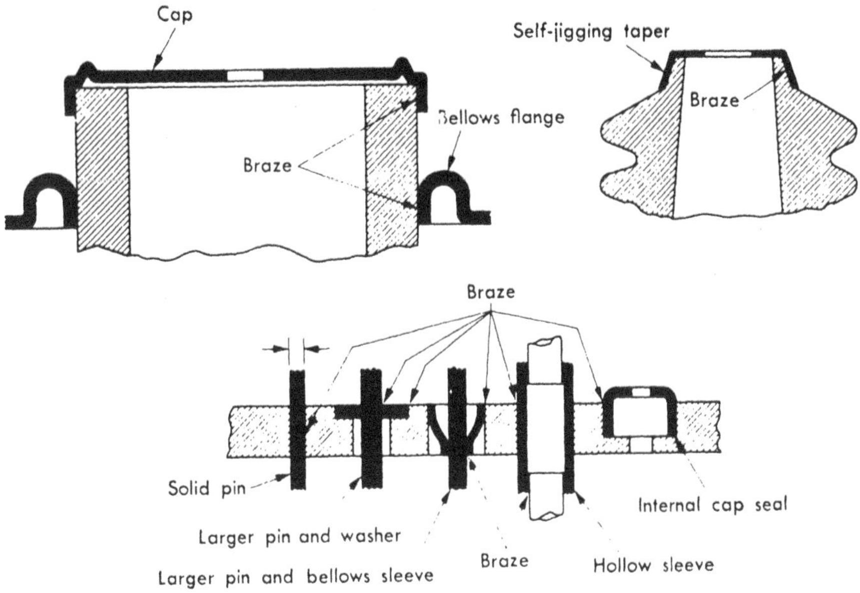

Figure 3.47 External, Tapered, and Internal Ceramic-to-Metal Seals

(From *Handbook of Electron Tube and Vacuum Techniques* by Fred Rosebury, Copyright 1965 by Addison-Wesley Publishing Company, Inc. Reproduced with permission.)

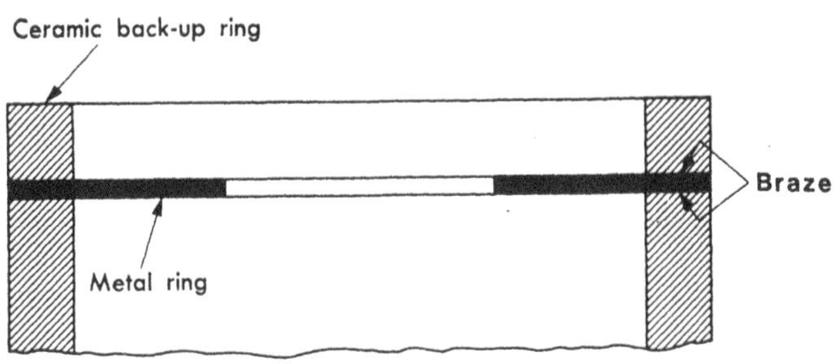

Figure 3.48 Butt Ceramic-to-Metal Seal

(From *Handbook of Electron Tube and Vacuum Techniques* by Fred Rosebury, Copyright 1965 by Addison-Wesley Publishing Company, Inc. Reproduced with permission.)

No.	Brazing -Material Constituents and Composition, wt. %	Temp, C		
		mp	Liquidus	Solidus
1	Copper (100)	1083		
2	Gold (100)	1063		
3	Pd (10)-Ag (90)		1065	1000
4	Au (35)-Cu (62)-Ni (3)		1030	975
5	B (2.9)-Cr (7)-Si (4.5)-Fe (3)-Ni (82.6)		1000	971
6	Au (40)-Cu (60)		980	950
7	Au (82)-Ni (18)	950		
8	Pd (10)-AG (58)-Cu (32)		852	825
9	Ag (72)-Cu (28)	780		
10	In (10)-Cu (27)-Ag (63) *		730	685

METAL COUPLES WHICH CAN BE BRAZED WITH THE ABOVE FILLERS

Metal 2 \ Metal 1	W	Mo	Ni	Monel	Kovar	Stainless Steel	Cu
Copper	4, 7	6, 7, 8, 9	6, 7, 9	4, 9	6, 7, 9	7, 8	8, 9, 10
Stainless Steel	5, 7	3, 5	5, 7	1	2, 7, 9	1, 8	
Kovar	4, 7	7	2, 6, 7	4, 9	2, 9		
Monel	4, 7	4	4, 9	4, 9			
Nickel	4, 7	4, 7	4, 6, 9				
Molybdenum	4, 7, 8	4, 7, 8					
Tungsten	7, 8						

* Used as step braze after No. 9.

Figure 3.49 Some Braze Materials Used in the Construction of Microwave Tubes

(From *Handbook of Thin Film Technology* by Maissel and Glang, Copyright 1970 by McGraw-Hill, Inc. Reproduced with permission.)

Many braze materials are available for use in the construction of microwave tubes. Some of these are listed in Fig. 3.49 along with various metal couples that can be joined.

During recent years, in many high-power klystrons and TWTs, the use of alloys containing silver has been avoided. This is because, at high temperatures, sufficient vaporization of the silver may occur to degrade tube performance.

After assembly and processing of a metal-to-ceramic vacuum tube has been completed, it is necessary to separate the tube from the pumping station. (An appendage pump may remain as an integral part of the tube). The technique universally used to separate the tube from the pumping station is illustrated in Fig. 3.50. Soft copper tubulation is used to make the vacuum connection to the pumping station. This tubulation is molecularly clean on the internal surfaces because it has been baked along with the microwave tube (at 450° C or higher).

By squeezing the tubulation with a special tool, the internal mating surfaces of the copper can be forced to flow together to form a cold weld. In its crudest form, the tool is similar to a manually operated bolt cutter with the cutting edges replaced by cylindrical steel bars.

Because the manual squeeze-off (pinch-off) operation with the tool that is much like a bolt-cutter is highly operator-dependent, an automated system using a tool similar to the hydraulic tool shown in Fig. 3.51 is normally employed.

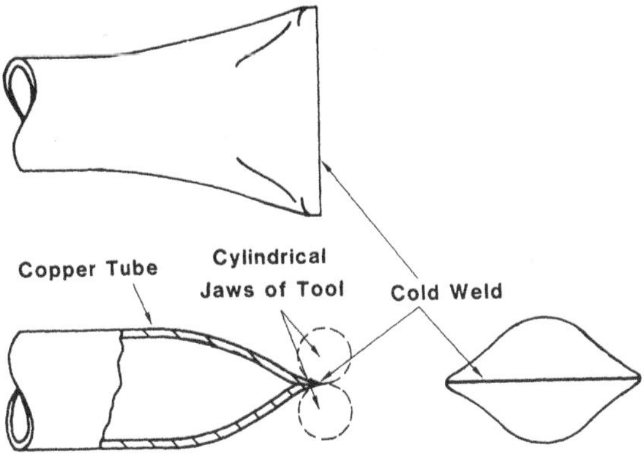

Figure 3.50 Technique Used to Squeeze-Off (Pinch-Off) Copper Tubulation from Tube to Pumping Station

(From *Power Travelling-Wave Tubes* by J.F. Gittins, Published 1965 by American Elsevier Publishing Company, Inc.)

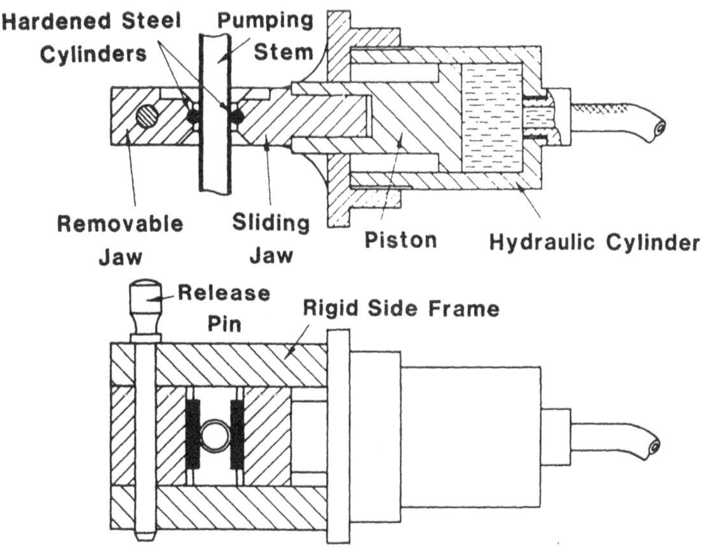

Figure 3.51 Hydraulic Squeeze-Off Tool

(From *Power Travelling-Wave Tubes* by J.F. Gittins, Published 1965 by American Elsevier Publishing Company, Inc.)

REFERENCES

[3.1] For additional reading, the following are recommended:

"Vacuum Physics and Technology," Vol. 14, *Methods of Experimental Physics*, New York: Academic Press, 1979.

S. Dushman and J.M. Lafferty, *Scientific Foundations of Vacuum Techniques*, New York: John Wiley and Sons, 1962.

M. Pirani and J. Yarwood, *Principles of Vacuum Engineering*, New York: Reinhold, 1963.

R.W. Roberts and T.A. Vanderslice, *Ultrahigh Vacuum and its Applications*, Englewood Cliffs, New Jersey: Prentice-Hall, 1963.

[3.2] F. Rosebury, *Handbook of Electron Tube and Vacuum Techniques*, Reading, Massachusetts: Addison-Wesley, 1965.

[3.3] W. H. Kohl, *Materials and Techniques for Electron Tubes*, New York: Reinhold, 1960.

[3.4] M. Knoll, *Materials and Processes of Electron Devices*, Berlin: Springer-Verlag, 1959.

Chapter 4

Magnetic Focusing

Magnetic fields exert forces on moving electrons as shall be described in subsequent chapters. As a result, magnetic fields are used for focusing electron beams in nearly all microwave tubes.

Three magnetic quantities are of primary interest:

- Magnetic field intensity, H, is one of the elementary fields in nature
- Magnetic flux density, B, is the density of magnetic flux lines produced by the magnetic field
- Magnetic permeability, μ, is the property of a medium that determines how much flux is produced by a certain field level. Flux density is related to field intensity by

  $$B = \mu H$$

The units commonly used for H, B, and μ are given below along with conversion factors.

Quantity	*MKS Unit (a)*	*CGS Unit (b)*	*Conversion*
H	Amperes/meter	Oersteds	$a = 79.577b$
B	Tesla	Gauss	$a = 10^{-4}b$
μ	Henrys/meter	Gauss/Oersted	$a = 1.257 \times 10^{-6}b$

The value for μ in air (free space) is

$$\mu = 4\pi \times 10^{-7} \quad \text{H/m}$$

It is common to use the term relative permeability, μ_r, which relates the permeability of a material to that of free space.

Magnetic focusing fields may be generated by

- Electromagnets
- Permanent magnets

In an electromagnet, current passing through a wire produces a magnetic field surrounding the wire. In a permanent magnet, electron motion within the material produces the magnetic field.

Electromagnets and permanent magnets are discussed in the following sections [4.1].

4.1 ELECTROMAGNETS

A wire carrying a current produces a magnetic field as shown in Fig. 4.1 through 4.4. The magnitude of the field may be determined from the following relation, known as Ampere's law:

$$\oint H \cdot dl = I$$

That is, the integral of the magnetic field around a closed path is the current contained within that path. For a circular path, as shown in Fig. 4.1, the integration is easily performed and the result is

$$H = \frac{I}{2\pi a}$$

where a is the radius of the path.

The direction of the field lines in relation to the direction of current flow can be remembered with the aid of the familiar right-hand rule illustrated in Fig. 4.2

4.1.1 Solenoids

For microwave tube applications, the current-carrying wire is frequently formed into a solenoid, as shown in Figs. 4.3 and 4.4, to produce a reasonably constant axial magnetic field near the center.

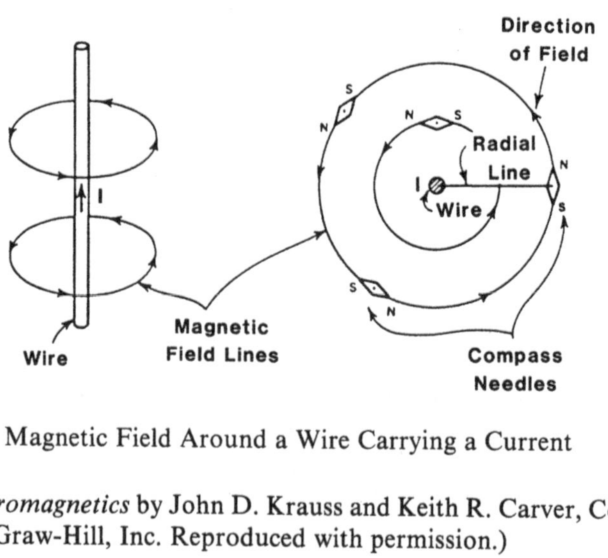

Figure 4.1 Magnetic Field Around a Wire Carrying a Current

(From *Electromagnetics* by John D. Krauss and Keith R. Carver, Copyright 1973 by McGraw-Hill, Inc. Reproduced with permission.)

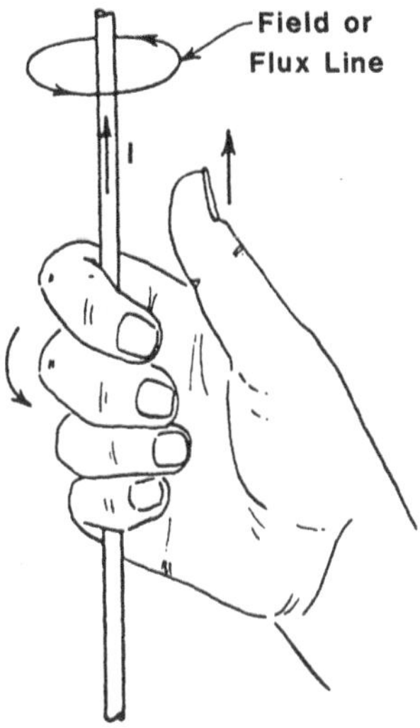

Figure 4.2 Right-Hand Rule Relating Direction of Magnetic Field or Flux Lines (Fingers) to Direction of Current *I* (Thumb)

(From *Electromagnetics* by John D. Krauss and Keith R. Carver, Copyright 1973 by McGraw-Hill, Inc. Reproduced with permission.)

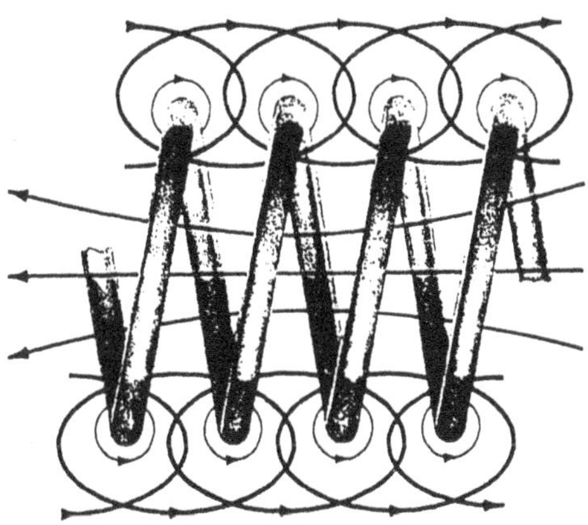

Figure 4.3 Portion of Solenoid Showing Partial and Total Flux Linkages

(From *Engineering Electromagnetics* by William H. Hayt, Jr., Copyright 1981 by McGraw-Hill, Inc. Reproduced with permission.)

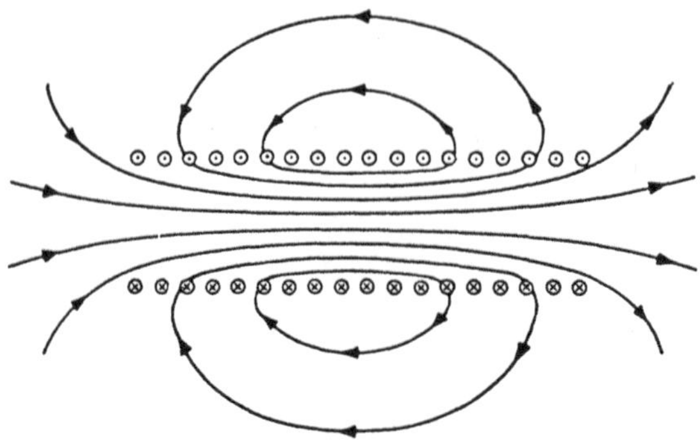

Figure 4.4 Magnetic Flux Pattern for a Single-Layer Solenoid

Very close to each turn of the solenoid, the flux lines are nearly circular. However, at greater distances, the magnetic flux in the region between adjacent turns cancels, so the flux lines around each pair of turns are nearly elliptical. At still greater distances, there is further cancellation of opposing flux lines and so the pattern of Fig. 4.4 occurs.

In the center of the solenoid, the flux lines are parallel and the flux density is constant. Near the ends of the solenoid, the flux lines diverge and the flux density decreases as indicated by the dashed line in Fig. 4.5. This large decrease in field near the ends of the solenoid is not acceptable for most microwave tube applications.

To enhance uniformity of the field on axis where the electron beam is to flow, magnetic pole pieces, as shown in Fig. 4.6, are used. The effect of the pole pieces is to strengthen the field near the ends of the solenoid as shown by the solid curve in Fig. 4.5. The magnetic behavior of the pole pieces is described in the following section.

Figure 4.7 shows a typical solenoid for a TWT or klystron. Individual coils of the solenoid are often separated to provide space for RF feed-throughs, cooling lines, *et cetera*.

Normally, a solenoid is surrounded by a magnetic shield, or magnetic bars, as shown in Fig. 4.7, to provide a return path for the magnetic flux lines. These magnetic shields or bars do not enhance the magnetic flux density on the axis of the solenoid. Instead, they serve to confine the magnetic flux lines surrounding a solenoid to a localized region near the solenoid. This helps to minimize the possibility that a magnetic object such as a tool will accidentally be pulled toward the tube being focused by the solenoid, thus causing damage to, or destruction of, the tube. The magnetic shields or bars

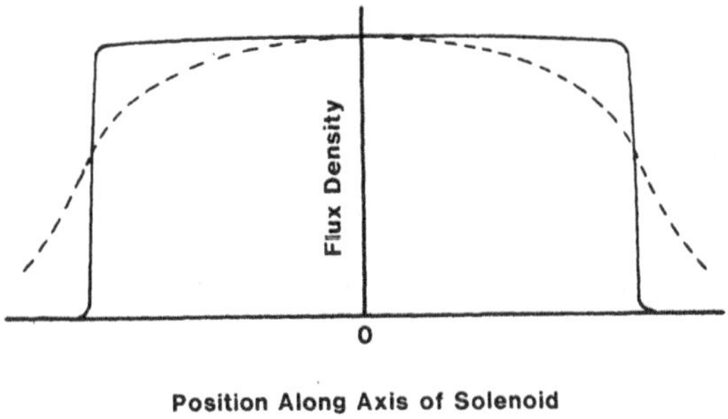

Figure 4.5 Solenoid Flux Density Pattern Without (Dashed Curve) and With (Solid Curve) Magnetic Pole Pieces

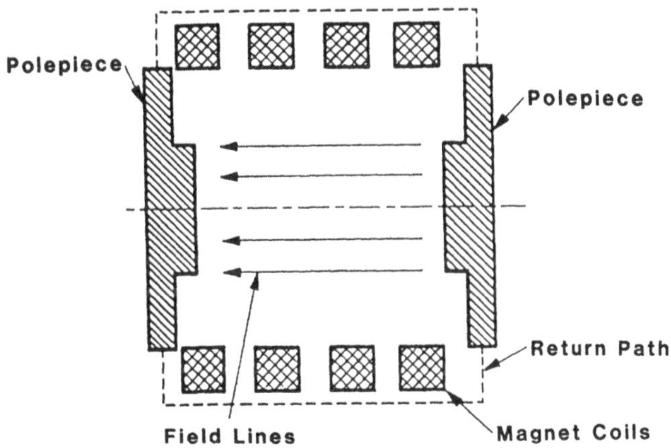

Figure 4.6 Focusing Solenoid Consisting of Coils and Magnetic Pole Pieces to Improve Field Uniformity

also help to minimize interactions of the solenoid with nearby electrical circuits.

In most cases, the solenoid used for focusing the beam of a tube is removable from the tube. Sometimes, however, when weight and power utilization must be minimized, the solenoid is made integral with the tube as shown in Fig. 4.8.

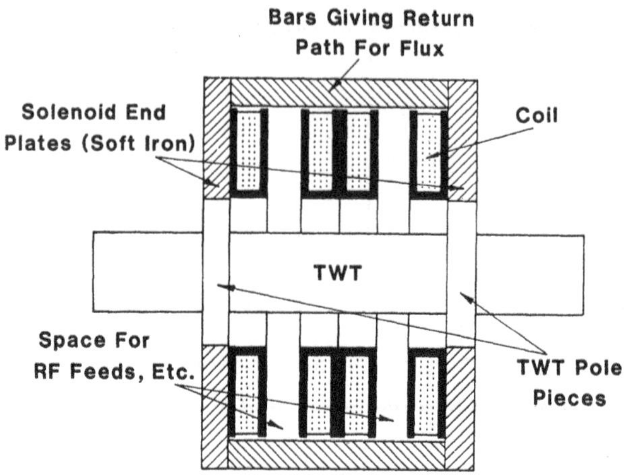

Figure 4.7 Typical Configuration of a Solenoid and a Microwave Tube

(From *Power Travelling-Wave Tubes* by J.F. Gittins, Published 1965 by American Elsevier Publishing Company, Inc.)

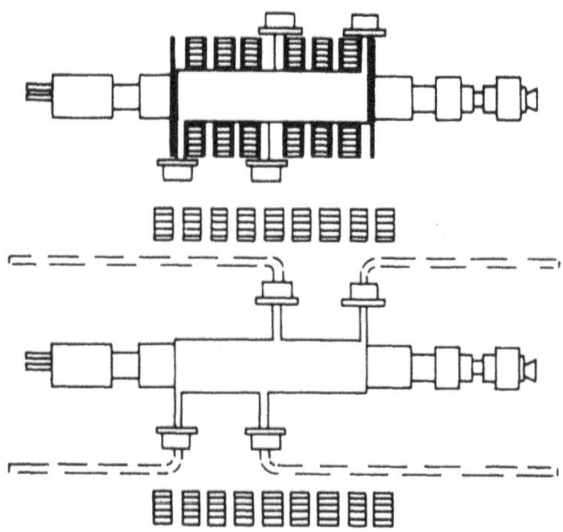

Figure 4.8 Illustration of Size and Weight Reduction Resulting from the Use of an Integral Solenoid

(From *Power Travelling-Wave Tubes* by J.F. Gittins, Published 1965 by American Elsevier Publishing Company, Inc.)

4.2 PERMANENT MAGNETS

Magnetic materials can be divided into 3 categories. These are

- $\frac{\mu}{\mu_0} < 1$ diamagnetic
- $\frac{\mu}{\mu_0} > 1$ paramagnetic
- $\frac{\mu}{\mu_0} >> 1$ ferromagnetic

These categories are discussed in the following paragraphs.

4.2.1 Diamagnetism

Diamagnetism is present in all materials. The establishment of magnetic flux is opposed by the material so that the relative permeability is less than unity. This effect is explained by the laws of Faraday and Lenz. When a magnetic field is applied, an electric field is produced. This electric field acts on the electron orbits in a material to produce microscopic reaction currents. The reaction currents produce a magnetic field that opposes the applied field.

Diamagnetism is an extremely small effect. In most materials, it is overridden by paramagnetism or ferromagnetism.

4.2.2 Paramagnetism

Paramagnetic materials have a relative permeability slightly greater than unity. In these materials, the application of an external magnetic field aligns permanent magnetic dipoles of atoms, ions or molecules with the applied field, thus enhancing the field. This alignment action is illustrated in Fig. 4.9.

4.2.3 Ferromagnetic Materials

In ferromagnetic materials, a highly specialized condition exists which greatly aids the applied magnetic field. This condition involves the electronic structures of atoms and the distances between atoms.

The most stable configuration of the electrons is when their spin magnetic moments are parallel. The result is that complete alignment of the spins occurs within regions called magnetic domains.

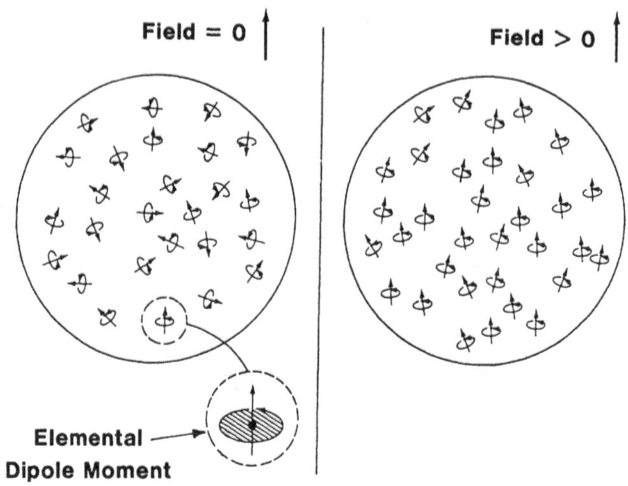

Figure 4.9 Alignment of Magnetic Dipole Moments by an Applied Magnetic Field

The application of an external field both aligns domains (Fig. 4.10) and shifts domain boundaries (Fig. 4.11) to produce an enormous field [4.2].

Table 4.1 contains a listing of the relative permeabilities of materials. Notice that diamagnetic and paramagnetic materials have relative permeabilities that are only a small fraction of a percent different than unity. For ferromagnetic materials, on the other hand, the relative permeability can be many orders of magnitude greater than unity.

4.2.4 Characteristics of Ferromagnetic Materials [4.3]

Ferromagnetic materials are highly nonlinear. The relation between B and H must be determined experimentally. This is done with a circuit like that shown in Fig. 4.12.

From Ampere's law, the magnetic field, H, in a magnetic sample is proportional to the current, I. From Faraday's law, the voltage, V, induced in a winding is proportional to the time rate of change of magnetic flux linking the winding. The integral of voltage is, therefore, proportional to flux and magnetic flux density, B.

A signal proportional to B is applied to the vertical input of an oscilloscope. A signal proportional to H is applied to the horizontal input. The result displayed on the oscilloscope is illustrated in Fig. 4.13.

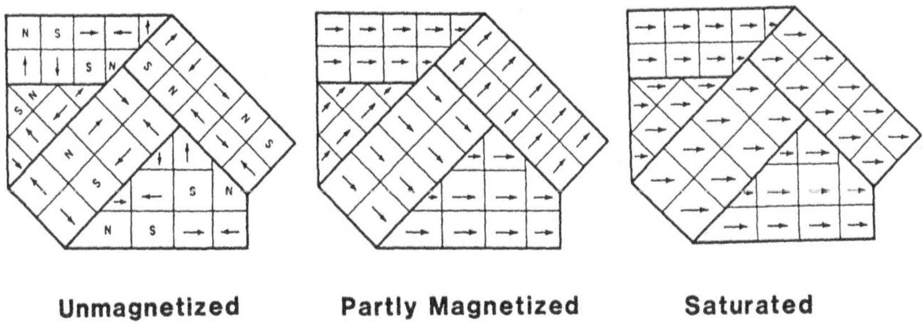

Figure 4.10 Illustration of Successive Stages of Magnetization with Field Increasing from (a) to (c)

(From *Electromagnetics* by John D. Krauss and Keith R. Carver, Copyright 1973 by McGraw-Hill, Inc. Reproduced with permission.)

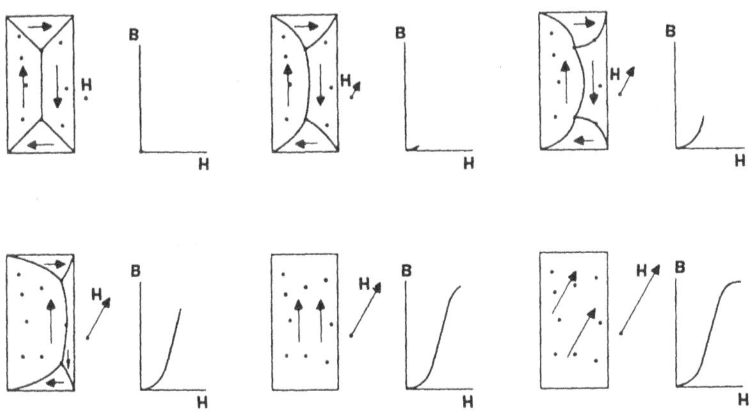

Figure 4.11 Simplified Representation of the Part Played by Domain Boundaries in the Process of Magnetization

When current is increased, the right-hand portion of the hysteresis curve is followed. When current is decreased, the left-hand portion is followed. At very large current excursions, the shape of the hysteresis curve is relatively independent of current and a loop known as the saturation loop is traversed.

There are several important characteristics of magnetic materials that may be illustrated with the hysteresis loop. Some of these are shown in Fig. 4.14.

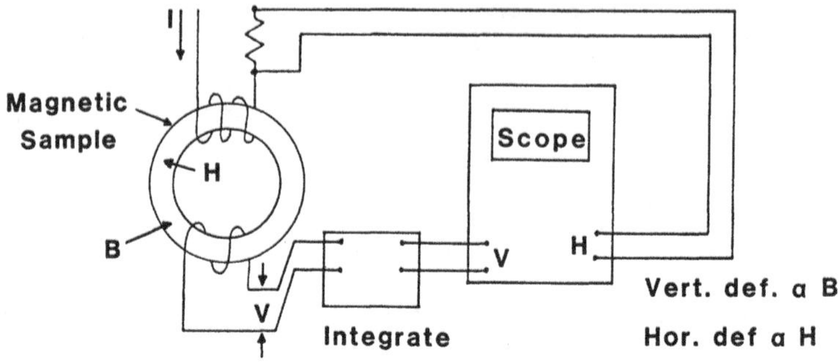

Figure 4.12 Circuit for Determining the Magnetic Characteristic of Materials

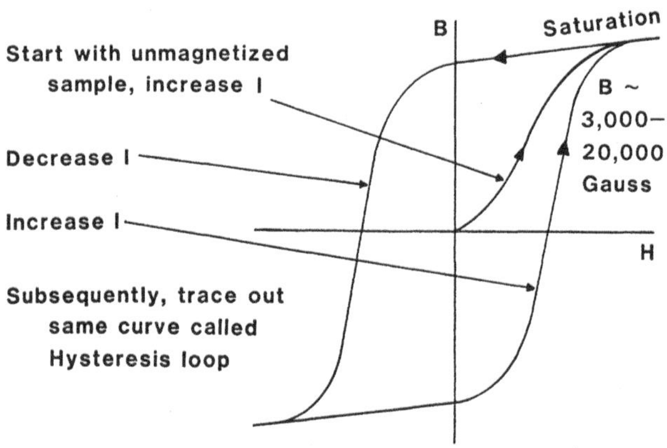

Figure 4.13 Hysteresis Curve Resulting from Use of Circuit Shown in Fig. 4.12

When a magnetic field is first applied to a demagnetized sample, the initial magnetization curve is followed. When H is zero, there is a residual flux density, B_r, in the material called the remanence or retentivity. The flux density, B, is reduced to zero when a negative field, $-Hc$, called the coercive force, is applied.

At lower current levels, smaller loops are traversed. In fact, as shown in Fig. 4.15, a material can be damagnetized by gradually reducing the alternating magnetic field to zero. The tips of the hysteresis loops form the "normal magnetization curve" as shown in Fig. 4.16. This is very similar in shape to the initial magnetization curve. This curve is a reproducible characteristic for a magnetic material.

TABLE 4.1

Relative Permeabilities of Some Diamagnetic, Paramagnetic, and Ferromagnetic Materials
(From *Electromagnetics* by John D. Krauss and Keith R. Carver, Copyright 1973 by McGraw-Hill, Inc. Reproduced with permission.)

Substance	*Group type*	Relative permeability (μ)
Bismuth	Diamagnetic	0.99983
Silver	Diamagnetic	0.99998
Lead	Diamagnetic	0.999983
Copper	Diamagnetic	0.999991
Water	Diamagnetic	0.999991
Vacuum	Nonmagnetic	1*
Air	Paramagnetic	1.0000004
Aluminum	Paramagnetic	1.00002
Palladium	Paramagnetic	1.0008
2–81 Permalloy powder (2 Mo, 81 Ni)†	Ferromagnetic	130
Cobalt	Ferromagnetic	250
Nickel	Ferromagnetic	600
Ferroxcube 3 (Mn-Zn-ferrite powder)	Ferromagnetic	1,500
Mild steel (0.2 C)	Ferromagnetic	2,000
Iron (0.2 impurity)	Ferromagnetic	5,000
Silicon iron (4 Si)	Ferromagnetic	7,000
78 Permalloy (78.5 Ni)	Ferromagnetic	100,000
Mumetal (75 Ni, 5 Cu, 2 Cr)	Ferromagnetic	100,000
Purified iron (0.05 impurity)	Ferromagnetic	200,000
Supermalloy (5 Mo, 79 Ni)‡	Ferromagnetic	1,000,000

* By definition.
† Percentage composition. Remainder is iron and impurities.
‡ Used in transformer applications with continuous tape-wound (gapless) cores.

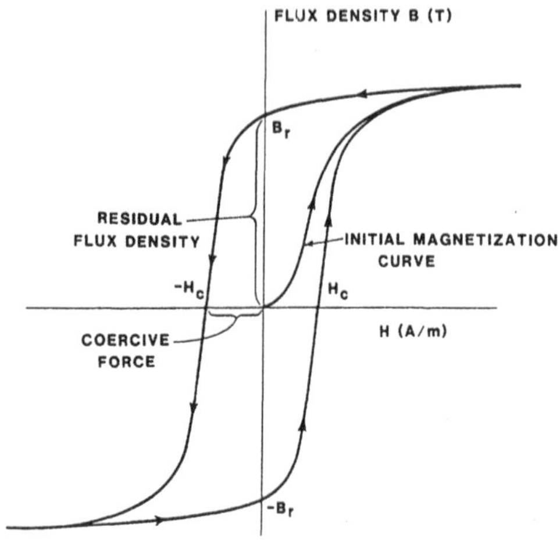

Figure 4.14 Hysteresis Loop Showing Initial Magnetization Curve, Residual Flux Density, and Coercive Force

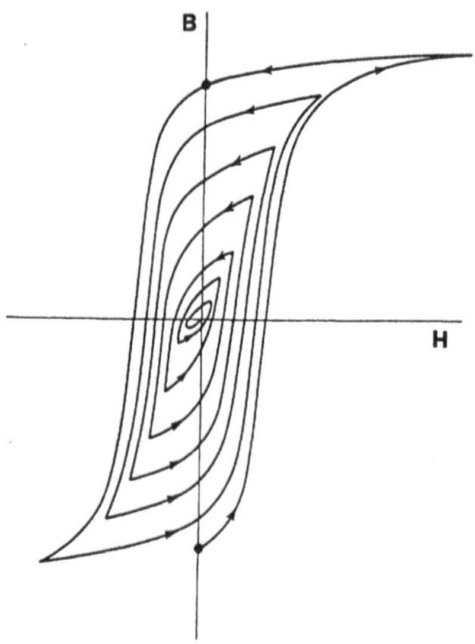

Figure 4.15 Demagnetization by Reversals of Magnetic Field

(From *Electromagnetics* by John D. Krauss and Keith R. Carver, Copyright 1973 by McGraw-Hill, Inc. Reproduced with permission.)

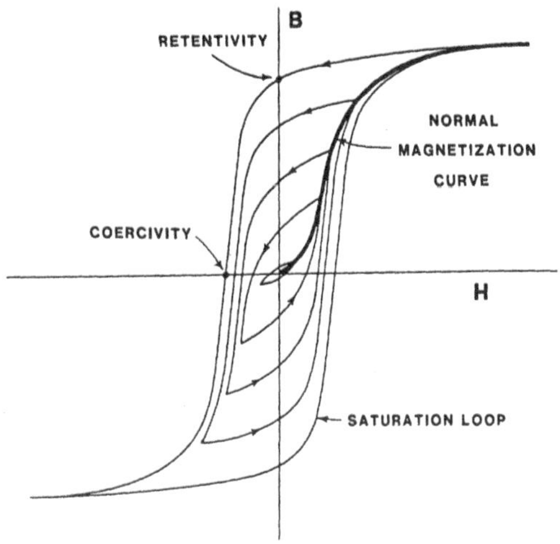

Figure 4.16 Normal Magnetization Curve with Relation to Hysteresis Loop

The relative permeability quoted for a material is based on the slope of the normal magnetization curve as shown in Fig. 4.17. The maximum value is normally used.

It is significant to note that, at very low values of the magnetic field, the permeability is low. In fact, for most magnetic materials, the permeability at low values of H is so low that they will not shield against stray magnetic fields. This is sometimes of importance in the electron gun of a tube when it is necessary to shield the cathode from stray magnetic flux. At high values of H, the permeability is also low. In fact, when a material is driven well into saturation, the permeability approaches the permeability of free space. This must be kept in mind in the design of magnetic pole pieces to be used in regions of high magnetic field. The concentration of flux lines at corners of pole pieces, for example, can cause saturation and distortion of the field pattern.

4.2.5 Energy Product

The energy per unit volume, W_m, stored in a magnetic field is given by

$$W_m = \int H \cdot dB$$

For magnetic materials, several values of energy are of interest:

1. The energy dissipated as a material is repeatedly magnetized and demagnetized. The area enclosed by a hysteresis loop represents the energy expended in one complete cycle. Loops for soft (small area) and hard (large area) materials are given in Fig. 4.18.
2. The energy required to bring a specimen to saturation from an originally unmagnetized condition (shown by the singly cross-hatched areas in Fig. 4.19). A magnetically "hard" material (Fig. 4.19b) requires more energy than a magnetically "soft" material (Fig. 4.19a)
3. The maximum amount of energy that can be stored in the space near a magnetic material having the most favorable shape. This energy is represented, as shown in Figs. 4.20, 4.21, and 4.22, by the maximum *BH* product (energy product) in the second quadrant (known as the demagnetization curve) of the hysteresis loop.

The third energy quantity listed above is the one of interest in selecting permanent magnets. High B_r is desirable, but high H_c is also important so that a magnet will not be easily demagnetized.

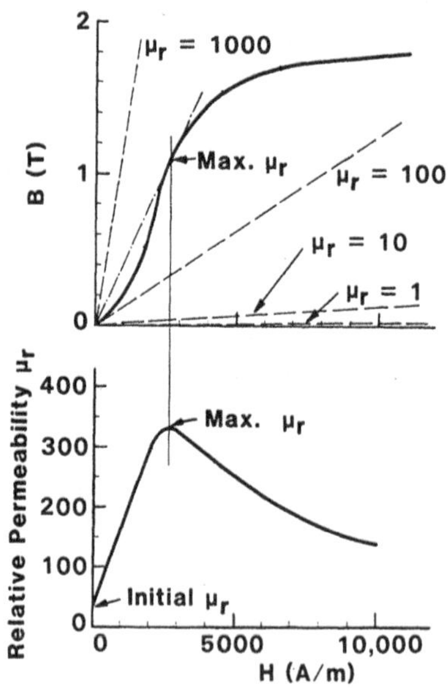

Figure 4.17 Relation of Relative Permeability to Normal Magnetization Curve

(From *Electromagnetics* by John D. Krauss and Keith R. Carver, Copyright 1973 by McGraw-Hill, Inc. Reproduced with permission.)

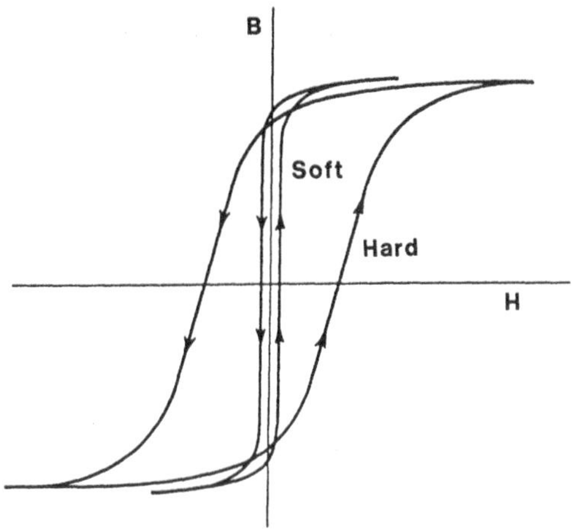

Figure 4.18 Hysteresis Loops for Soft and Hard Magnetic Materials

(From *Electromagnetics* by John D. Krauss and Keith R. Carver, Copyright 1973 by McGraw-Hill, Inc. Reproduced with permission.)

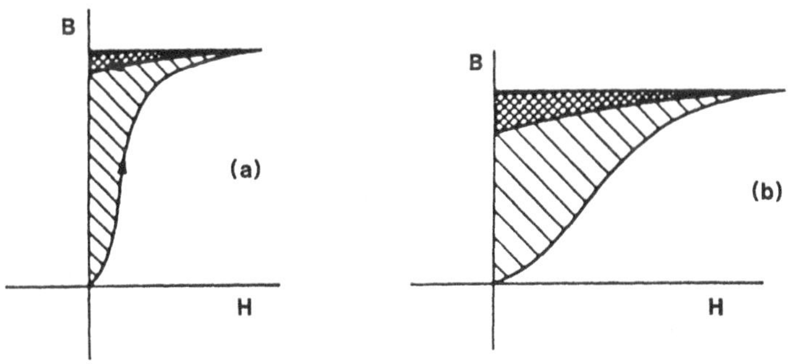

Figure 4.19 Energy Density Areas for (a) Soft and (b) Hard Magnetic Materials

(From *Electromagnetics* by John D. Krauss and Keith R. Carver, Copyright 1973 by McGraw-Hill, Inc. Reproduced with permission.)

In Fig. 4.21, curve 1 represents a material having a high retentivity but a low coercivity, while curve 2 represents a material which is just the reverse. Curve 3 represents a material which is a compromise between the two and is thus the best permanent magnet.

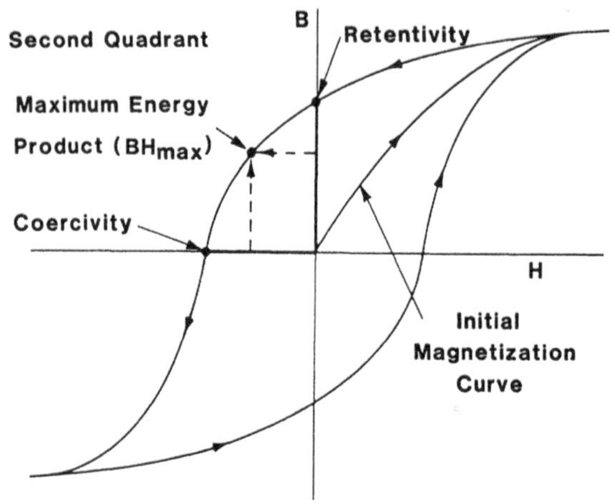

Figure 4.20 Hysteresis Curve Showing Maximum Energy Product

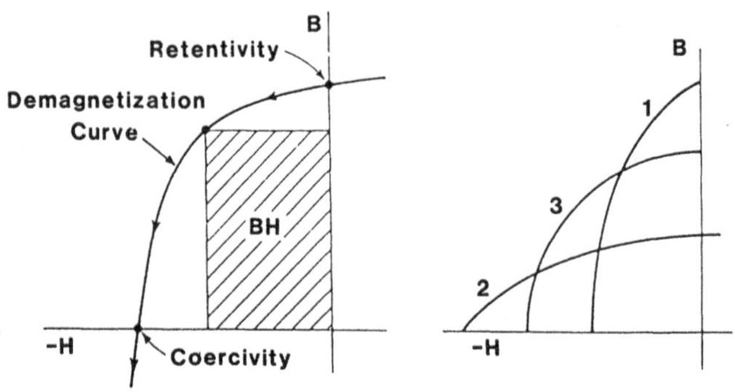

Figure 4.21 Demagnetization Curves for Permanent Magnets Having Various Characteristic Shapes

(From *Electromagnetics* by John D. Krauss and Keith R. Carver, Copyright 1973 by McGraw-Hill, Inc. Reproduced with permission.)

Alnico magnets have been commonly used for microwave tube applications for many years. Demagnetization and *BH* product curves are given for alnico V in Fig. 4.22.

During the past decade, samarium cobalt (SmCo) has been used as the magnet material for a large percentage of the tube developments requiring permanent magnets. Demagnetization curves for SmCo and for several other materials, shown in Fig. 4.23, indicate the advantage of SmCo over the others.

The energy product for SmCo is approximately four times that of alnico V. The coercive force is over an order of magnitude greater than that of alnico V, with the result that the SmCo magnets are extremely stable.

After temperature stabilization, the thermal effects in SmCo are reversible up to the stabilization temperature and are a function of the thermal coefficient, which is .04%/°C. Operation above 300°C is not recommended.

4.2.6 Permanent-Magnet Configurations

The configurations of permanent magnets used for focusing the beams in microwave tubes vary widely depending on tube type and application. Some examples are shown in Fig. 4.25 through 4.29.

The cylindrical and barrel shaped magnets in Fig. 4.25 and 4.26 are large and heavy, but have the advantages over electromagnets of requiring no power and *always* providing a magnetic field. This is in contrast to electromagnets and their power supplies, which may fail at critical times, and may result in damage to tubes if they are improperly protected or interlocked.

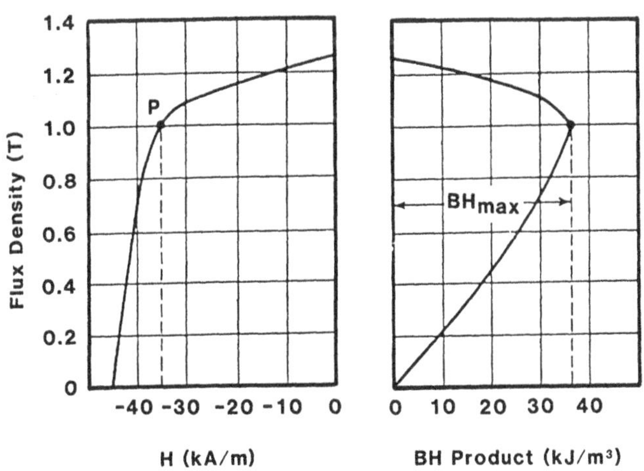

Figure 4.22 Demagnetization and BH Product for Alnico V (Maximum *BH* Product Occurs at *P*)

(From *Electromagnetics* by John D. Krauss and Keith R. Carver, Copyright 1973 by McGraw-Hill, Inc. Reproduced with permission.)

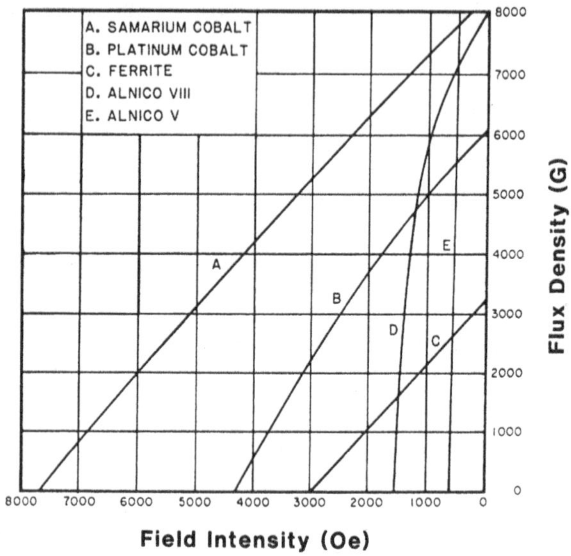

Figure 4.23 Comparison of Rare Earth Cobalt Magnets with Conventional Magnets

(From Rare Earth Cobalt Magnets, Varian Brochure, 1972.)

Periodic-permanent-magnet (ppm) structures are frequently used on tubes in lightweight applications. A ppm structure with axially charged magnets is shown in Fig. 4.27. A structure with radially charged magnets is shown in Fig. 4.28. In either case, magnetic lenses are formed which periodically focus the electron beam traveling along the center of the structure. Periodic magnetic focusing of an electron beam is discussed in Chapter 7.

A problem with the ring-shaped magnets illustrated in Fig. 4.27 and 4.28 is that each must be split in order to be placed on the tube. This is one of the factors leading to nonuniformities in the magnetic fields produced by these structures. To obtain acceptable tube performance, it is often necessary to compensate for the nonuniformities by placing magnetic shims on the outside of the magnetic assembly.

A technique for avoiding some of the problems of ring-shaped magnets is illustrated in Fig. 4.29. Several (often four) pill-shaped magnets are placed between magnetic pole pieces and surround the axis of the tube. With this technique, the magnets can be magnetized to the desired level and then placed on the tube without being split. Field uniformity is increased (as compared with the ring-shaped magnets). In addition, magnet cost is reduced because magnets of one size can be used on tubes having a wide range of sizes.

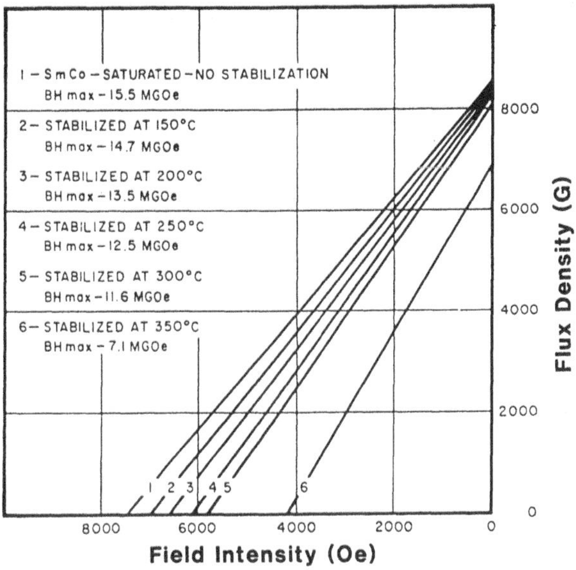

Figure 4.24 Effect of Thermal Stabilization of Energy Product of Samarium Cobalt (SmCo)

(From Rare Earth Cobalt Magnets, Varian Brochure, 1972.)

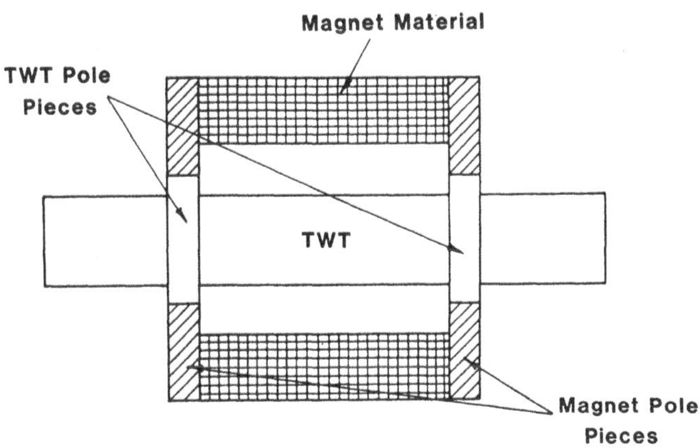

Figure 4.25 Cylindrical Pot Permanent Magnet

(From *Power Travelling-Wave Tubes* by J.F. Gittins, Published 1965 by American Elsevier Publishing Company, Inc.)

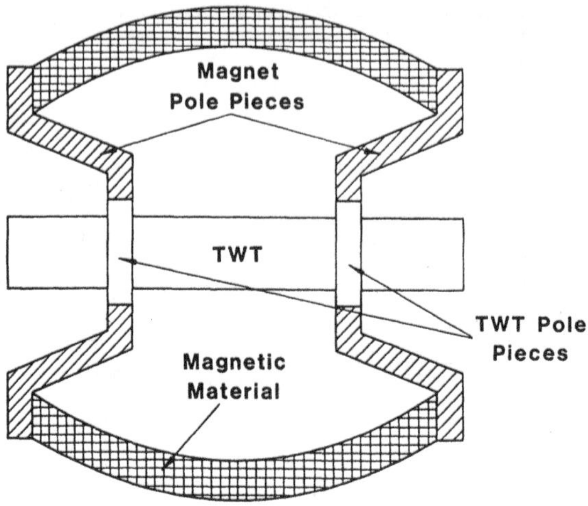

Figure 4.26 Barrel Permanent Magnet with Re-entrant Pole Pieces

(From *Power Travelling-Wave Tubes* by J.F. Gittins, Published 1965 by American Elsevier Publishing Company, Inc.)

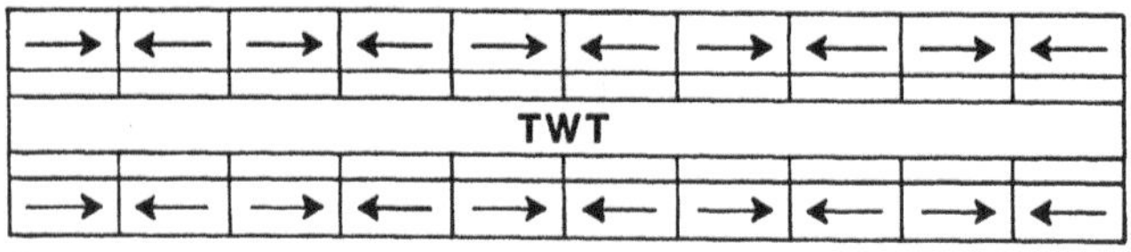

Figure 4.27 Periodic-Permanent-Magnet (PPM) Structure Using Axially Charged Magnets

(From Hughes Aircraft Co., TWT and TWTA Handbook.)

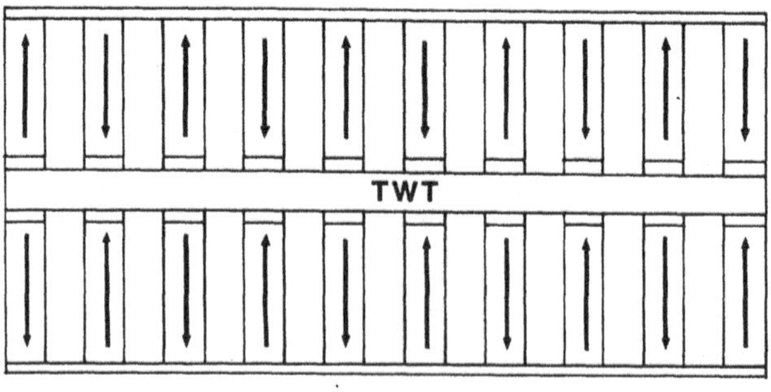

Figure 4.28 PPM Structure Using Radially Charged Magnets

(From Hughes Aircraft Co., TWT and TWTA Handbook.)

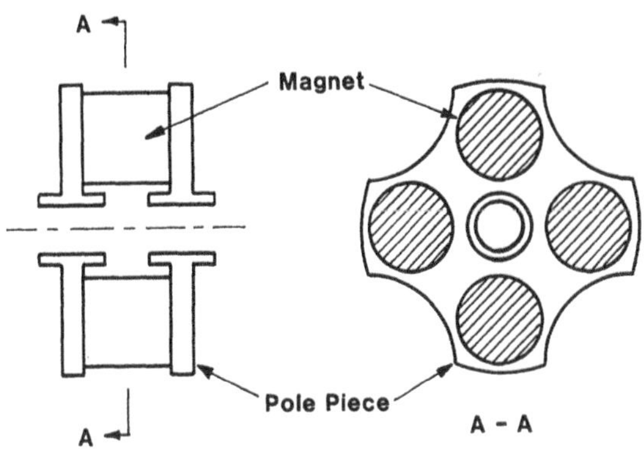

Figure 4.29 PPM Configuration Using Pill-Shaped Magnets

REFERENCES

[4.1] For additional reading, the following are recommended:

J. D. Krauss and K. R. Carver, *Electromagnetics.* New York: McGraw-Hill, 1973.

W. H. Hayt, Jr., *Engineering Electromagnetics.* New York: McGraw-Hill, 1981.

L. M. Magid, *Electromagnetic Fields, Energy, and Waves.* New York: McGraw-Hill, 1972.

C. G. Skitek and S. V. Marshall, *Electromagnetic Concepts and Applications.* Englewood Cliffs, New Jersey: Prentice-Hall, 1982.

[4.2] E. C. Snelling, *Soft Ferrites, Properties and Applications.* London: Iliffe Books Ltd.

[4.3] For additional reading, the following is recommended:

R. J. Parker and R. J. Strudders, *Permanent Magnets and Their Applications.* New York: John Wiley and Sons, 1962.

Chapter 5

Cathodes

The cathode is the source of electrons for the electron beam in every microwave tube. The current density of electron emission from the cathode ranges from milliamperes to tens of amperes per square centimeter of cathode area.

Two mechanisms for the emission of electrons from the cathode are almost universally used. These are:

- Thermionic emission
- Secondary emission

In the large majority of microwave tubes, such as klystrons and traveling wave tubes, only thermionic emission is used. In crossed-field tubes, such as magnetrons and some crossed-field amplifiers, both thermionic and secondary emission are used. In some crossed-field amplifiers, only secondary emission is used.

In this chapter, those aspects of cathodes of primary importance to microwave tube engineers and users are discussed.

J.R. Pierce [5.1] listed the primary characteristics that an ideal cathode should have:

1. would emit electrons freely, without any form of persuasion such as heating or bombardment. Electrons would leak off from it into vacuum as easily as they pass from one metal to another.
2. would emit copiously, supplying an unlimited current density.
3. would last forever, its electron emission continuing unimpaired as long as needed.
4. should emit electrons uniformly and traveling at practically zero velocity.

Of course, real cathodes have none of these ideal characteristics. For example, real cathodes must be heated to a temperature in the vicinity of 1000°C in order to produce appreciable electron emission. At that temperature, the current density is not unlimited, but instead is on the order of amperes per square centimeter at most.

Because of the necessity for a high operating temperature, some of the key constituents of a cathode evaporate, which leads to their depletion and, eventually, to the end of the life of the cathode.

The reason for the fourth characteristic of an ideal cathode is not as obvious as the other three. First of all, it should be noted that in real cathodes, electron emission fluctuates on a microscopic scale because the condition of the cathode surface changes rapidly with time. Secondly, electrons are emitted with finite random velocities in random directions relative to the cathode surface. There are at least two important reasons why an ideal cathode "should emit electrons uniformly and traveling at practically zero velocity." These are:

1. fluctuations in electron emission and variations in electron emission velocities lead to noise currents in the electron beam, which, in turn, produce noise on the output signal from the tube.
2. variations in electron emission velocities and in the directions with which electrons leave the cathode surface lead to problems in focusing the electrons into a well defined beam. This problem is particularly severe in very small, high-density beams.

5.1 THERMIONIC EMISSION [5.2, 5.3]

At temperatures above absolute zero, some electrons have sufficient energy to escape from the surface of a solid. As temperature is increased, the number of electrons with sufficient energy to escape increases. In addition to temperature, the nature of the surface of the solid has an extremely strong effect on the rate at which electrons are emitted. Electron emission resulting from the heating of a surface is referred to as thermionic emission.

The fundamental aspects of thermionic emission may be understood with the aid of Fig. 5.1, which is the classical energy level diagram for electrons near the interface between a solid and vacuum. In the solid, the parabolic curves represent electron energy levels adjacent to atoms. The energy levels merge to form a conduction band. The difference in energy between the top of the conduction band in the solid and the vacuum level adjacent to the solid is known as the work function and is commonly denoted by $e\phi$.

Electron emission from the solid results from electrons in the solid having sufficient momentum directed toward the surface to overcome the potential barrier (work function) and to escape from the solid.

The thermionic emission current can be predicted by considering the density of the electron energy states and the probability of their occupation (see Fig. 5.2). By multiplying the density of states by the probability of occupation, the electron energy distribution as a function of energy is obtained.

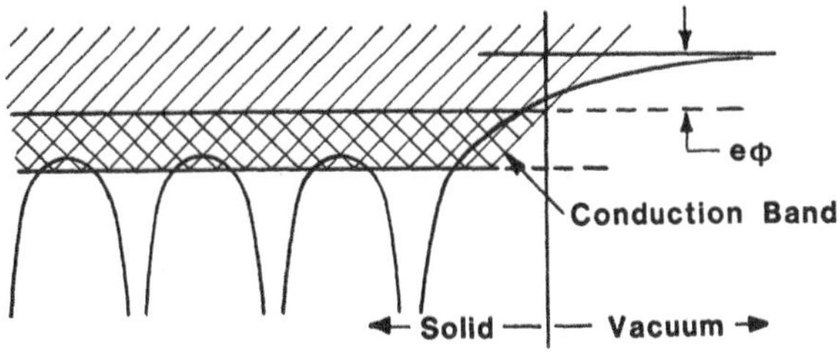

Figure 5.1 Energy Level Diagram for Electrons Near the Surface of a Metal

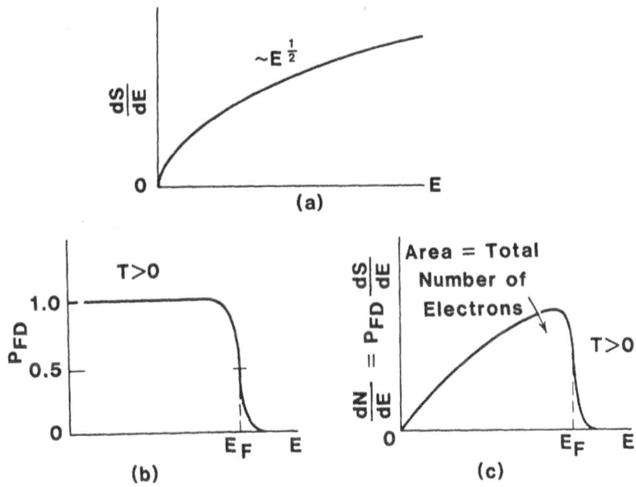

Figure 5.2 Density of States (a), Fermi-Dirac Probability Factor (b), and Energy Distribution Function (c) *versus* Energy

Only electrons in the high energy "tail" of the distribution curve have sufficient energy to overcome the work function barrier and escape from the emitter. The very strong dependence of electron energy in the high energy "tail" on temperature is shown in the Fig. 5.3.

When the theoretical analysis of thermionic emission is carried out, the resulting emission current density, J, is

$$J = \left(\frac{4\pi mek^2}{h^3}\right) T^2 e^{-\frac{e\phi}{kT}}$$

or

$$J = A_0 T^2 e^{-\frac{e\phi}{kT}} \quad \text{A/m}^2$$

which is known as the Richardson-Dushman equation. The various constants are:

m = electron mass
e = electron charge
k = Boltzmann constant
h = Planck's constant
ϕ = work function
T = temperaturue
$A_0 = 1.20 \times 10^6 \text{ A/m}^2 \text{deg}^2$

The most significant aspect of this equation is the exponential variation of current density with work function and temperature. The variation of current density with T^2 is negligible by comparison. This can be readily demonstrated by considering the relative effects of a small change in temperature. For example, for a typical operating temperature of 1000° C, a 1% change in T produces roughly a 70% change in the value of the exponential term, but only a 2% change in T^2.

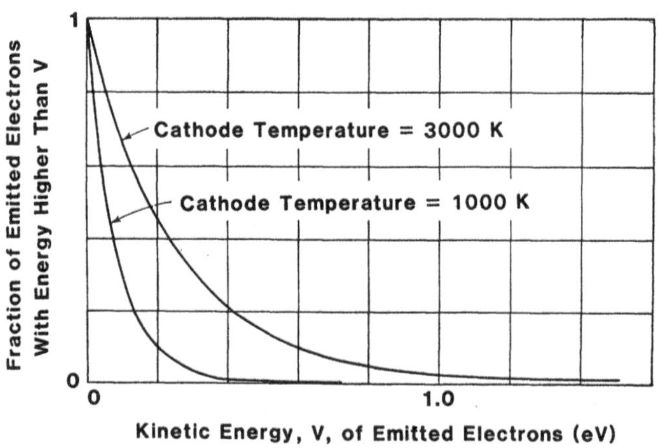

Figure 5.3 Kinetic Energy Distribution of Thermionically Emitted Electrons

(From *Fundamentals of Electronic Motion,* Willis W. Harman, Copyright 1953 by McGraw-Hill, Inc. Reproduced with permission.)

To compare theory with experiment, the Richardson-Dushman equation is first written in the following form:

$$\ln \frac{J}{T^2} = \ln A_0 - \frac{e\phi}{kT}$$

This suggests that a plot of J/T^2 *versus* $1/T$ should be a straight line. The work function, $e\phi$, should be the slope of the line and $\ln A_0$ should be the intercept of the extrapolation of the line to infinite temperature ($1/T = 0$). The data are plotted in the form shown in Fig. 5.4, which is known as a Richardson polt.

In practice, the experimental value of A_0 turns out to be on the order of 1/4 to 1/2 of the predicted value. Two reasons are:

1. The work function has been assumed to be independent of temperature. In fact, $e\phi$ may vary with temperature. A first-order approximation sometimes made is that the incremental change in $e\phi$ varies linearly with temperature. Thus,

   $$\phi = \phi_R + \alpha T$$

   where $e\phi_R$ is the Richardson work function and α is the temperature coefficient of change in work function. Now, the Richardson-Dushman equation may be rewritten as

   $$J = \left[A_0 e^{-\frac{e\alpha T}{kT}} \right] T^2 e^{-\frac{e\phi_R}{kT}}$$

   where the quantity in brackets is the corrected value of A_0.

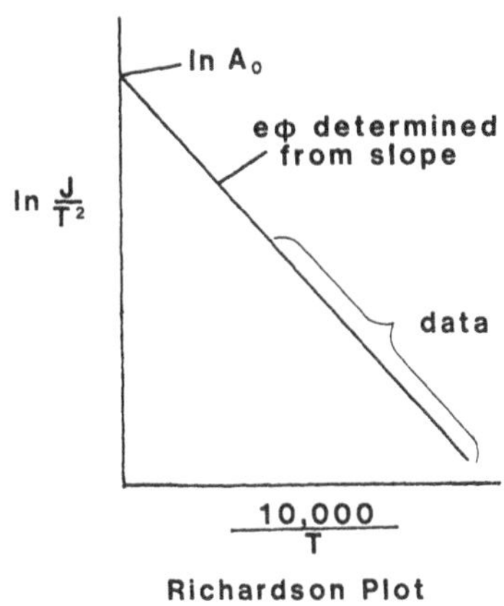

Figure 5.4 Richardson Plot for Determining Work Function and Emission Constant A_o

2. The same value of $e\phi$ has been assumed for all parts of the emitting surface. In fact, $e\phi$ may be different for different crystal surfaces. Suppose, for example, that half of the emitter area has work function $e\phi_1$ and the other half has work function $e\phi_2$. In addition, suppose that $e\phi_2 > e\phi_1$. Then, because of the exponential dependence of emission on work function, practically all of the emission will be from the $e\phi_1$ areas. The emitting area is, therefore, only half of the nominal cathode area and the measured value of A_0 is only half of the theoretical value.

From the Richardson-Dushman equation, it is clear that to obtain high values of emission, the work function must be low and the temperature must be high. Unfortunately, as shown in Table 5.1, those materials with low work functions also have low melting temperatures. Cesium, with a work function of 1.7 eV and a melting temperature close to room temperature (28° C), is an extreme example. Barium also has a low work function, but has a moderate melting temperature (729° C). As a result, barium is commonly used in thermionic cathodes. Still, modern thermionic cathodes operate at temperatures above 1000° C, so the barium is in the molten state during cathode operation. As a result, barium evaporation at a relatively high rate accompanies the operation of these thermionic cathodes.

An indication of the rate of barium evaporation at 1000° C can be obtained by referring to the vapor pressure curves for metals given in Chapter 3. The vapor pressure for bulk barium at 1000° C is nearly 10 Torr. Not only does barium evaporation lead to a finite cathode life, but it also results in the necessity for careful design of the cathode structure so that the evaporated barium does not contaminate critical surfaces such as insulators.

Figure 5.5 shows theoretical emission densities as functions of temperature and work function (dashed lines). Also shown are measured emission densities for several cathodes, many of which contain metals listed in Table 5.1. It is significant that, for the most part, the experimental curves are located very near to where they would be expected, based on listed room temperature work function values.

Some variations of work function with temperature are evident (in the barium-on-nickel and the barium-on-tungsten cathodes, for example). Also, the effect of the base cathode material on the apparent work function of the emissive material can also be seen in some cases. (Compare barium-on-nickel and barium-on-tungsten for example). This will be discussed in Sec. 5.1.5 on the theory of cathode operation.

5.1.1 Schottky Effect

In the derivation of the Richardson-Dushman equation, the effect of an electric field at the cathode surface was not taken into consideration. As a

TABLE 5.1
Room Temperature Work Functions for Various Metals
(From *Fundamentals of Electronic Motion,* Willis W. Harman, Copyright 1953 by McGraw-Hill, Inc. Reproduced with permission.)

Metal	*Work function* *eV*	*Melting temperature* °C
Aluminum	3.0	660
Barium	1.8	720
Carbon	4.8	>3500
Cesium	1.7	28
Copper	4.3	1083
Gold	4.8	1063
Lithium	2.2	186
Molybdenum	4.2	2620
Nickel	5.0	1455
Platinum	6.3	1773
Potassium	1.9	62
Rubidium	1.8	38
Silver	4.6	960
Sodium	2.0	97
Strontium	2.0	800
Tantalum	4.1	2850
Thorium	3.5	1845
Tungsten	4.6	3370

result, neither voltage (from an anode adjacent to the cathode) nor electric field appear in the Richardson-Dushman equation.

In fact, it is found that when an electric field is applied to an emitting surface, emission increases. The reason for this is evident when the potential energy of an electron near an emitting surface is examined. As shown in Fig. 5.6, the applied field reduces the energy barrier that an electron must overcome in order to escape from the emitter.

In effect, the work function is reduced by $e\Delta\phi$ to become

$$e\phi - e\Delta\phi$$

where

$$\Delta\phi = \left(\frac{eE}{4\pi\epsilon_0}\right)^{\frac{1}{2}}$$

and E is the applied electric field.

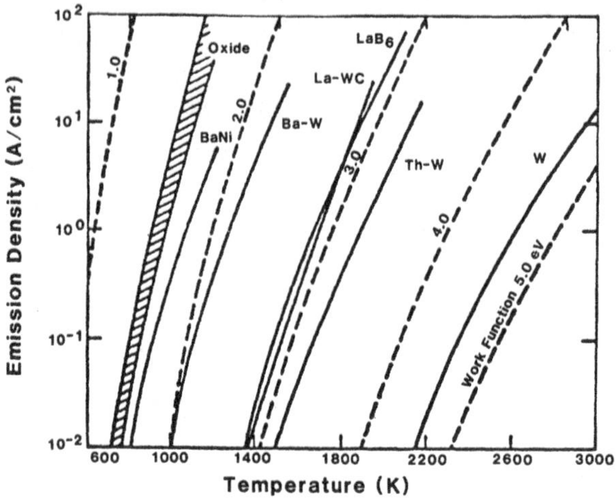

Figure 5.5 Emission Density as a Function of Temperature for Several Cathodes and Work Functions

(From *Methods of Experimental Physics,* Vol. 4, Part A, Section 1.1.1 by G.A. Haas, Copyright 1967 by Academic Press. Reproduced with permission.)

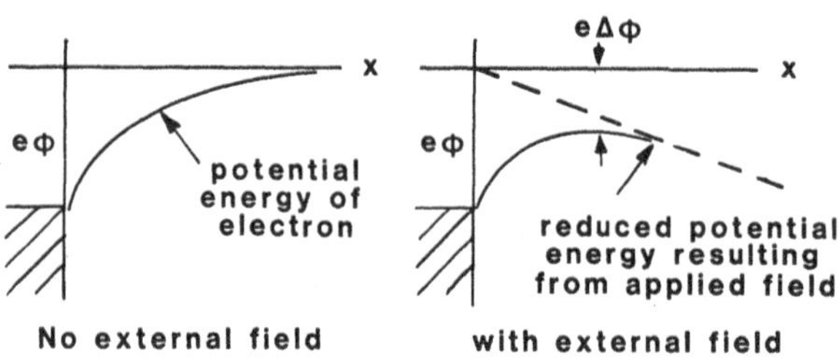

Figure 5.6 Illustration of the Schottky Effect

When the reduced work function is placed in the Richardson-Dushman equation, the modified current density can be written as

$$J = J_0 e^{\left(\frac{eE}{4\pi\epsilon_0}\right)^{\frac{1}{2}}} \tag{5.1}$$

where J_0 is the Richardson-Dushman current density and is sometimes referred to as the "zero field" current density [5.4]. J_0 is a useful parameter for comparing performances of different cathodes.

In practice, to determine J_0 experimentally, it is necessary to extrapolate from values measured at relatively high electric fields. This is because, at low electric field levels, electrons near the cathode surface alter the applied electric field. This is known as the electron space charge effect and will be discussed in detail in subsequent paragraphs.

To determine J_0, it is first noted that (5.1) can be written as

$$\ln J = \ln J_0 + \left(\frac{eE}{4\pi\epsilon_0}\right)^{\frac{1}{2}}$$

Thus, a plot of ln J *versus* E should be a straight line with an intercept, when extrapolated to $E = 0$, of the value of ln J_0. This is illustrated in Fig. 5.7. Notice that, at low values of electric field, the measured value of current density falls below the extrapolated value because of the effect of the charge of the electrons in the region near the cathode surface. Some error in J_0 may result from the change of the true work function by the applied electric field.

5.1.2 Field Emission

As the electric field applied to the surface of a cathode is increased to the $10^9 - 10^{10}$ V/m (0.1 – 1.0 V/Angstrom) level, it is found that electron emission increases very rapidly. Furthermore, the increase in emission is almost independent of cathode temperature.

The electron energy profile is like that for the Schottky effect (Fig. 5.6), except that the applied field is much larger. As a result, as indicated in Fig. 5.8, the width of the potential barrier at the surface of the cathode is very narrow. It is so narrow, in fact, that some electrons can pass through the barrier even though they do not have sufficient kinetic energy to overcome

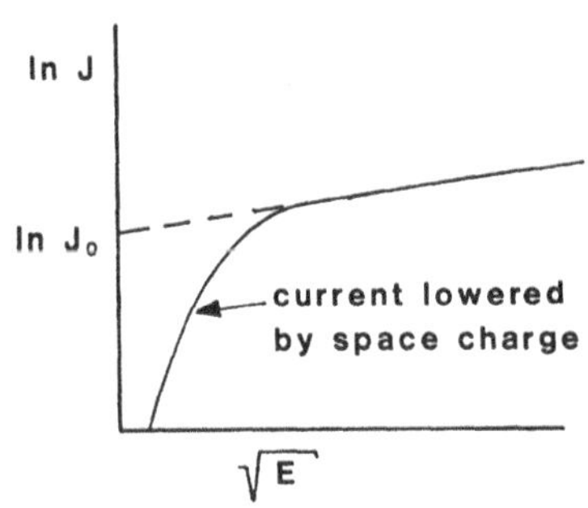

Figure 5.7 Illustration of Extrapolation to Find Zero Field Current Density

the barrier. This effect is known as the tunneling effect and results from the quantum mechanical nature of the electron.

The resulting electron emission is called field emission. Field emission increases exponentially with applied voltage. Because extremely large emission densities are possible at all temperatures, including room temperature, there have been many attempts to develop field emission cathodes for use in microwave tubes. Unfortunately, these attempts have not been successful. The primary reason is that to achieve the high electric field levels necessary for substantial emission, it has been necessary to fabricate the cathodes from arrays of very small needle-shaped emitters. The needle shape is used to enhance the electron emission so that emission occurs at the tip of each emitter, as illustrated in Fig. 5.9.

Emitted electrons ionize gas molecules to form positive ions, some of which impinge on and damage the tips of the emitter. As a result, emission is degraded.

5.1.3 Space-Charge Effect [5.5, 5.6]

The effect of the negative charge of an electron is, of course, to reduce the potential that is present in the absence of the electron. In the region near an emitting cathode where many electrons are present, the reduction in potential can be appreciable.

This is shown for a parallel-plane diode in Fig. 5.10. In the absence of electrons, the potential increases linearly from the cathode to the anode.

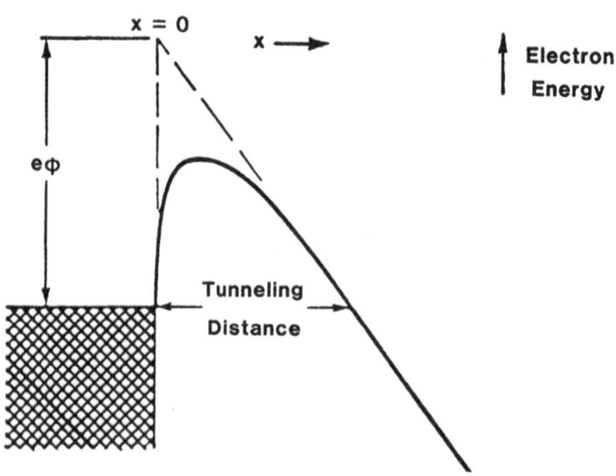

Figure 5.8 Electron Energy Profile with a Very Strong Electric Field

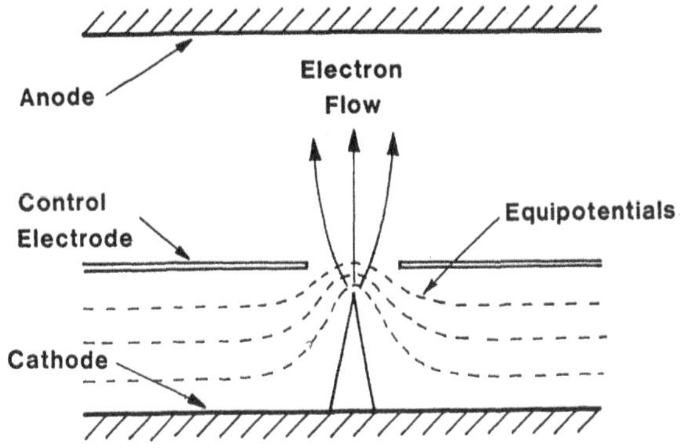

Figure 5.9 Illustration of Field Enhancement and Electron Emission from a Needle-Shaped Emitter

When electrons are present, the potential profile is depressed as indicated by the dashed line.

As the electron emission rate is increased (by increasing temperature, for example) the potential is further decreased. As indicated by Fig. 5.11, there is a limit to how far the electron emission rate can be increased. This limit occurs when the density of electrons adjacent to the cathode surface becomes so large that it tends to depress the potential below zero. If the potential were to become negative, electrons would be forced back to the cathode.

For (A) in Fig. 5.11, the E field at the cathode surface pulls additional electrons from the surface and depresses the potential. For (B) in Fig. 5.11, the E field at the cathode surface forces electrons back to the cathode surface and increases the potential.

An equilibrium condition occurs when, as shown in Fig. 5.12, the potential adjacent to the cathode surface is zero; that is, when the electric field at the cathode surface is zero.

If the potential tends to become positive, more electrons flow from the cathode, thereby depressing the potential. If the potential tends to become negative, fewer electrons flow from the cathode, thereby raising the potential.

When the electric field at the cathode surface is forced to zero by the electron cloud near the cathode surface, the emission is said to be space charge limited. When a sufficient number of electrons is present so that emission is space charge limited, then cathode temperature, surface condition, *et cetera* have no effect on emission. This is indeed a very fortunate situation because it eliminates the necessity for extreme uniformity of temperature across the surface of a cathode. It also eliminates the necessity for precise control of the voltage and current to the heater in the cathode. When the voltage is increased in a diode with a space-charge-limited cathode,

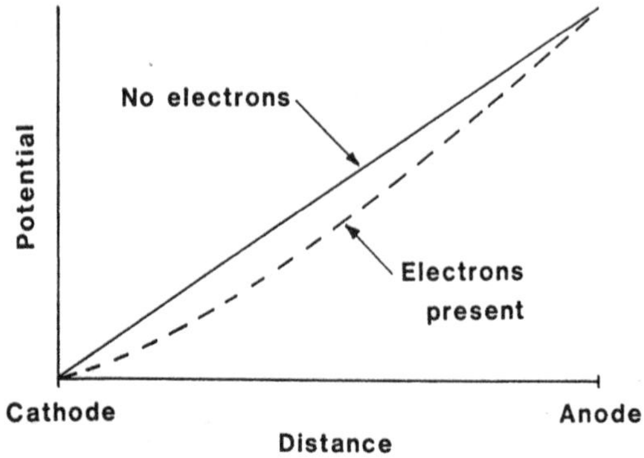

Figure 5.10 Potential Distribution With and Without Electrons in a Parallel-Plane Diode

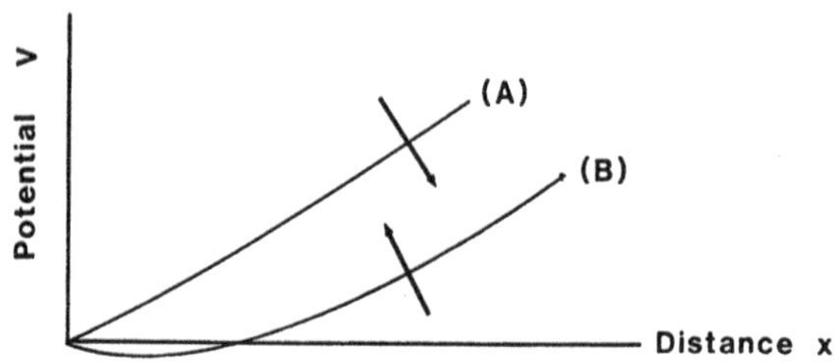

Figure 5.11 Potential Profiles Near Surface of the Cathode:

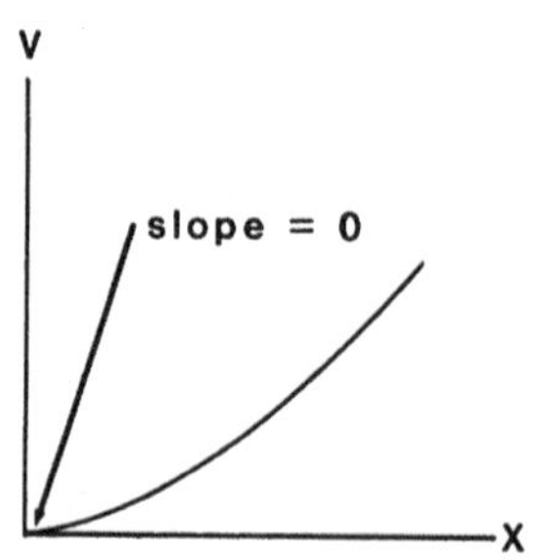

Figure 5.12 Equilibrium Occurs When the Electric Field at the Cathode Surface is Zero

the potential at all points between the cathode and anode tends to increase. As a result, additional electrons flow from the cathode to cause the electric field at the cathode surface to remain at zero. It is evident, therefore, that diode current is dependent on anode voltage.

The relation between voltage and current in a space charge limited diode can be determined in a straightforward manner. This is done by combining relationships for potential distribution as a function of charge density (Poisson's law), current density as the product of charge density and electron velocity, and electron velocity as a function of voltage (energy consideration).

The result is known as the Child-Langmuir law for the flow of electrons in a space charge limited (SCL) diode. For the parallel-plane diode, this law is

$$J = \frac{4}{9} \epsilon_0 (2\eta)^{1/2} \frac{V^{3/2}}{x^2} \tag{5.2}$$

where

J = current density
ϵ_0 = dielectric constant of free space
η = electron charge to mass ratio
x = distance from cathode
V = potential at x

For a diode with a cathode-to-anode spacing $x = d$, (5.2) is commonly written as

$$I = PV^{3/2} \tag{5.3}$$

where $I = JA$, with A being the area of the cathode, and P is known as the perveance of the diode, which is

$$P = \frac{4}{9} \frac{A\epsilon_0}{d^2} (2\eta)^{1/2}$$

$$= 2.33 \times 10^{-6} \frac{A}{d^2}$$

for the parallel-plane diode.

It is important to note that perveance is a function only of the geometry of the diode. SCL diodes with geometries other than planar (cylindrical, spherical, *et cetera*) all obey $I = PV^{3/2}$, where the expression for perveance is the one appropriate for the geometry used.

In practice, SCL diodes behave exactly as predicted. This is readily demonstrated by plotting current as a function of voltage using 3/2 power graph paper. As shown in Fig. 5.13 the resulting plot is a straight line for SCL operation. The slope of the line is the perveance.

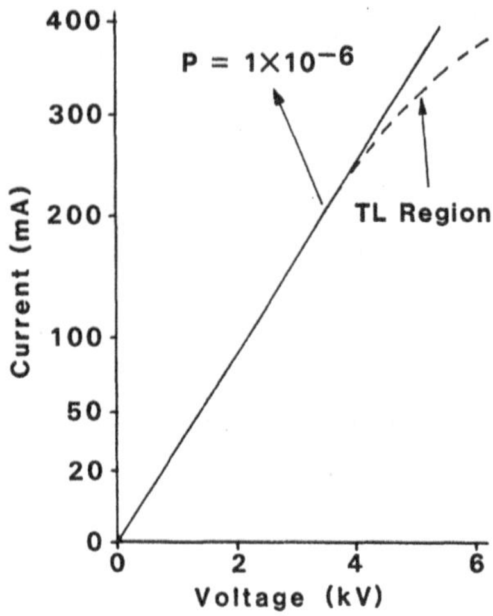

Figure 5.13 Current *versus* Voltage Plotted on 3/2 Power Graph Paper

When the voltage is increased to a high enough value so that the cathode cannot supply sufficient current to maintain the SCL condition, the cathode is said to be TL (temperature limited). The TL region is readily apparent in the 3/2 power *I*-*V* graph because the plotted results depart from a straight line.

In Fig. 5.14, current is shown plotted as a function of temperature for a thermionic cathode. In the region of TL operation, current increases with temperature. At high temperatures, current is independent of temperature, indicating SCL operation. As voltage is increased, the temperature at which the transistion from TL to SCL operation occurs is increased. As indicated in Fig. 5.15, the TL portion of a current *versus* temperature curve is translated to higher temperatures as a function of time.

It is necessary to select an operating temperature for a cathode that remains above the knee of the curve during the life of the cathode. Of course, the rate at which the TL curve moves increases with increasing temperature, so an excessive temperature will reduce cathode life.

In Fig. 5.15, there is a slight upward slope with increasing temperature in the SCL region. This is commonly observed and results from movement of the cathode toward the anode as a result of thermal expansion of elements of the cathode support structure.

The various mechanisms dominating current flow in a thermionic diode are summarized in Fig. 5.16. At low anode voltages, current is space charge

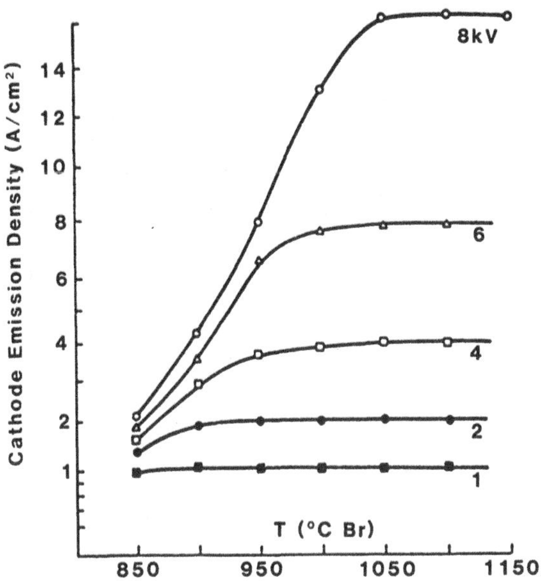

Figure 5.14 Emission as a Function of Temperature and Anode Voltage

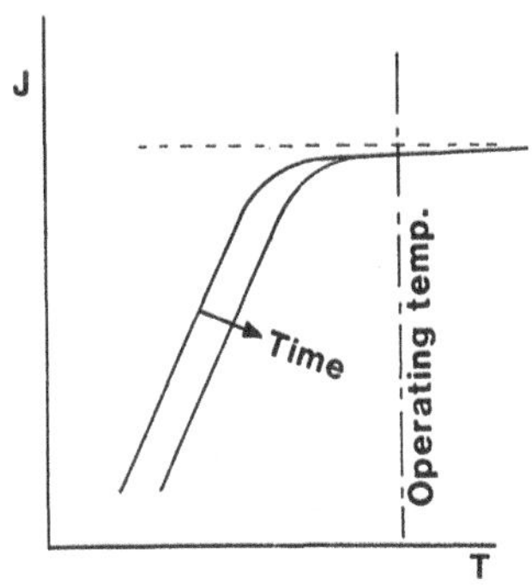

Figure 5.15 Movement of TL Curve with Time

limited. As voltage is increased, the emission limit of the cathode is reached and current becomes temperature limited. Current continues to increase slowly with voltage because of the Schottky effect. At high voltages current increases rapidly because of field emission.

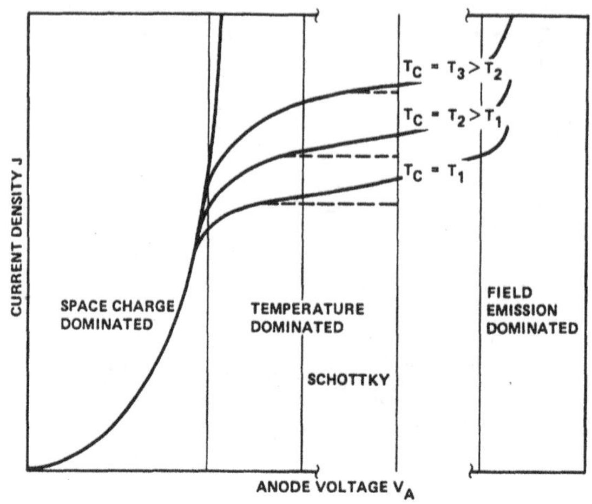

Figure 5.16 Summary of Mechanisms Dominating Current Flow Through a Thermionic Diode

5.1.4 Evolution of Thermionic Cathodes

The first thermionic cathodes used in quantity were those used in early radio tubes. They were made from pure tungsten, although other cathode materials were known which were capable of producing adequate emission at lower temperatures than tungsten.

Because of the high work function (~ 4.6 eV), the operating temperature was high. For example, 2200°C yields about 0.3 A/cm^2.

Apparently, tungsten was used in early tubes rather than other cathodes because it was foolproof. A tungsten cathode always had plenty of emission. Other cathodes were temperamental and were sometimes "poisoned" by mysterious agents.

For some applications, a small percentage of thoria was mixed with the tungsten. Thorium would diffuse to the surface of the tungsten and reduce the work function, thereby permitting the use of a lower operating temperature. Thoriated tungsten cathodes are still in common use in high-power switch tubes, transmitter tubes, and microwave-oven magnetrons. The cathode surface is sometimes carburized to reduce the evaporation rate of the thorium. The life expectancy of high-power switch tubes containing thoriated tungsten cathodes is plotted in Fig. 5.17.

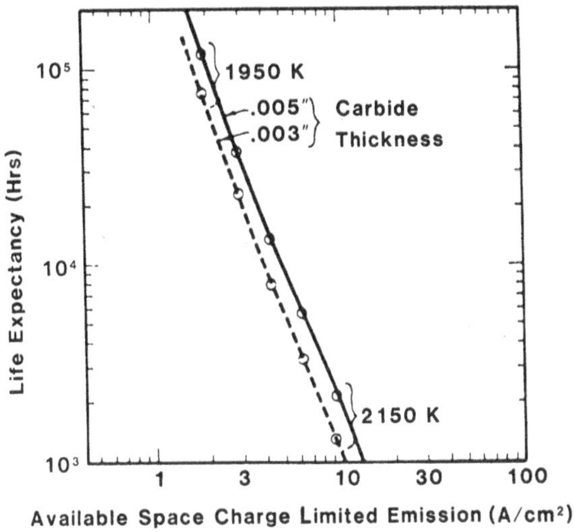

Figure 5.17 Life Expectancy for a Switch Tube with a Thoriated Tungsten Cathode (Actual Tube Life ≈ 1/2 of Life Expectancy)

(From T.E. Yingst, *et al.*, *Proc. IEEE*, March 1973.)

Early cathodes were all directly heated; that is, the filament, which was heated by passing a current through it, was also the cathode. Thus, as a result of the heater current, there was a voltage variation along the surface of the cathode. Because batteries were used as power sources, the filament voltage was dc, and so there was no modulation of the emission current.

The development of ac radios and the need to heat the filaments with ac current brought about the need for a cathode, the surface of which was all at the same potential. The indirectly heated cathode resulted, whereby the heater was separate from the electron emitting surface. Directly and indirectly heated cathodes are illustrated in Fig. 5.18.

In the indirectly heated cathode, tungsten is used for the filament. Because of heat losses and the relatively large cathode area, the cathode surface had to operate at a temperature substantially below the heater temperature. This provided the requirement for a low work function emitting surface.

The oxide-coated cathode, which had been discovered long before radio tubes by Wehnelt in 1904, provided the solution to the problem because of its very low work function and low operating temperature.

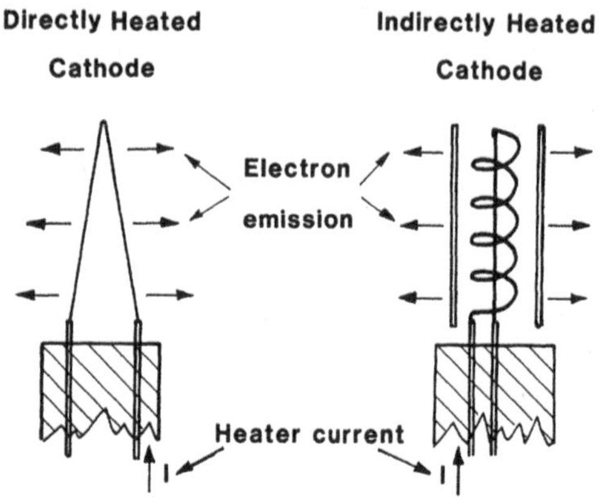

Figure 5.18 Comparison of Directly and Indirectly Heated Cathodes

Oxide-coated cathodes served the industry well from the 1920s through World War II and into the early 1950s. They are still prevalent today in applications requiring relatively modest cathode emission and long life.

Early oxide-coated cathodes were described by J.R. Pierce [5.1] as follows:

> The early history of oxide coated cathodes seemed without rhyme or reason. It was found that if a wire intended to emit electrons was heated and rubbed with sealing wax and the sealing wax then burned off, a truly phenomenal cathode was produced. Further work traced the emission to alkaline earth oxides, and the compound from which they came was humorously described as a barium resinate. The sealing wax technique died hard, and for several years the process of coating filaments consisted in stringing fine platinum ribbon out in long lengths, heating it with an electric current, and rubbing a resinous compound containing barium and strontium over it repeatedly. The narrow ribbon was then cut into lengths, put into a crude vacuum tube of the day, the tube was pumped and baked until a "hard vacuum" was achieved, and the filament was activated. This activation was a wishful process consisting of glowing the filament brightly, applying voltage, and drawing current to the plate until, in happy circumstances the plate got red-hot, giving off enough gas to produce a blue glow and further increase the activity and the current. After a lucky period of this treatment the filament became active; that is, it would emit electrons at a lower and reasonable temperature, at which it would last for a number of thousands of hours.

The art and science of oxide cathodes is now highly developed. A mixture of barium, strontium, and calcium carbonates is commonly used. These are applied to a metal base (usually nickel) and then activated in vacuum. Activation consists of heating and reducing the carbonates to oxides.

As oxide-coated cathodes were improved, dc emission increased from mA/cm^2 to hundreds of mA/cm^2. Pulsed emission densities are dependent on pulse length and reportedly can be as high as tens of A/cm^2.

The search for better cathodes than oxide-coated cathodes resulted from their low dc emission and susceptibility to damage. The L cathode, illustrated in Fig. 5.19, was developed by Philips about 1950. In this cathode, the emissive material was sandwiched between the heater and a porous tungsten pellet. During operation, emissive material diffused through the porous tungsten and formed a low work function emitting surface.

The L cathode provided good emission and long life. Some of the difficulties encountered in its use were the following:

1. There was leakage of barium vapor between the tungsten and molybdenum.
2. The temperature uniformity of the cathode surface was poor because of poor heat transfer through the cavity containing the emissive mix.
3. Outgassing and activation were difficult.

From 1950 to 1955, the evolution of the impregnated dispenser cathode illustrated in Fig. 5.20 occurred. The Philips B-type cathode developed by Levi in 1953 is still the most widely used cathode in the microwave tube industry.

The emissive mix is impregnated into the porous tungsten cathode pellet. The most common mix consists of five parts BaO, three parts CaO and two parts Al_2O_3 and is frequently referred to simply as 5:3:2. Other variations, such as 3:1:1 and 4:1:1, are in widespread use.

The B-type cathode is rugged, relatively impervious to damage, and can readily provide an emission density of several A/cm^2. At these high emission densities, the operating temperature is 1100° C or higher, and so life is limited to a few thousand hours.

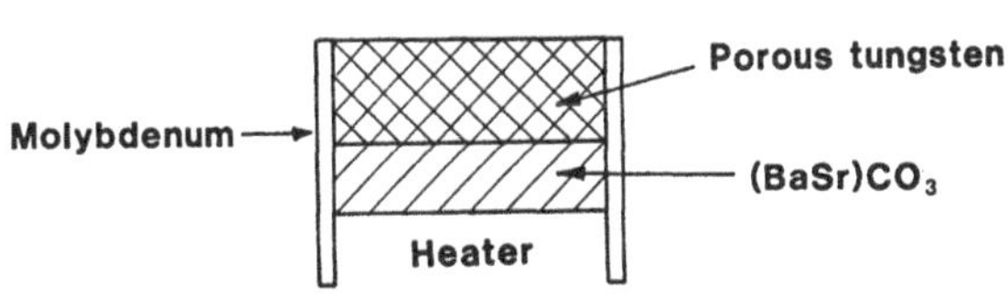

Figure 5.19 Essential Features of Philips L Cathode

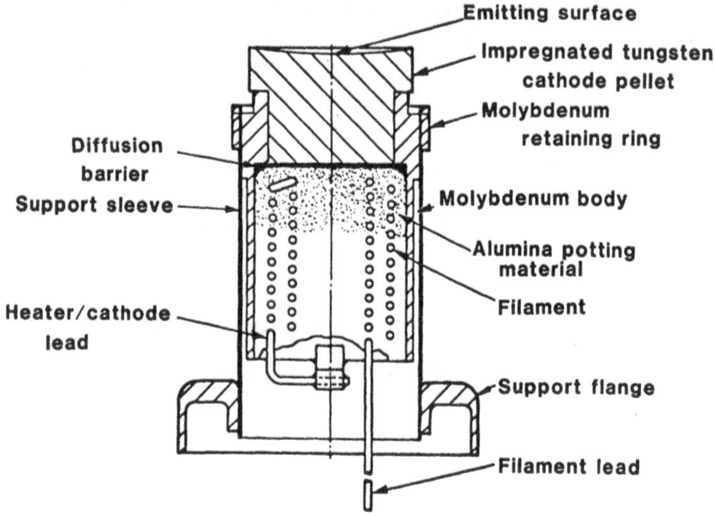

Figure 5.20 Typical B-type Impregnated Dispenser Cathode

The search for a cathode capable of an emission density comparable to or higher than the B-type cathode, but with a lower operating temperature and longer life, led to the development of the M-type cathode in the 1960s. Basically, the M-type cathode is a B-type with a thin film several thousand Angstroms thick of osmium, iridium, or rhenium applied to the surface. The effect of the thin film is to reduce the cathode operating temperature about 90° C (dependent on film metal) for the same emission density as is obtained from a B cathode.

Unfortunately, when higher current densities than a few A/cm^2 are required, the temperature is sufficient to cause the thin coating to diffuse into the tungsten matrix and the M-type cathode becomes a B-type.

Some tube manufacturers have been reluctant to use M-type cathodes because of reported problems with nonuniform production and life deterioration resulting from degradation of the thin enhancing film (by diffusion or ion bombardment, for example).

A recently observed difficulty with M-type cathodes is in obtaining the expected emission as soon as the cathode reaches the operating temperature. In one case, emission is delayed about four minutes. This emission delay is intolerable in fast-warm-up applications.

Since about 1975, the mixed metal matrix (MMM) cathode has received considerable attention in an effort to overcome the problems in life deterioration because of degradation of the surface of the M-type cathode. In the MMM cathodes, the enhancing metal (iridium or osmium) is mixed in powder form with tungsten and then pressed and sintered to form a porous matrix.

When iridium is used, it is found that initial performance is excellent (4 A/cm^2 at <1000°C), but emission decays rapidly after 500 hours of operation. When osmium is used, initial performance is comparable to a B-type cathode, but after 400 hours the cathodes stabilize to a value that is about 50% better than the M-type.

More recently, new types of cathodes have been investigated. In one, a tungsten-osmium alloy is deposited on the surface of a porous tungsten matrix. In another, a thin foil of tungsten with an array of laser drilled or ion etched pores is used to permit the dispensing of the emissive material.

These new types of cathodes are the subject of ongoing research, which it is hoped will lead, not only to the development of significantly improved cathodes, but also to a detailed understanding of the physics of cathode operation.

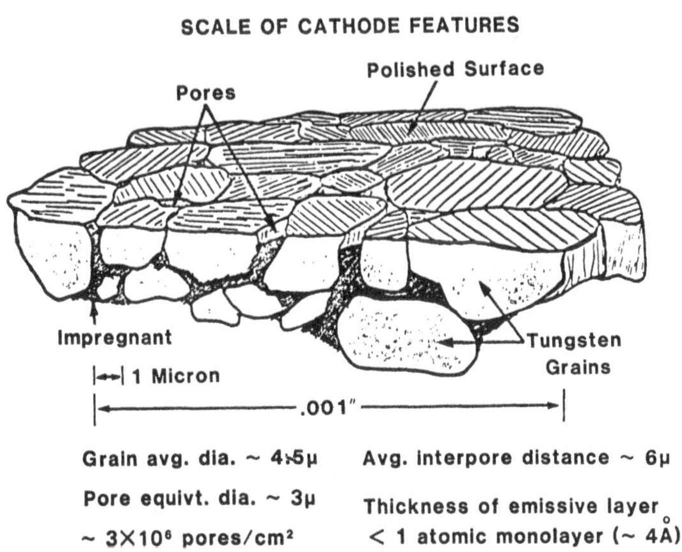

Figure 5.21 Scale of Features of an Impregnated Cathode

(From M.C. Green, Technical Report RADC-TR-81-211, July 1981.)

5.1.5 Impregnated Dispenser Cathode Operation [5.7, 5.8]

Figure 5.21 indicates the scale of the features of the surface of a typical B-type, M-type, or MMM cathode. In particular, it should be noticed that all dimensions are typically a few microns.

When the cathode is heated to the operating temperature, barium is released from the interior of the tungsten matrix by reactions of the emissive mix with the tungsten. The freed barium then migrates to the surface of the emitter. Some of the barium spreads out over the emitter surface and eventually evaporates.

The fraction of the surface that is covered depends on the rate of barium evaporation from the surface. As shown simplistically in Fig. 5.22, there are islands surrounding the pores in the emitter surface that are covered by barium.

The sizes of the islands depend on the temperature and age of the cathode. As temperature is increased, the rate at which barium migrates to the surface increases, as do the sizes of the islands. As the cathode ages and the pores in the cathode surface are depleted, the rate of migration of barium to the surface decreases, and so the islands become smaller.

In those portions of the emitter surface covered by barium, the barium atoms interact with, and reduce the work function of, the surface. In those regions not covered by barium, the work function is that of the base emitter material.

Because of the large differences between the work functions of the areas covered by barium and those not covered by barium, some portions of the cathode operate in the space-charge-limited condition and others operate in the temperature-limited condition. In fact, Longo [5.9, 5.10] has shown

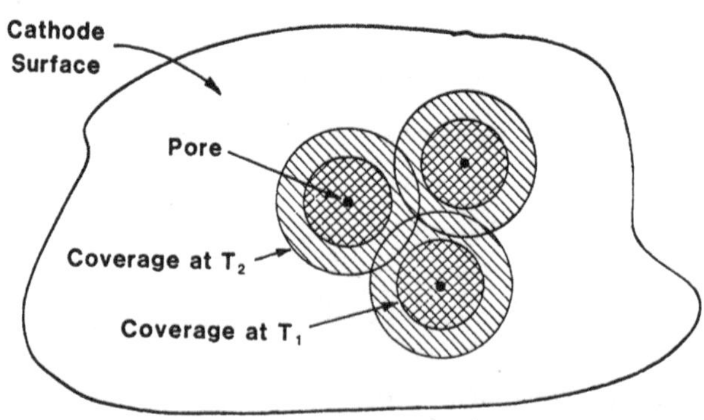

Figure 5.22 Simplified View of Barium Coverage of Cathode Surface at Two Temperatures T_2 and T_1 where $T_2 > T_1$

empirically that the following simple equation describes the observed emission from an impregnated dispenser cathode over a wide range of voltages and temperatures. Thus, we have

$$\frac{1}{J_{OP}} = \frac{1}{J_{SC}} + \frac{1}{J_{TL}} \tag{5.4}$$

In this equation, J_{OP} is the observed current density, J_{SC} is the current density that would be observed if the entire cathode was operating space-charge-limited, and J_{TL} is the current density that would be observed if the entire cathode was operating temperature-limited. Shown in Fig. 5.23 is a comparison of (5.4) with measured results.

Using (5.4) and assuming the mechanism by which barium migrates to the surface of the cathode (gas flow under conditions known as Knudsen flow), Longo shows that current decay from an impregnated dispenser cathode may be expressed as

$$\frac{\Delta J}{J} = Ae^{\frac{a}{\sqrt{t}}}$$

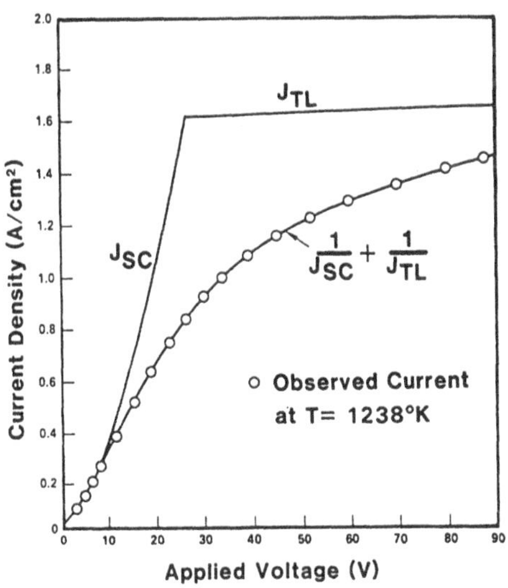

Figure 5.23 Comparison of Longo's Empirical Relation with Measured Results

(From R.T. Longo, *Technical Digest*, 1978 IEDM.)

In this equation, $\Delta J/J$ is the percentage change in current, t is time, and A and a are functions that are independent of time. Equation (5.5) is shown to be valid for long operating times in Fig. 5.24

5.1.6 Life Considerations

In the great majority of cathode applications, long life is highly desirable. Long life normally implies that the life should be as long as possible to help minimize tube replacement costs, or to help limit redundancy costs (in satellite applications for example).

In a very limited number of applications (decoys and space defense, for example) cathode life can be very short. Except for the very fast warm-up times and long shelf lives that these applications also usually require, there does not appear to be a cathode problem in these applications.

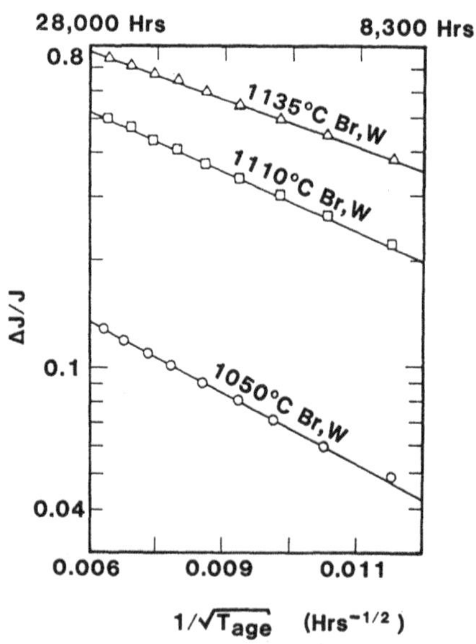

Figure 5.24 Comparison of Life Test Data with Predictions of Surface Depletion Model

(From R.T. Longo, *Technical Digest*, 1978 IEDM.)

Cathode life can be indicated in many ways. For example, Figs. 5.25 and 5.26 show Miram plots for a standard B-type cathode and a MMM cathode. In Miram plots, cathode emission data are always plotted in the same way to facilitate comparison of cathodes. The current density is normalized with respect to the space-charge-limited value. Also, values of anode voltage are normally chosen so that emission densities are separated by factors of 2. From the 3/2 power relationship between current and voltage, this means that anode voltages are separated by factors of $2^{2/3}$ or 1.59.

As Figs. 5.25 and 5.26 show, in the TL region, the curves are translated to higher and higher temperatures as cathodes are operated. The key to optimizing cathode life is to choose an operating temperature that remains slightly above the knee of the curve during the life of the cathode. It must be kept in mind that the rate at which the TL curve moves increases with increasing temperature. Thus, the operating temperature should not be any further above the knee than is absolutely necessary.

For the person interested in long-life applications of tubes, data of the form given in Fig. 5.27 are useful. This figure shows the relation between cathode life and the operating current density that can be expected if that life is to be achieved.

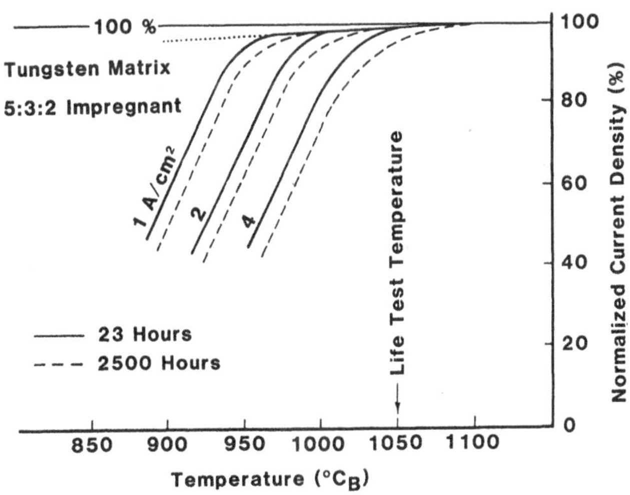

Figure 5.25 Miram Plot for Standard Cathode

(From M.C. Green, Technical Report RADC-TR-81-211, July 1981.)

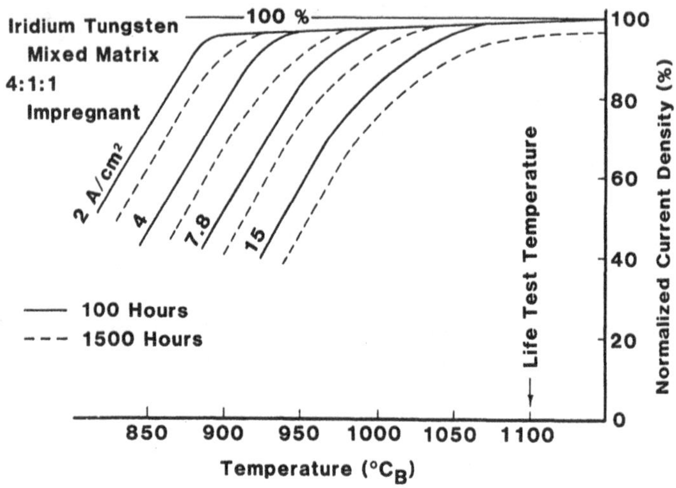

Figure 5.26 Miram Plot for Mixed Matrix Cathode

(From M.C. Green, Technical Report RADC-TR-81-211, July 1981.)

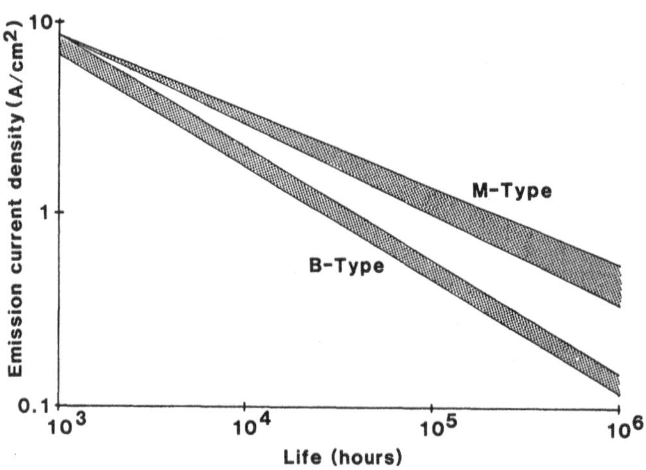

Figure 5.27 Operating Current Densities that Can be Expected as a Function of Desired Cathode Life

5.1.7 Dispenser Cathode Surface Physics

The basic theory of thermionic emission discussed at the beginning of this chapter is well established and understood. What is not well understood are the work functions of surfaces with adsorbed atoms of various materials. As a result, cathodes have historically been developed more by trying various combinations of materials and techniques, rather than by design based on a thorough knowledge of surface physics.

Fortunately, however, the modern state of cathode theories is significantly advanced from that given by Pierce in 1946 [5.1].

> It is clear that as far as science goes, the state of the oxide coated cathode is confused indeed. Not only does theory fail to help us make good cathodes; it seems not to fit all the evidence. Maybe cathodes work sometimes one way and sometimes another, which isn't as ridiculous as it may sound. At any rate, while some physicists are uninterested and others confused, cathode making continues to be an art practiced, it seems, by artists with a good deal of witch doctor in their make-up. When something goes wrong, this or that demon is exorcised, and eventually things go right again. And, unlikely as it seems, cathodes continue to improve.

Still, while cathode theories have become very advanced, it is not yet possible in theory to design a new type of cathode having a predetermined emission density, operating temperature, and life.

Consider, for example, the apparently contradictory nature of the information provided in Fig. 5.28. M-type cathodes are fabricated by applying a very thin coating of osmium, iridium, or rhenium to the tungsten surface of a B-type cathode. The emission enhancement factors for the M-type cathode compared with the B-type are from about 2.7 for rhenium to about 3.7 for osmium. Emission is enhanced even though the work functions or rhenium, osmium, and iridium are significantly above that for tungsten.

The explanation for the emission enhancement observed in the M cathode is not straightforward. Detailed measurements [5.7] show that barium and oxygen atoms form a sandwich-like layer on the surface of the cathode. Oxygen atoms form a monolayer on the metallic cathode surface. It appears that for each oxygen atom, there is a barium atom in a layer on top of the oxygen. The barium and oxygen atoms are oriented by their interaction with the metallic cathode surface (see Fig. 5.29) to produce an overall reduction in work function that is dependent on the electronic nature of the metallic cathode surface.

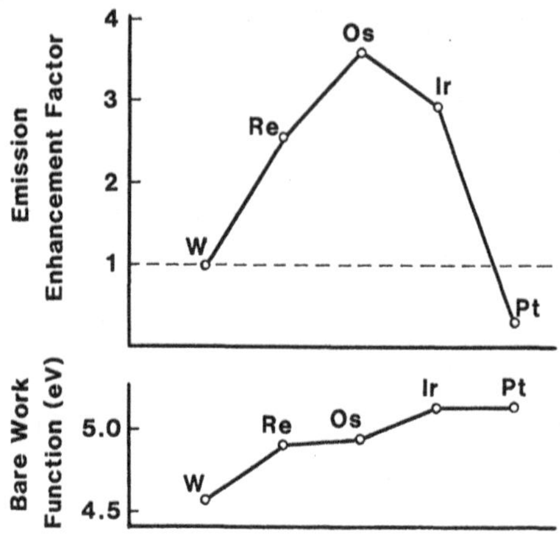

Figure 5.28 Emission Enhancement and Work Function for M-type Cathode Enhancement Metals

(From M.C. Green, Technical Report RADC-TR-81-211, July 1981.)

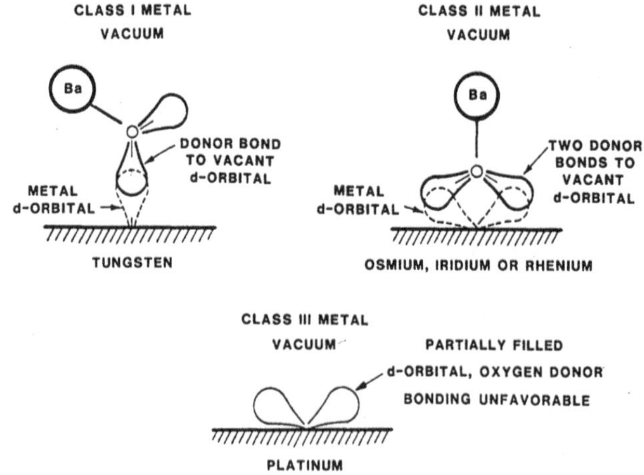

Figure 5.29 Bonding of Barium and Oxygen Atoms to Metallic Surface of Cathode

(From M.C. Green, Technical Report RADC-TR-81-211, July 1981).

The development of a detailed understanding of the exact nature of the interactions at the surface of the dispenser cathode is the subject of cathode research at many organizations. It is hoped that, once the interactions are understood, it will be possible to engineer cathodes.

5.2 SECONDARY EMISSION

When electrons bombard a surface, they may cause other electrons to be emitted from the surface. The bombarding electrons are called primary electrons and the emitted electrons are called secondary electrons. Secondary emission is of extreme importance in the cathodes of crossed-field tubes where the electric and magnetic field configurations cause many primary electrons to bombard the cathode.

Often, a small percentage of the cathode emission is thermionic and the remainder is secondary. A good example is the Philips YJ1183 spin-tuned magnetron in which it is estimated that the thermionic emission is 3 A/cm^2 and the secondary emission if 23 A/cm^2 for a total cathode emission of 26 A/cm^2.

For many cathode-surface materials, a high secondary emission yield is accompanied by a relatively short cathode life. For example, electron back-bombardment leads to dissociation of barium oxide, which is commonly the emitting surface in a magnetron or crossed-field amplifier. The only cathodes free from the deleterious effects of back-bombardment are those fabricated from pure metal or from carburized thoriated tungsten.

In some crossed-field amplifiers, cold cathodes are used and, thus, all of the emission is secondary. Emission is initiated by the relatively small number of free electrons that are always present (because electrode surfaces are not at absolute zero, interaction of electrodes with background radiation, *et cetera*). Some of these electrons are driven into the cathode by the fields.

In the secondary emission process, when a primary electron strikes a solid, it gives up energy by repelling electrons within the solid. The electrons within the solid thereby gain energy and may or may not be emitted, depending on how much energy they receive, how close they are to the surface, and the direction of their motion.

The amount of secondary emission depends on the secondary emission ratio and on the impact energy and angle of incidence of the primary electrons.

The secondary emission ratio for several common materials is shown in Fig. 5.30. Materials used in microwave tubes normally have relatively high secondary ratios as indicated in Figs. 5.31 and 5.32. An oxide-coated cathodes may have an exceedingly high secondary emission ratio (as high as 10[5.3]).

The importance of the angle of incidence of the primary electrons is indicated in Figs. 5.32 and 5.33.

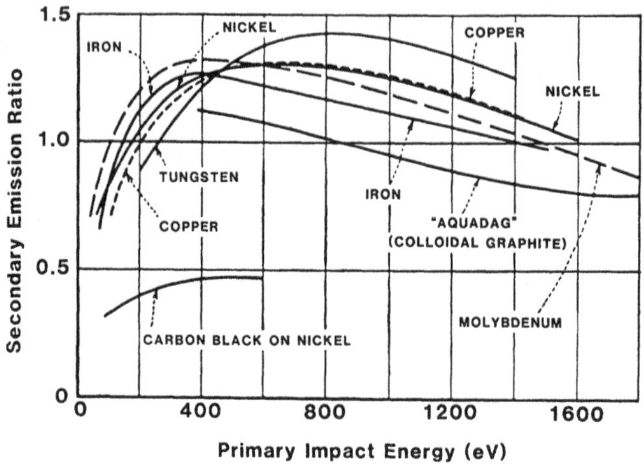

Figure 5.30 Secondary Emission Ratio for Several Common Materials

(From *Fundamentals of Electronic Motion* by Willis W. Harman, Copyright 1953 by McGraw-Hill, Inc., Reproduced with permission.)

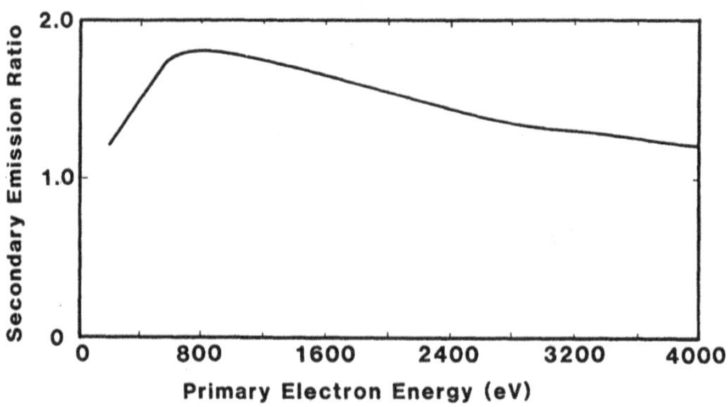

Figure 5.31 Secondary Emission Ratio for Platinum

(From John F. Skowron, *Proc. IEEE*, March 1973.)

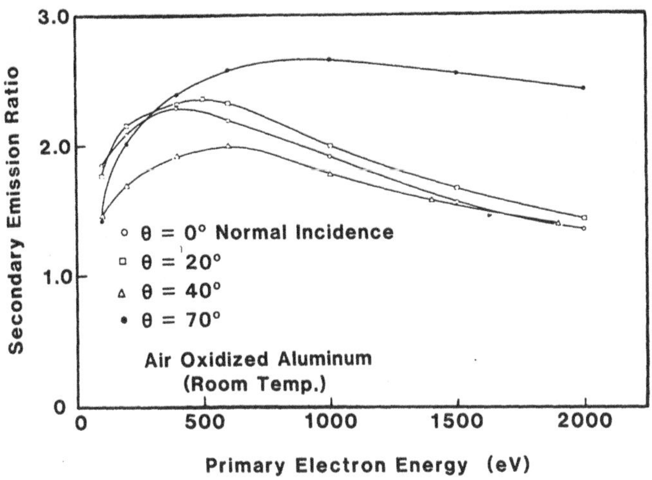

Figure 5.32 Secondary Emission Ratio for Oxidized Aluminum

(From John F. Skowron, *Proc. IEEE*, March 1973.)

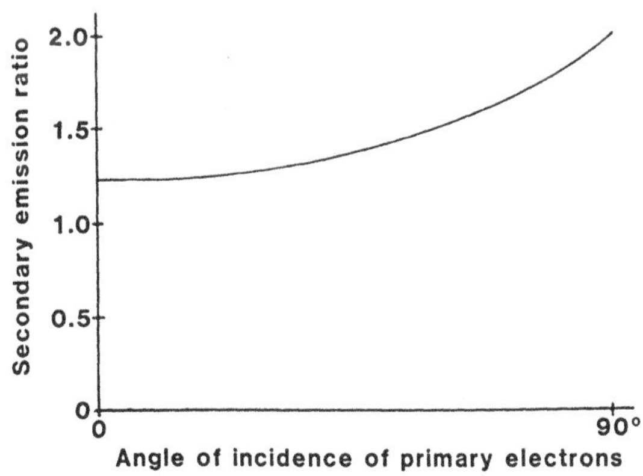

Figure 5.33 Typical Variation of Secondary Emission Ratio with Primary Electron Angle of Incidence

Materials used as secondary emission cathodes in microwave tubes include pure platinum, oxidized beryllium or aluminum, and conventional B-type cathodes.

A typical distribution for the energies of emitted secondary electrons is shown in Fig. 5.34. Perhaps the most significant aspect of this energy distribution is the high values of electron energies (compared with thermionically emitted electrons). Thermionic electrons have energies of a small fraction of an electron volt. Secondary electrons have energies two orders of magnitude higher. This is partially the cause of the relatively high noise output of crossed-field devices.

The secondary emission coefficient of a surface is very sensitive to the presence of contaminants. Thus, for example, emission from a platinum cathode will remain constant as long as the surface remains uncontaminated, but will usually decrease if coated by foreign materials [5.11].

In the case of impregnated cathodes, secondary emission depends on the effective surface material, which can change with time due to depletion, erosion, or change in chemical composition under electron bombardment.

At peak power levels of less than 200 kW, cathodes with metal oxide surfaces of beryllium or aluminum are often used. Special provisions are often made to provide a low background pressure of oxygen to replenish the oxide surface as it is depleted due to various processes. Oxygen is provided by heating cupric oxide to a controlled temperature where it decomposes and releases oxygen.

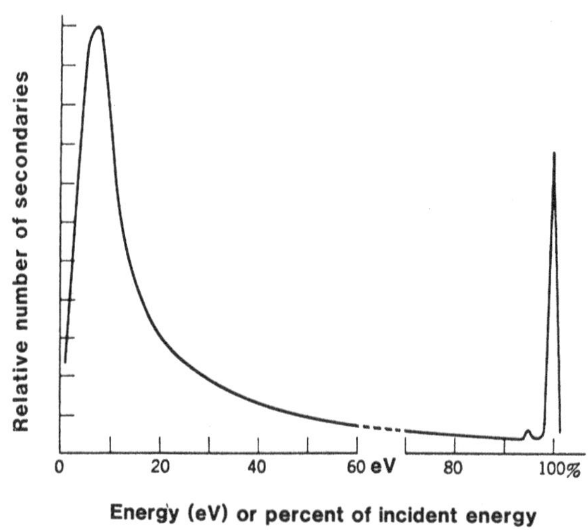

Figure 5.34 Distribution of Energies of Secondary Electrons

REFERENCES

[5.1] Material extracted from an article by J.J. Coupling, a pseudonym of J.R. Pierce, which appeared in the November 1946 edition of *Science Fiction*, a Street and Smith publication

[5.2] Additional suggested reading: G. Herrman and P.S. Wagener, *The Oxide-Coated Cathode*. London: Chapman and Hall, 1951.

[5.3] Additional suggested reading: R.L. Sproull, *Modern Physics*. New York: John Wiley and Sons, 1956.

[5.4] J.L. Cronin, "Modern dispenser cathodes" *IEE Proc.* vol. 128, pt. 1, no. 1, Feb. 1981, pp. 19-32.

[5.5] Additional suggested reading: C.L. Hemenway, R.W. Henry, and M. Caulton, *Physical Electronics*. New York: John Wiley and Sons, 1967.

[5.6] Additional suggested reading: M. Chodorow and C. Susskind, *Fundamentals of Microwave Electronics*, New York: McGraw-Hill, 1964.

[5.7] Additional suggested reading: M.C. Green, "Dispenser cathode physics," Final Technical Report, RADC-TR-81-211, July 1981.

[5.8] Additional suggested reading: M.C. Green, "Dispenser cathode design" Final Technical Report, RADC-TR-83-99, May 1983.

[5.9] R.T. Longo, "Long Life, High Current Density Cathodes," *Technical Digest*, 1978 International Electron Devices Meeting, Dec. 1978, pp. 152-158.

[5.10] R.T. Longo, "A Study of Thermionic Emitters in the Regime of Practical Operation," *Technical Digest*, 1980 International Electron Devices Meeting, Dec. 1980, pp. 467-470.

[5.11] Technical Manual, Installation, Operation, Maintenance, Care and Handling Instructions, General, *Microwave Tubes, Magnetron Tubes, Electron Tubes*, by Varian Associates. Air Force Publication Number T.O. 00-25-251, 1 October 1979.

Chapter 6

Electron Guns for Linear-Beam Tubes

In linear-beam microwave tubes, the electron beam essentially flows in a straight line. In nearly all linear-beam tubes, an axial magnetic field (aligned with the axis of the beam) is used to focus the beam. Linear-beam tubes are frequently referred to as "O-type" devices. (The names O-type and M-type (for crossed-field tubes) originated in France and come from the French designations for linear-beam and crossed-field tubes.)

Linear-beam tubes are divided into two major categories: klystrons and traveling wave tubes (TWTs). In addition, there are hybrid devices in which klystron and TWT technologies are combined.

The electron guns for all types of linear-beam tubes have operating principles that are essentially identical. The only major differences are in size and operating voltage and current, which result from differences in power requirements for the tubes. This chapter contains a discussion of the operating principles of the electron guns used on nearly all linear-beam tubes. These guns are named after J.R. Pierce and are known as Pierce guns [6.1, 6.2].

6.1 PIERCE GUNS

The electron gun in a microwave tube is used to form the electrons from the cathode into a beam suitable for interaction with a microwave circuit. Electron gun designers must cope with the following two basic problems:

1. The electrostatic repulsion forces between electrons tend to cause the beam to diverge.
2. The current density required in the electron beam is normally far greater than the emission density of the cathode.

6.1.1 Pierce Electrodes

Dealing first with the repulsion force problem, it is instructive to examine the problem of attempting to cause electrons to travel in parallel paths as they leave a cathode as indicated in Fig. 6.1. Space charge forces will be considered, but it will be assumed that no magnetic field is present.

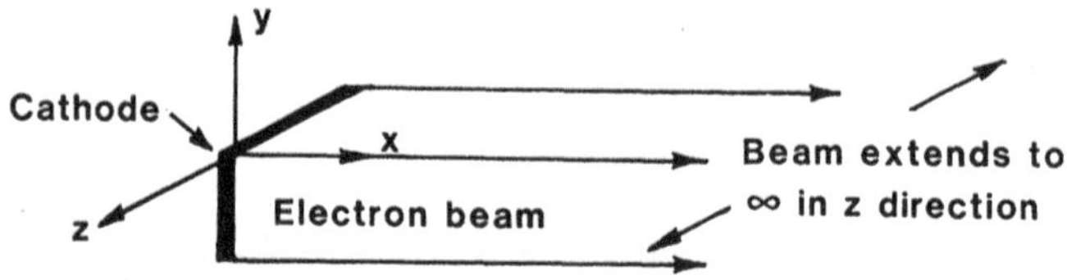

Figure 6.1 Parallel Electron Flow from a Planar Cathode

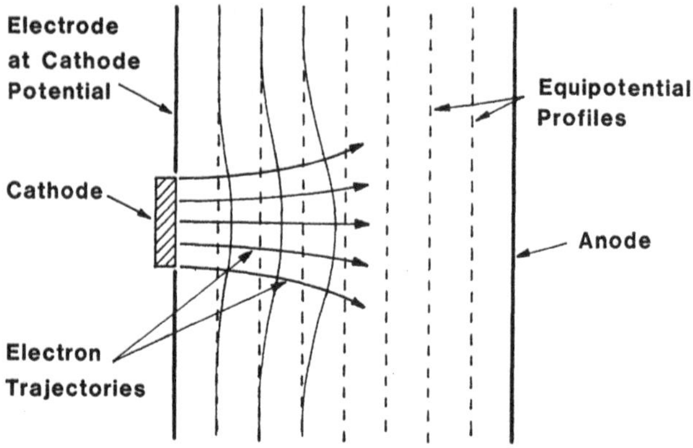

Figure 6.2 Equipotential Profiles without (Dashed Lines) and with (Solid Lines) Electrons

First off all, the equipotential profiles in a parallel-plane diode containing no electrons are parallel and spaced equal distances apart for equal voltage increments. This is shown by the dashed lines in Fig. 6.2.

When electrons are present, the equipotential profiles are deflected to the right as indicated by the solid lines. Electrons tend to move perpendicular to the equipotential profiles, so the electron trajectories diverge as the electrons move away from the cathode. This is simply another way of saying that the electrons are pushing on each other and causing their trajectories to diverge.

If an electrode arrangement could be devised that would straighten the equipotential profiles even though electron space charge forces were present, then the electrons would travel in parallel paths.

Intuitively, it can be seen that if the electrodes adjacent to the cathode are tipped toward the anode, as indicated in Fig. 6.3, then the equipotential profiles are straightened.

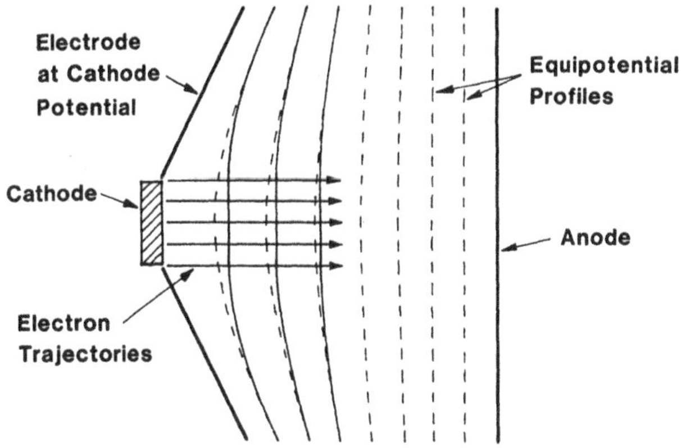

Figure 6.3 Straightening of Equipotential Profiles in Electron Beam

In the absence of electrons, the profiles curve toward the cathode. Electron charge causes them to move toward the anode.

The mathematical solution that leads to a definition of the electrode configuration for parallel electron flow is similar to the solution for the problem of space-charge-limited flow discussed in Chapter 5. The main difference now is that the equations must be solved inside and outside the electron stream, and then the solutions must match at the edge of the beam. The resulting electrode and equipotential profiles for parallel electron flow are given in Fig. 6.4.

Notice that planar electrodes at cathode potential and at an angle of 67-1/2 degrees with respect to the beam edge coupled with a curved anode produces parallel electron flow. The 67-1/2 degree electrodes are frequently referred to as "Pierce electrodes."

Figure 6.5 contains an illustration of an electron gun containing 67-1/2 degree Pierce electrodes for producing parallel (rectilinear) electron flow. Notice that grid wires are shown across the aperture in the positive electrode. This is to maintain the electrode potential within the beam and, thus, prevent divergence of the beam after it passes through the positive electrode. Beam divergence caused by an anode aperture will be discussed in subsequent paragraphs.

6.1.2 Convergent Flow

Thus far, the problem of overcoming space charge forces in a parallel-flow electron gun has been solved. The next step is to examine a technique

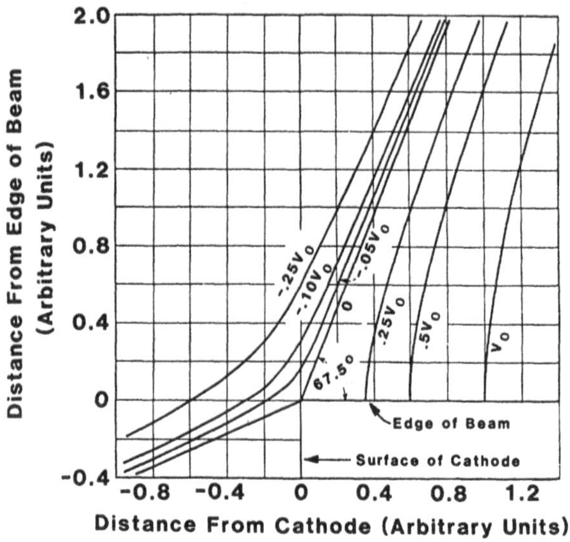

Figure 6.4 Equipotential Profiles for Parallel Electron Flow in a Planar Two-Dimensional Diode

(From *Theory and Design of Electron Beams* by J.R. Pierce, Copyright 1954 by D. Van Nostrand, Inc., Reproduced with permission.)

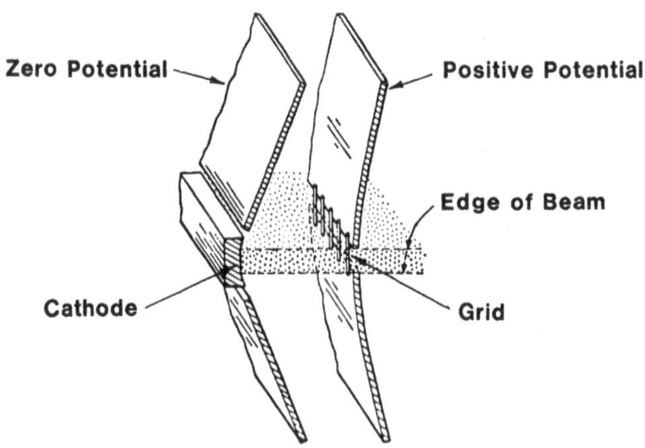

Figure 6.5 Electron Gun for Producing Rectilinear (Parallel) Electron Flow

(From *Theory and Design of Electron Beams* by J.R. Pierce, Copyright 1954 by D. Van Nostrand, Inc., Reproduced with permission.)

for obtaining an electron beam density greater than the cathode emission density.

In the spherical diode of Fig. 6.6, under space-charge-limited flow conditions, current as function of voltage is given by

$$I = PV^{3/2}$$

where

$$P = \frac{29.33 \times 10^{-6}}{\alpha^2}$$

and where α^2 was first calculated by Langmuir and Blodgett and is given in Fig. 6.6. This is the current for the entire sphere of solid angle 4π steradians.

Next, consider the conical segment of a spherical diode as shown in Fig. 6.7. If this diode could be made to function as shown, electron flow from the cathode to the anode would be convergent so that the beam density at the anode would be larger than the cathode emission density. This, of course, is what is desired.

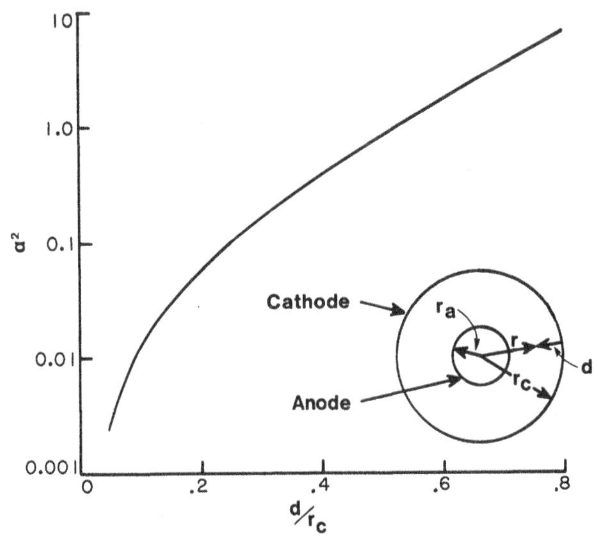

Figure 6.6 Langmuir and Blodgett's α^2 as a Function of Distance from Cathode

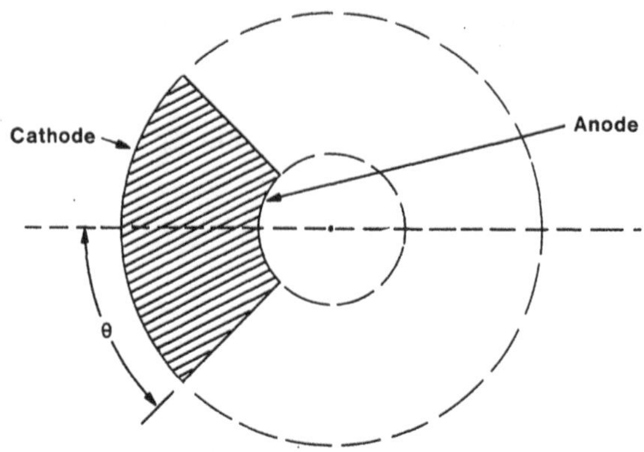

Figure 6.7 Conical Diode of Half Angle θ

For a cone of half angle θ the current I_o related to the total current in a sphere is

$$I_o = \frac{1 - \cos\theta}{2} I$$

so

$$I_o = 14.67 \times 10^{-6} \frac{1 - \cos\theta}{\alpha^2} V^{3/2}$$

Figure 6.8 shows perveance as a function of cone angle and normalized cathode-anode spacing. It will be shown in subsequent paragraphs that the extremely high values of perveance indicated in Fig. 6.8 cannot be achieved in real electron guns when anode effects are considered.

The problem with the conical diode is that space charge forces will cause the beam to diverge just as they did for the planar diode. Equipotential contours, which should be spherical in shape, are deflected by the electron space charge potential as indicated in Fig. 6.9.

To retain radial electron flow when the remainder of the sphere is removed, it is necessary, as was the case for rectilinear flow in a parallel plane diode, to use electrodes that cause the equipotential profiles to have the shape which they had in the absence of the electrons (spherical in this case).

Conceptually, as was the case for the planar diode, it can be seen that by tipping the electrodes adjacent to the cathode forward towards the anode, the equipotential contours can be made more nearly spherical.

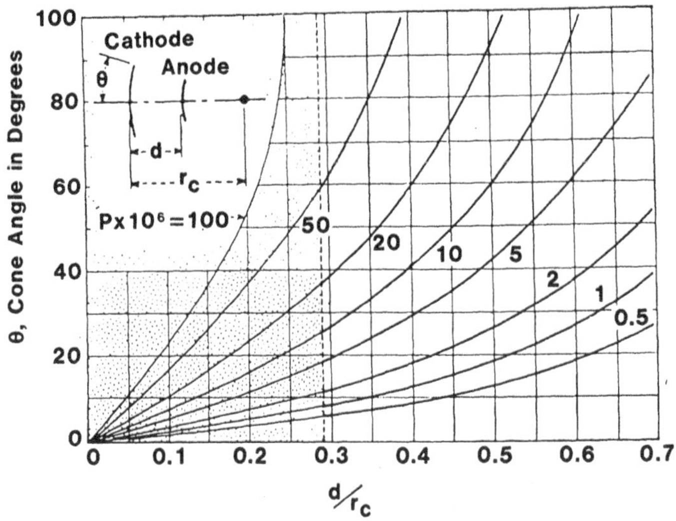

Figure 6.8 Perveance as a Function of Cone Angle and Cathode-Anode Spacing

(From *Theory and Design of Electron Beams* by J.R. Pierce, Copyright 1954 by D. Van Nostrand, Inc., Reproduced with permission.)

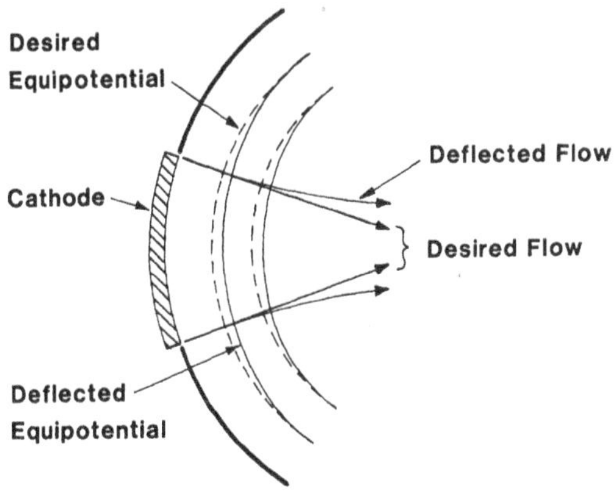

Figure 6.9 Deflection of Equipotential Contours in Conical Diode by Electron Space Charge Potential

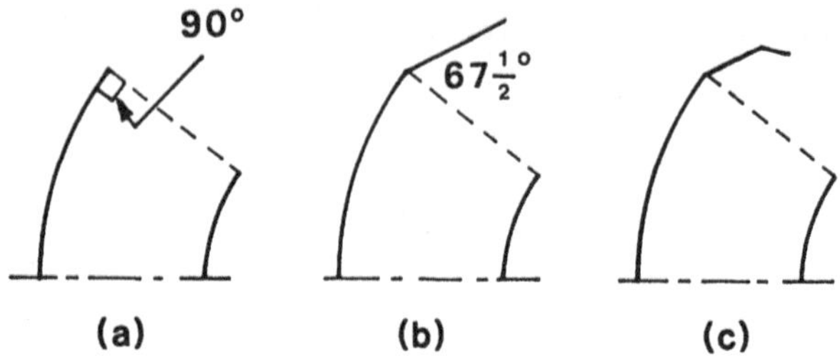

Figure 6.10 Use and Modification of Pierce Electrodes to Produce Spherical Equipotential Profiles

In fact, as indicated in Fig. 6.10(a), close to the cathode, electron flow appears to be two-dimensional, so an electrode placed at a 67 1/2 degree angle with respect to the edge of the beam (Fig. 6.10(b)) will have the desired effect on the equipotential profiles.

Near the center of the cathode and away from the cathode, the potential profiles will still not be spherical, and so the shape of the focusing electrode must be modified (Fig. 6.10(c)) to achieve proper electron flow. This modification of the focus electrode shape will be discussed in later paragraphs in which high-perveance Pierce guns are described.

6.1.3 Anode Aperture

To remove electrons from the conical diode, it is necessary to cut a hole in the anode. The resulting distortion of the equipotential profiles, especially that which occurs near the anode aperture, is illustrated in Fig. 6.11.

It should be noticed in Fig. 6.11, that, for a perveance of 3×10^{-6}, two potential distributions are shown. One, with Pierce electrodes at 67 1/2 degrees, shows distortion of the equipotential profiles from the ideal spherical shape in the region close to the cathode. With a modified-focus electrode, the equipotential profiles in much of the region between the cathode and anode appear to be spherical in shape.

Attempts to increase perveance by reducing anode-to-cathode spacing may be completely offset by the reduction in perveance resulting from the anode aperture.

In addition, it becomes increasingly difficult to maintain spherical potential profiles near the cathode. The result, as shown in Fig. 6.11, is that the electric field in the center of the cathode is less than it is near the outer edge of the cathode. As a result, current is drawn predominantly from the annulus of the cathode as indicated in Fig. 6.12.

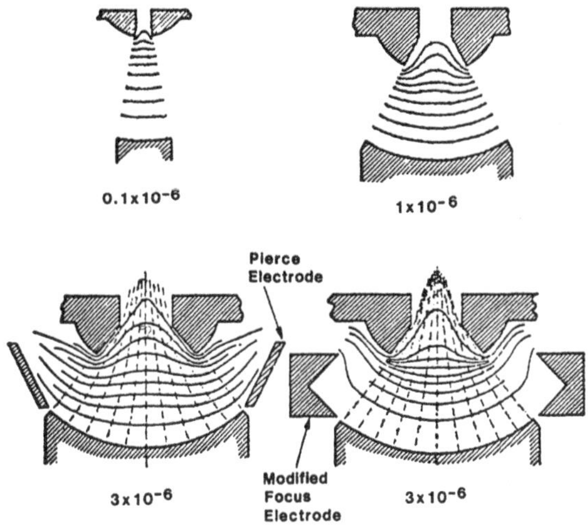

Figure 6.11 Effect of Anode Aperture on Equipotential Profiles for Various Values of Perveance

(From *Focusing of Charged Particles*, Vol. II, Chapter 3.2 by G.R. Brewer, Copyright 1967 by Academic Press. Reproduced with permission.)

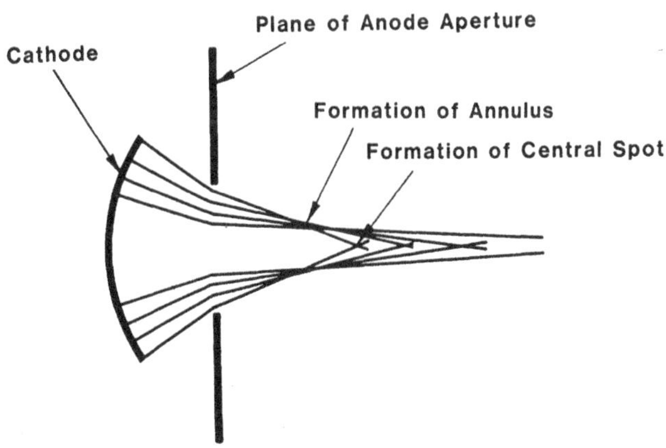

Figure 6.12 Concentration of Electron Emission and Beam Focusing in High-Perveance Guns

(From *Power Travelling-Wave Tubes* by J.F. Gittins, Published 1965 by American Elsevier, Inc.)

While a discussion of focusing is somewhat premature at this point, it should be noted in Fig. 6.12 that the electrons from the outer edges of the cathode are focused to a spot relatively close to the plane of the anode aperture. Electrons from regions closer to the center of the cathode are focused at greater distances from the anode-aperture plane. The results of these nonuniformities in focusing are further illustrated by the guns shown in Figs. 6.13 and 6.14. These guns have a cylindrical projection from the anode into the focusing electrode. Still, the resulting current distributions in the beams are extremely nonuniform.

As a result of the problems illustrated in Figs. 6.12 to 6.14, the design of Pierce guns with perveances above 2 or 3×10^{-6} is very difficult.

Under ideal conditions, as electrons emerge from the cathode-to-anode region of the gun, electron trajectories do not cross and the current throughout the beam at any axial position is constant. The distortion of the equipotential profiles near the anode aperture (illustrated in Fig. 6.11) results in the formation of a divergent electrostatic lens. The defocusing effect of this lens is given in Fig. 6.15.

After the beam leaves the anode lens region, it enters a region where only the space charge effects exist. The resulting beam behavior is illustrated in Fig. 6.16. The beam radius reaches a minimum value and then increases as a result of the space charge forces.

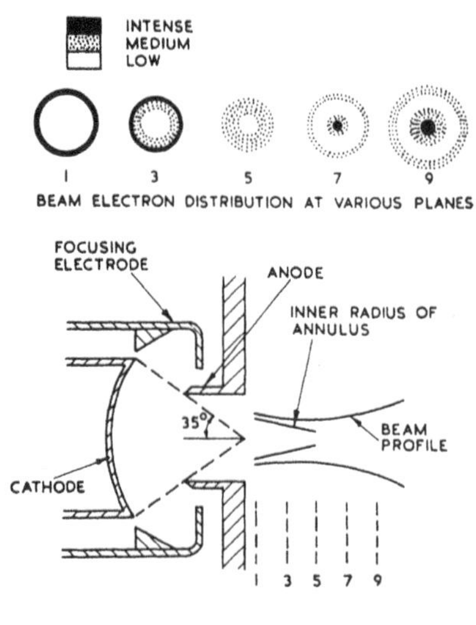

Figure 6.13 Gun Designed by Mathias According to Muller's Charts

(From *Power Travelling-Wave Tubes* by J.F. Gittins, Published 1965 by American Elsevier, Inc.)

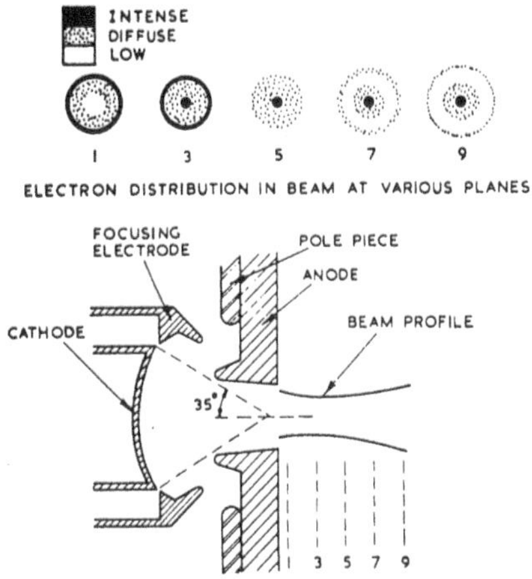

Figure 6.14 Mathias and King's High-Voltage Version of Muller Type Gun

(From *Power Travelling-Wave Tubes* by J.F. Gittins, Published 1965 by American Elsevier, Inc.)

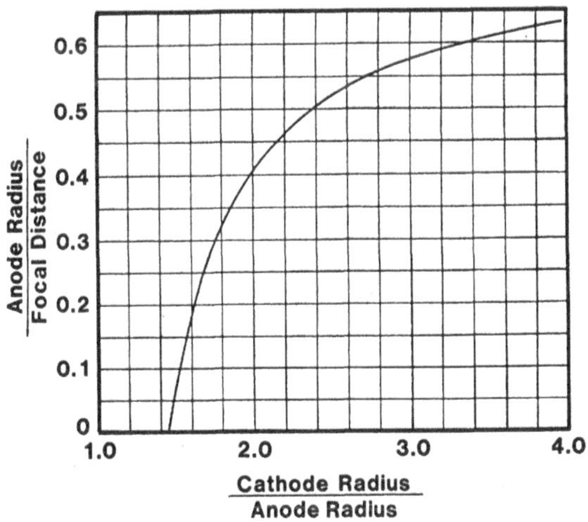

Figure 6.15 Defocusing Effect of Anode Aperture

(From *Theory and Design of Electron Beams* by J.R. Pierce, Copyright 1954 by D. Van Nostrand, Inc., Reproduced with permission.)

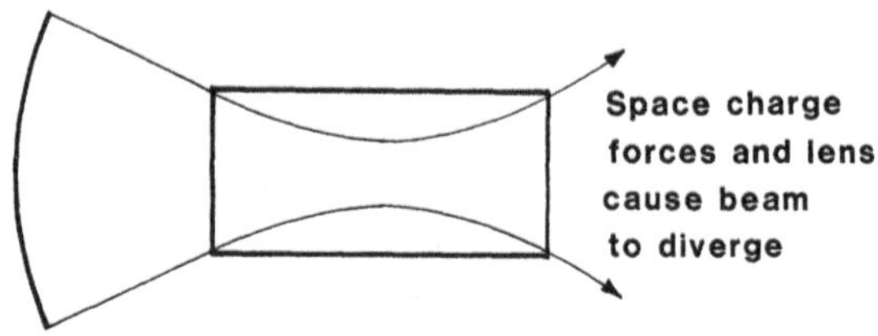

Figure 6.16 Beam Behavior as a Result of Anode Lens Action and Space Charge Forces

6.1.4 Beam Minimum

As the beam approaches or leaves the position of minimum radius, its shape is a function of perveance and follows the universal beam-spread curves given in Fig. 6.17.

To properly inject the electron beam from the gun into the magnetic focusing structure of a linear-beam tube, it is essential to have accurate knowledge of the minimum beam radius and the axial position at which the minimum radius occurs.

The minimum beam radius is given in Fig. 6.18 for the case when thermal velocities of electrons are ignored. In practice, only the upper left-hand

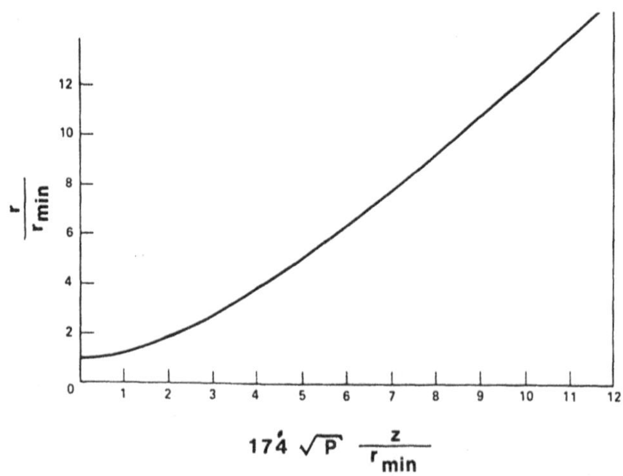

Figure 6.17 Universal Beam Spread Curve Showing Normalized Beam Radius as a Function of Normalized Axial Position and Perveance, P

portion of the curve is used. The axial position of the minimum beam diameter is given in Fig. 6.19.

Beam profiles resulting from the use of the design procedure outlined in the preceding paragraphs are illustrated in Fig. 6.20. It is important to note the extreme sensitivity of the beam radius to relatively small changes in anode location. This is also evident from Fig. 6.18 where, for example, a 5% reduction in normalized cathode radius (from 2.1 to 2.0) very nearly results in a 50% increase in normalized beam radius. This clearly illustrates the importance of precisely locating the cathode with respect to the anode.

The effects of the initial velocities imparted to electrons as they leave the cathode have been ignored in the previous paragraphs describing beam formation. The effects of thermal velocities are indicated in Fig. 6.21. This figure illustrates the shape of a beam under the influence of space charge forces so that the beam edge has the shape of the universal beam-spread curve shown in Fig. 6.17. Electrons injected with a transverse thermal velocity component result in the beam edge having the shape shown by the dashed curves. The effect of the thermal velocity component on the minimum radius can be significant.

The regions in which thermal effects predominate in determining the minimum beam diameter are indicated in Fig. 6.22. It comes as no surprise that an electron gun having a large cone angle, which implies that the beam is being compressed to a small radius, is the one in which thermal effects are important.

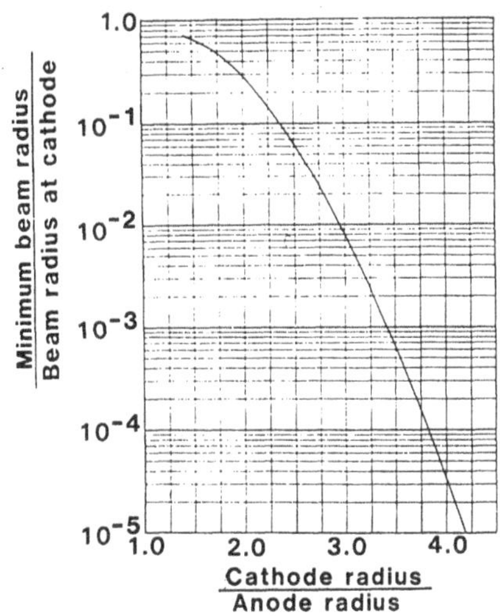

Figure 6.18 Minimum Beam Radius

(From *Theory and Design of Electron Beams* by J.R. Pierce, Copyright 1954 by D. Van Nostrand, Inc., Reproduced with permission.)

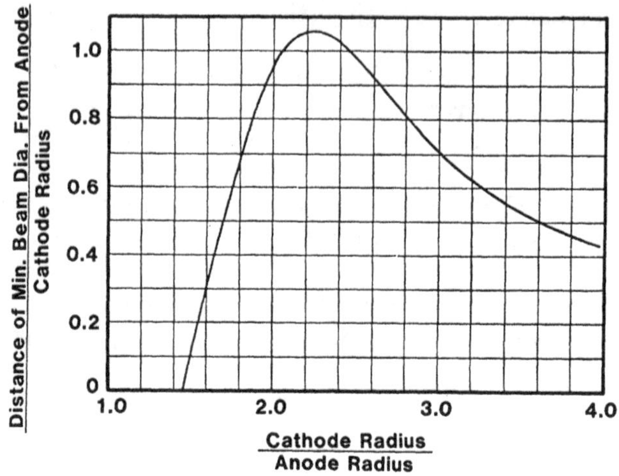

Figure 6.19 Axial Position of Minimum Beam Diameter

(From *Theory and Design of Electron Beams* by J.R. Pierce, Copyright 1954 by D. Van Nostrand, Inc., Reproduced with permission.)

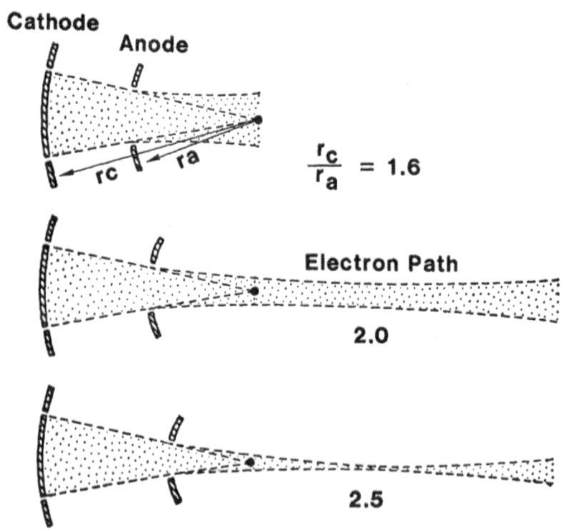

Figure 6.20 Emerging Electron Stream for Various Ratios of Cathode Radius to Anode Radius

(From *Theory and Design of Electron Beams* by J.R. Pierce, Copyright 1954 by D. Van Nostrand, Inc., Reproduced with permission.)

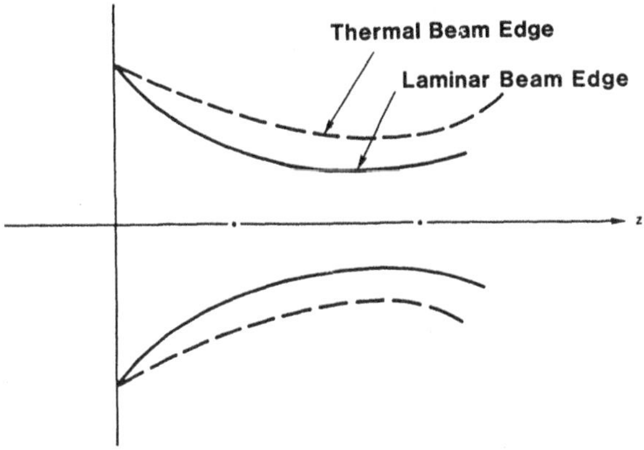

Figure 6.21 Effect of Thermal Velocities on Shape of Beam Under the Influence of Space Charge Forces

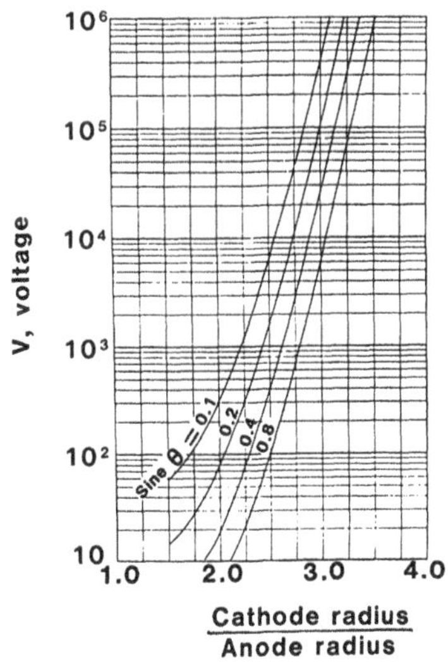

Figure 6.22 For Voltages Above the Curves, Space Charge Predominates in Determining the Minimum Beam Diameter; For Voltages Below the Curves, the Effect of Thermal Velocities Predominates

(From *Theory and Design of Electron Beams* by J.R. Pierce, Copyright 1954 by D. Van Nostrand, Inc., Reproduced with permission.)

6.2 BEAM CONTROL ELECTRODES

6.2.1 Grids

Grids are often placed in Pierce guns to control the current from the cathode. The primary reason for doing this is to permit the beam current to be controlled with a voltage much smaller than the cathode-to-anode voltage. This results in a reduction in the size and weight of the modulator for the tube.

Several types of grid structures are illustrated in Fig. 6.23. The simplest structure has a single grid element. The problem with this grid is that the cathode current intercepted by it can be very high. In some cases as much as 15% or more of the cathode current is intercepted. The results of the interception are excessive heating of the grid and the requirement for a relatively powerful grid modulator.

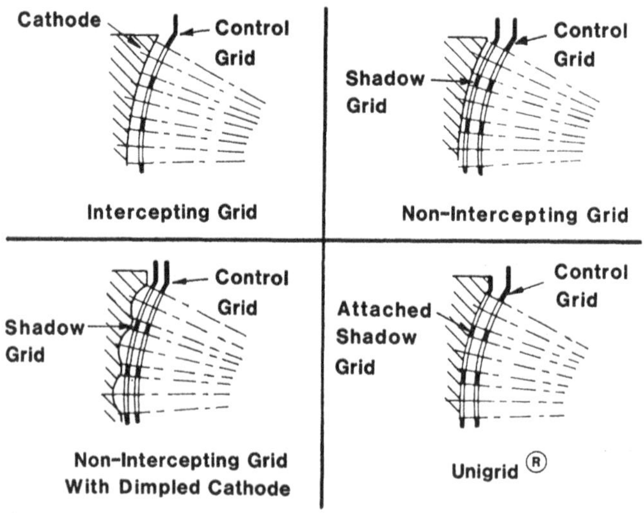

Figure 6.23 Various Grid Structures Used in Pierce Guns for Current Control

(From *Microwave Tube Manual* by Varian Associates. Air Force Publication Number T.0. 00-25-251, October 1979.)

Most grid structures contain a shadow grid between the control grid and the cathode surface. The shadow grid operates at or near cathode potential so that heating by intercepted current is small. The elements of the shadow grid are aligned with the elements of the control grid so that the control grid is shielded and intercepts very little current. Several shadow grid structures are shown in Fig. 6.23.

The fabrication of shadow-grid structures is difficult, especially in high-frequency tubes where dimensions are small. Alignment of the elements of the control and shadow grids is critical. Alignment must be maintained despite the fact that shadow grid operates at a higher temperature than the control grid because it is closer to or perhaps in contact with the cathode. The top surface of the structure is the control grid and the bottom surface is a diffusion barrier between the cathode and the insulator. With this design, alignment and close grid-cathode tolerances are easily obtained and maintained with the cathode at operating temperature. The control grid is fabricated from pyrolytic graphite and coated with a thin layer of boron nitride to reduce grid emission to an acceptable level.

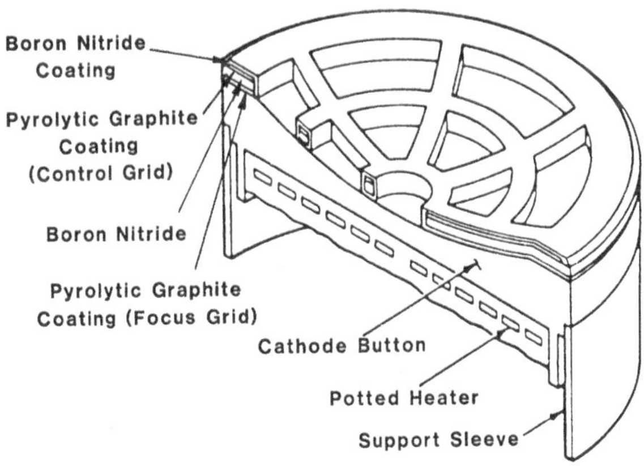

Figure 6.24 Bonded Grid and Cathode Assembly

(From G.V. Miram and E.L. Lien, *Technical Record* 1979 IEDM.)

While gridded guns facilitate beam control, the use of grids leads to the disturbance of electron trajectories in the gun and in the beam from the gun. As a result, the amount of beam current that impinges on the RF structure as the beam passes through a tube is increased. This can lead to a reduction in the power capability and efficiency of the tube.

A Pierce gun with an anode, focus electrode, control grid, and shadow grid is shown in Fig. 6.25. The focus electrode is used to help shape the equipotential profiles in the cathode-to-anode region.

As the grid-to-cathode voltage is varied in a gridded electron gun, cathode current varies as shown in Fig. 6.26. In addition, the beam current intercepted, usually by the RF structure of the tube (often called body current), varies. Of course, body current must be limited to prevent damage to the RF circuit.

Very sophisticated grid structures have been developed for some special microwave tube applications. An example is shown in Fig. 6.27. This particular grid structure was developed to make possible a tube that could operate efficiently at two different power levels. This type of operation is frequently referred to as dual-mode operation.

With the grid structure of Fig. 6.27, the beam current and diameter can be switched to change beam power level while maintaining constant beam voltage and proper focusing conditions.

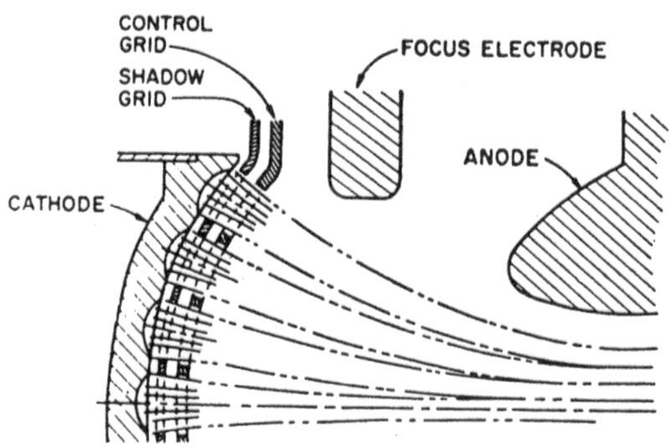

Figure 6.25 Shadow-Grid Gun

(From Armand Staprans, *et al., Proc. IEEE,* March 1973.)

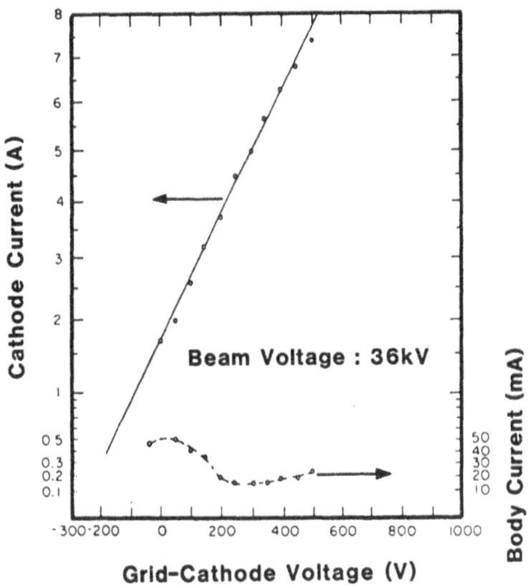

Figure 6.26 Shadow-Grid Electron Gun Characteristics

(From Armand Staprans, *et al.*, *Proc. IEEE*, March 1973.)

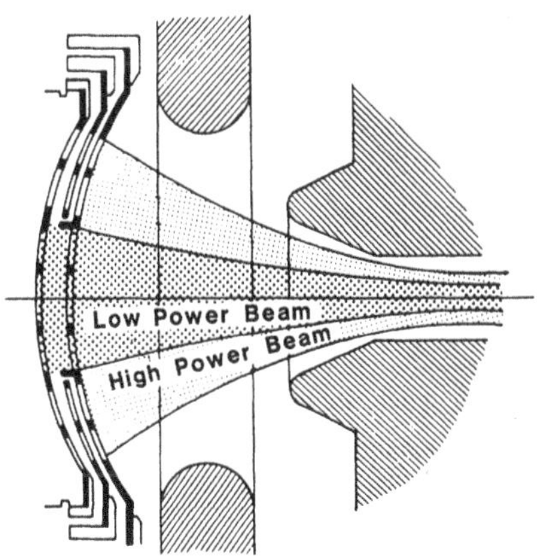

Figure 6.27 Dual-Mode Gun Design

(From *Microwave Tube Manual* by Varian Associates. Air Force Publication Number T.0. 00-25-251, October 1979.)

6.2.2 Control Focus Electrodes

If a sufficiently negative voltage is applied to the control electrode of a Pierce gun, the beam current can be reduced to zero. The voltage required to cut off the beam current can be reduced by locating an additional electrode in the center of the cathode as shown in Fig. 6.28.

The control focus electrodes can be used to turn on the beam by setting the voltage at the cathode potential. The beam is turned off by setting the voltage significantly below cathode potential. This system is not appropriate for the intermediate control of current because of the severe defocusing that occurs in the electron gun.

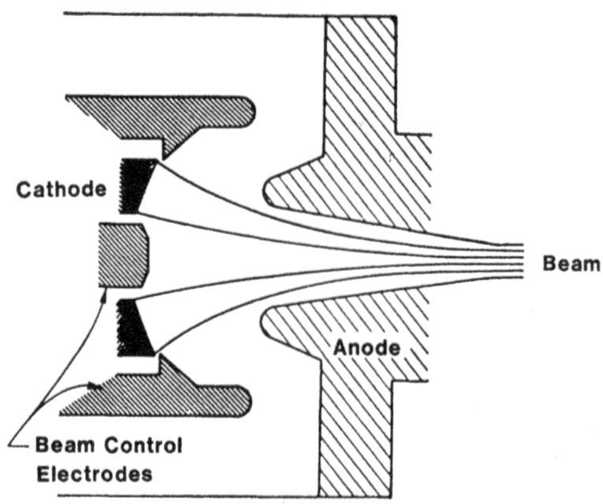

Figure 6.28 Control Focus Electrode Gun

(From Armand Staprans, *et al.*, *Proc. IEEE*, March 1973.)

6.2.3 Modulating Anode

In the modulating anode gun, the anode of the electron gun (the modulating anode) is separated from the body of the tube, a part of which is a second (final) anode. The configuration of a modulating anode gun is shown in Fig. 6.29.

Although the voltage range over which the modulating anode must be varied in order to vary beam current is large, the power required to drive the modulating anode is small because the current intercepted by this anode is very small. At intermediate voltage levels, perveance may be controlled with the modulating anode without severely affecting beam focusing.

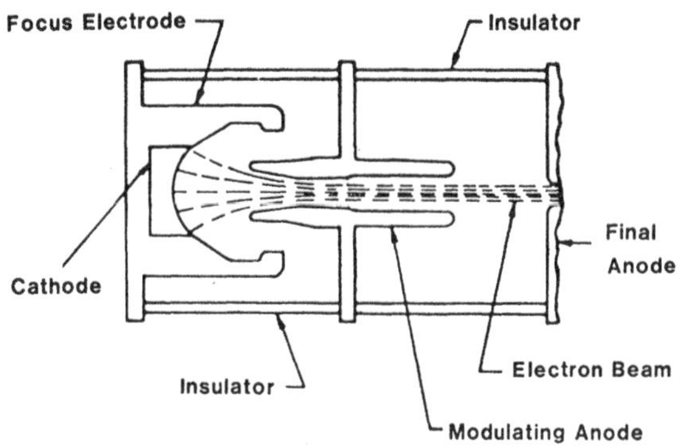

Figure 6.29 Modulating Anode Gun

(From Armand Staprans, *et al.*, *Proc. IEEE*, March 1973.)

6.2.4 Summary of Beam Control Electrode Characteristics

Table 6.1 contains a summary of the more important characteristics of beam control electrodes. Two amplification factors are given. The total amplification factor, μ, is the ratio of beam to control electrode voltage required for beam control. The cut-off amplification factor is the ratio of beam voltage to the negative control electrode voltage (with respect to the cathode) effectively required to reduce beam current to zero.

TABLE 6.1
Typical Characteristics of Control Electrodes
(From Armand Staprans, *et al.*, *Proc. IEEE*, March 1973.)

Type	μ	μ_c	*Capacity*	*Grid Interception*	*Focusing at Low Voltage*
Modulating Anode	1 to 3		50 pF	0	Good
Control Focus Electrode	2 to 10	2 to 10	100 pF	0	Poor
Intercepting Grid	50	100	50 pF	15%	Fair
Shadow Grid	30	300	50 pF	0.1%	Fair

As would be expected, the gridded guns have by far the highest amplification factors and the modulating anode gun has the lowest amplification factor.

The capacitances of all of the control electrode structures to the surrounding electrodes are essentially the same. Capacitance is important, of course, in the design of the modulator that must be used to drive the control electrode structure.

Finally, as would be expected, the interception of the intercepting grid structure is much higher than the other structures. Only the modulating anode gun has good focusing because the design of that gun is essentially the design of an ideal Pierce gun.

REFERENCES

[6.1] J.R. Pierce, *Theory and Design of Electron Beams,* 2nd ed. New York: Van Nostrand, 1954.

[6.2] Recommended for further reading: M. Chodorow and C. Susskind, *Fundamentals of Microwave Electronics.* New York: McGraw-Hill, 1964.

Chapter 7

Electron Beams for Linear-Beam Tubes

In Chapter 6, the Pierce type of electron guns used in linear-beam (0-type) tubes were discussed. In this chapter, the magnetic focusing of a beam from a Pierce gun is examined [7.1].

Before continuing with the discussion of magnetically focused beams, it is important to point out that numerous electrostatic focusing techniques have been examined in the forty-plus years that microwave tubes have been in existence. With the exception of the very few cases where the tube is designed to accommodate a beam following the universal beam-spread curve, in which case no focusing is required, only magnetic focusing is used at the present time.

The primary problem in focusing, which is illustrated in simplified fashion in Fig. 7.1, is that space charge forces are so high for the typical beams used in microwave tubes that rapid beam divergence occurs in the absence of focusing forces. With equal and opposite constraining (focusing) forces, beam divergence can be prevented.

In virtually all linear-beam (0-type) tubes, the focusing forces are provided by a magnetic field aligned with the axis of the electron beam. Frequently, a solenoid, like that illustrated in Fig. 7.2, is used to generate the magnetic field.

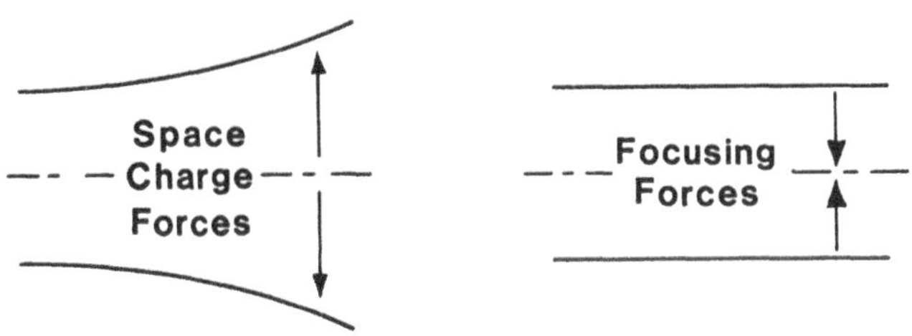

Figure 7.1 Illustration of the Effects of Space-Charge Forces and Focusing Forces on an Electron Beam

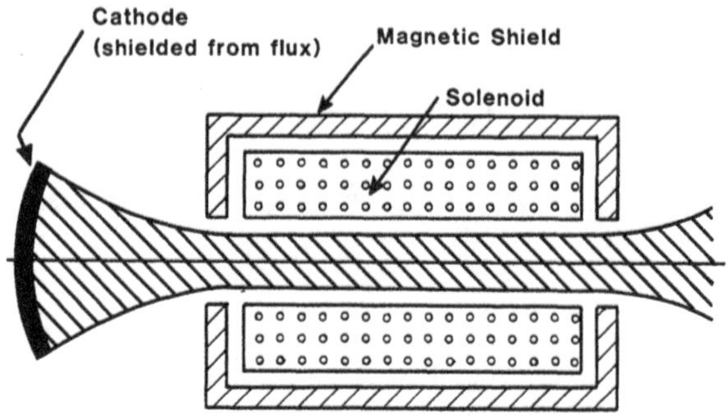

Figure 7.2 Illustration of Use of Solenoid to Focus the Beam in a Linear-Beam (O-type) Tube

7.1 FOCUSING WITH A SHIELDED CATHODE

To simplify the description of the focusing process, it is initially assumed that there is no magnetic flux through the cathode and that the beam first encounters magnetic flux lines as it enters the solenoid.

Magnetic focusing of the beam is initiated as the electron beam crosses radially directed components of magnetic flux lines (illustrated in Fig. 7.3). Consider, for example, an electron above the axis of the beam (and solenoid). That electron encounters a component of magnetic flux that is radially directed downward. The magnetic force on the electron is out of the paper. (The cross product of the velocity, u, and the flux density, B, is into the paper and the force on an electron is in the reverse direction from the $u \times B$ product).

Consider, next, an electron below the axis. That electron encounters a component of magnetic flux that is radially directed upward. The magnetic force on that electron is into the paper.

When the magnetic forces on all electrons are considered, the net effect is that the electron beam starts rotating in the clockwise direction as it enters the field with the flux lines in the directions shown in Fig. 7.3. If the directions of the flux lines were reversed, the direction of rotation of the beam would be counterclockwise.

As the electron beam moves further into the magnetic field, it encounters stronger and stronger axial components of magnetic flux. It is the rotational velocity component of the beam (established as the beam crossed radially directed flux lines upon entering the field) interacting with the axial components of flux that produces the magnetic focusing force.

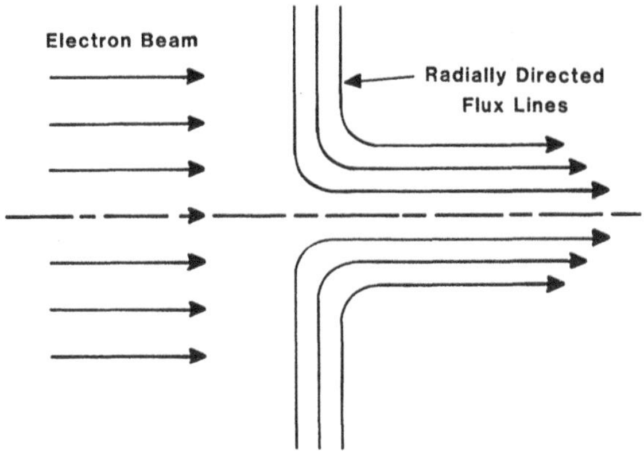

Figure 7.3 Configuration of Magnetic Flux Lines at Entrance of Electron Beam into Solenoid

To continue the illustration of the focusing process, consider again an electron above the beam axis. This electron has a velocity component out of the paper. The cross product of that velocity component with the axial magnetic flux is is a vector away from the axis so that the force on the electron is toward the axis. Similar considerations for other electrons show that they all experience a magnetic force toward the axis.

As each electron is followed, it is found that a helical trajectory about the axis of the solenoid is traversed as indicated in Fig. 7.4.

7.1.1 Brillouin Flow

If the space-charge forces tending to cause the beam to diverge are just balanced by equal and opposite magnetic forces tending to compress the beam, and if the electrons enter the magnetic field with no radial velocity component, then the beam diameter will remain constant as the beam continues to rotate and move through the magnetic field.

In the absence of flux through the cathode, the magnetic flux level that produces a magnetic force that exactly balances the space-charge force is called the Brillouin flux level and is commonly denoted by B_B. The resulting electron motion in the form of a smooth rotating beam is called Brillouin flow. When the beam exits from the magnetic field at the collector end of the tube, the sequence of interactions is just the opposite of that at the entrance of the beam into the field. That is, as the beam leaves the magnetic field, the radial compressive force disappears and rotation of the beam stops.

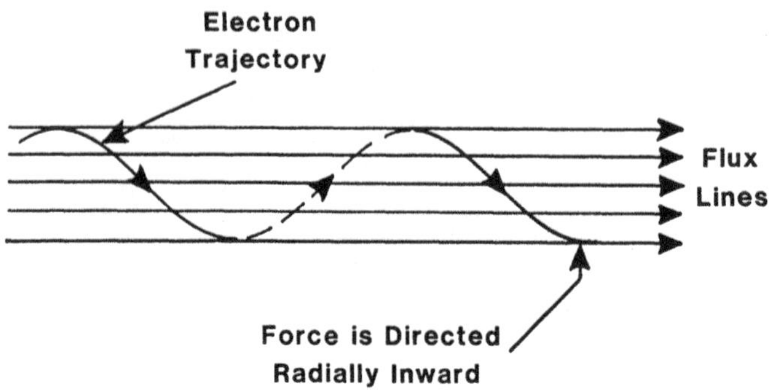

Figure 7.4 Electron Trajectory in Axial Field

As a result, the beam expands under the influence of space-charge forces as it enters the collector.

Now, as we have said, for Brillouin flow to occur, the beam must enter the magnetic field on axis with no radial electron motion and with a radius equal to the equilibrium radius (the radius at which space-charge forces are balanced by magnetic forces). This situation is illustrated in Fig. 7.5.

By equating the oppositely directed radial space-charge and magnetic forces, it is not difficult to show that the equilibrium radius, a, of the beam is

$$a = \frac{1}{B_z} \sqrt{\frac{2\,I}{\pi\,\eta\,\epsilon_0\,\mu_z}} \tag{7.1}$$

where

B_z is the axial flux density
I is beam current
u_z is axial beam velocity
η is electron charge to mass ratio
ϵ_0 is the dielectric constant of free space

There are no surprises in (7.1). As B_z increases, the magnetic forces increase, and so the equilibrium radius must decrease. As current increases, the charge density in the beam increases, and so the space charge forces and equilibrium radius must increase. As velocity increases, the space charge density and the space charge forces decrease, so the equilibrium radius must decrease.

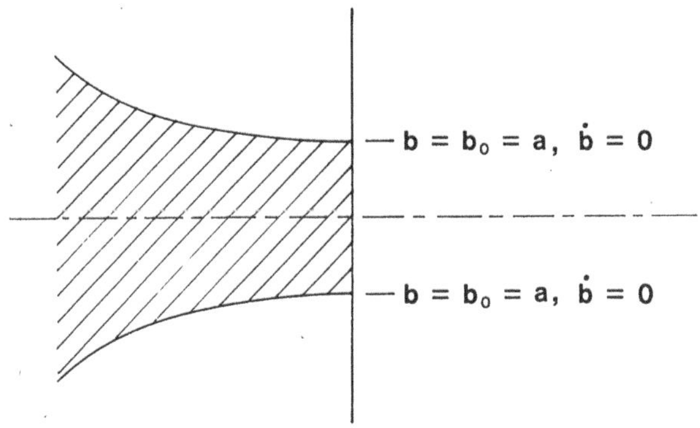

Figure 7.5 Entrance Conditions for Cylindrical Beam Formation (Brillouin Flow). [The initial beam radius is b_o, the equilibrium radius is a, and no radial velocities exist ($\dot{b} = 0$)].

7.1.2 Scalloping

In practice, when the electron gun is shielded from magnetic flux, it is extremely difficult to achieve the perfectly balanced force condition that is required for Brillouin flow. For example, consider the case when the electron beam enters the magnetic field with a radius smaller than the equilibrium radius as shown in Fig. 7.6.

At plane z_1 where the beam enters the magnetic field, the space-charge force is larger than the magnetic focusing force, so the beam starts to expand. When the beam reaches plane z_2, the radial forces are equal, but the electrons are moving radially outward. As a result, the electrons overshoot the equilibrium radius. By the time the electrons have reached plane z_3, the radial motion of the electrons has been stopped because the space-charge forces have become smaller than the magnetic forces.

In fact, this force imbalance causes the beam to become smaller again and, by plane z_4, the electrons are again crossing the equilibrium radius. Because the electrons are moving radially inward at plane z_4, they overshoot the equilibrium radius. By the time the beam reaches plane z_5, the inward radial motion has been stopped because the space-charge forces exceed the magnetic forces.

In a uniform axial field, the beam radius at plane z_5 is the same as it was at plane z_1 and the entire process that took place between planes z_1 and z_5

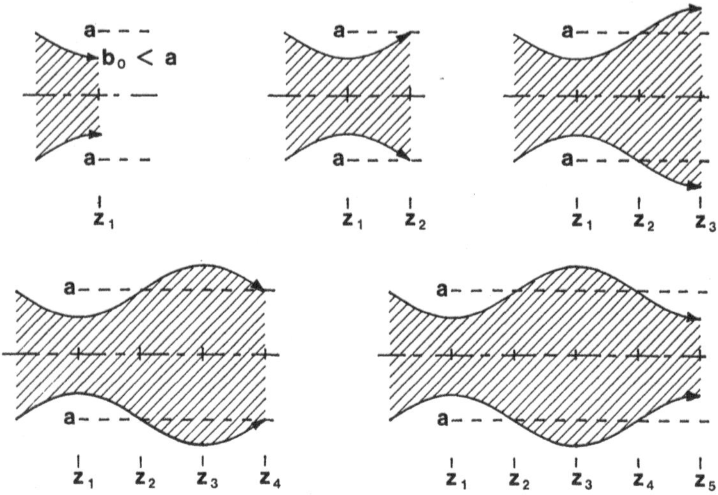

Figure 7.6 Beam Dynamics Resulting from Beam Entrance into Magnetic Field at Less than the Equilibrium Radius

starts over again. As a result, the beam radius continues to oscillate as the beam moves in the axial direction. The resulting beam shape is shown in Fig. 7.7. Beam scalloping is said to occur and the scallop wavelength is given by

$$\lambda_s = \frac{4\,\pi\,u_z}{\sqrt{2}\,\eta\,B} \qquad (7.2)$$

As the magnetic flux density is changed, the scallop wavelength is changed along with the equilibrium radius. The result is shown in Fig. 7.8. For $B < B_B$, the equilibrium radius is greater than the initial beam radius, b_0, and the beam scallops outward. When $B = B_0$, no scalloping occurs. When $B > B_b$, the equilibrium radius is less than b_0 and the beam scallops inward.

The preceding discussion shows that if the beam enters the magnetic field on axis with electrons traveling parallel to the axis, it is possible to adjust the magnetic field level to eliminate the scalloping.

If the electron gun is at the wrong axial position so that the beam is either diverging or converging as it enters the magnetic field, then no adjustment of the magnetic field can be made to eliminate the resulting scalloping.

At this point it is appropriate to discuss the beam behavior that results if the axis of the electron beam is not aligned with the axis of the magnetic field. This could happen for any one of several reasons. For example, the electron gun could be misaligned during tube fabrication. The gun axis could be either off center or tilted with respect to the axis of the tube. Other causes

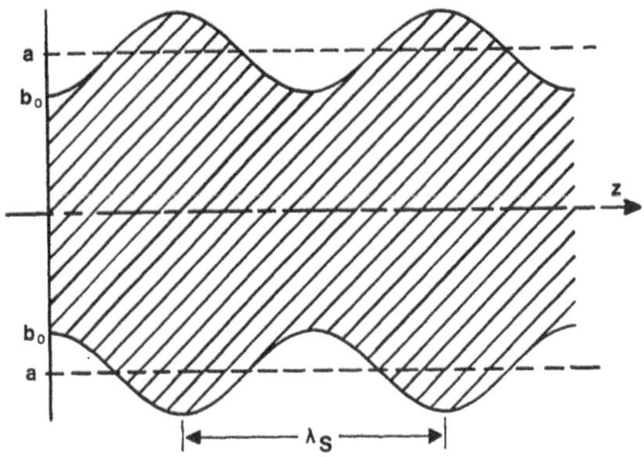

Figure 7.7 Beam Scalloping Resulting from Entrance at Less than the Equilibrium Radius

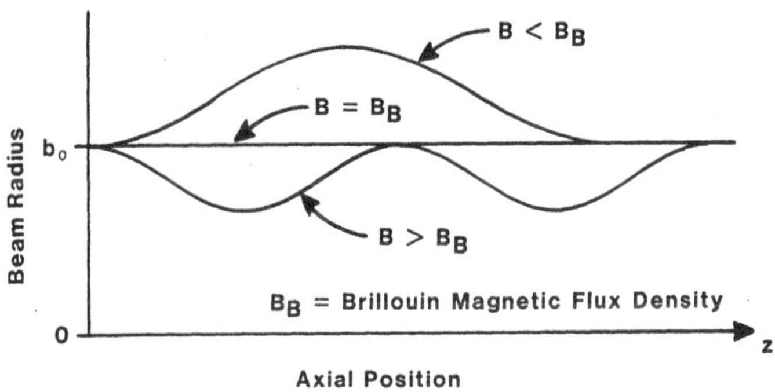

Figure 7.8 Scalloping for Various Magnetic Flux Density Levels

for misalignment include errors in cathode, grid, or control electrode positioning or nonuniformities in magnetic pole piece material.

The result of beam misalignment is that the entire beam spirals about the magnetic field axis at the cyclotron frequency. Figure 7.9 contains a summary of the problems that could be encountered in attempting to form a Brillouin beam flowing parallel to the tube axis. These problems are:

1. Flux density too small
2. Flux density too large
3. Beam is converging at entrance
4. Beam is diverging at entrance
5. Beam axis is offset from magnetic field axis
6. Beam axis is tilted with respect to magnetic field axis

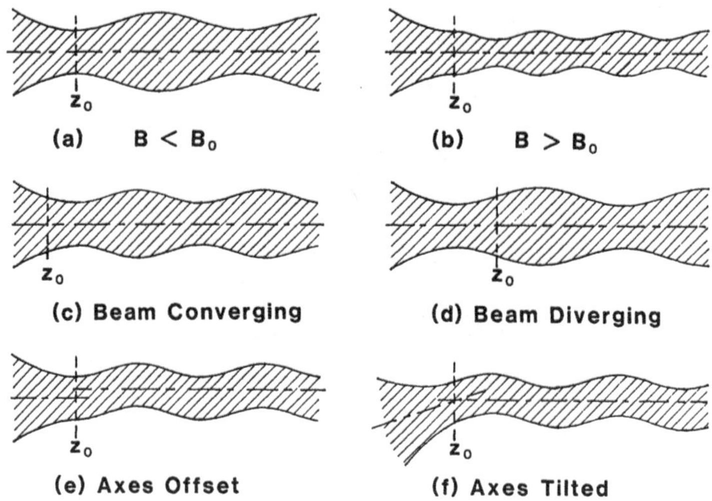

Figure 7.9 Problems in Forming a Brillouin Beam Flowing Parallel to the Tube Axis

With Brillouin flow, there is a very delicate balance between space-charge and magnetic forces. If the charge density is changed, then this balance is upset and a change in beam diameter results. The charge density is changed when the beam is used in a klystron or traveling wave tube and the RF fields cause the electrons to form into bunches.

Figure 7.10 shows the magnitude of the perturbations in beam diameter that result from changes in charge density produced by changes in current density. Notice that there is very nearly a one-to-one relationship between changes in current density and the resulting changes in beam diameter.

An electrode and pole-piece arrangement for launching a beam under Brillouin flow conditions is illustrated in Fig. 7.11. Once again, it is important to note that no flux links the cathode.

Detailed measurements made on Brillouin-focused electron beams show that, in fact, these beams behave very nearly as predicted by analysis [7.2]. In Fig. 7.12, a beam-scanning device is illustrated which was used for current density measurements in a beam from a magnetically shielded Pierce gun [7.3]. Results obtained by scanning transversely across the center of a beam at equally spaced axial positions are shown in Fig. 7.13. The inward scalloping of the beam when $B > B_B$ is clearly shown. When $B = B_B$, the beam diameter was essentially constant.

Ideally, each of the current profiles shown in Fig. 7.13 should be rectangular in shape. Two primary factors prevented the current from rising abruptly from zero to a constant value. One of these factors was the finite

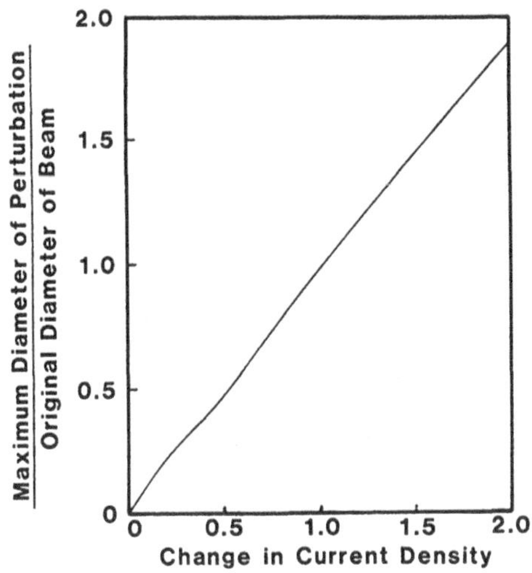

Figure 7.10 Magnitude of Perturbations in Beam Diameter Caused by Changes in Current Density for a Brillouin Flow Beam

(From *Power Travelling-Wave Tubes* by J.F. Gittins, Published 1965 by American Elsevier Publishing, Inc.)

diameter (0.10 in) of the aperture used for current sampling in the beam-scanning device. The other factor was the thermal velocities of the electrons.

From data of the type shown in Fig. 7.13, beam radius as functions of axial position and magnetic flux density were determined and are given in Fig. 7.14. At the predicted Brillouin flux density of 285 Gauss, the variation in beam diameter was less than 2%.

The high degree of susceptibility of the Brillouin beam to perturbations is illustrated in Figs. 7.15 through 7.17. An ac current was used in the heater of the electron gun and the voltage to the cathode of the electron gun was pulsed on at various times during the heater current cycle [7.4].

When the heater current was zero, the electron beam cross section of Fig. 7.15 was obtained. When the heater current was at a maximum value, cross sections like that of Fig. 7.16 were obtained. Under very specialized conditions, an image of the spiral-wound heater could be detected in the beam as shown in Fig. 7.17 [7.5].

The perturbations shown in Figs. 7.16 and 7.17 resulted from the linking through the cathode surface of magnetic flux lines generated by the heater current.

The phenomenon shown in Fig. 7.17 can be explained theoretically by tracing electron trajectories from the cathode as indicated in Fig. 7.18 [7.6].

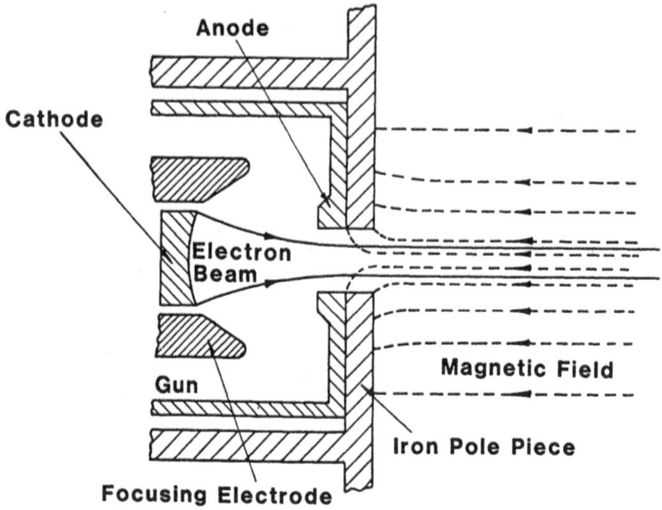

Figure 7.11 A Practical Arrangement for Launching Brillouin Flow

(From *Power Travelling-Wave Tubes* by J.F. Gittins, Published 1965 by American Elsevier Publishing, Inc.)

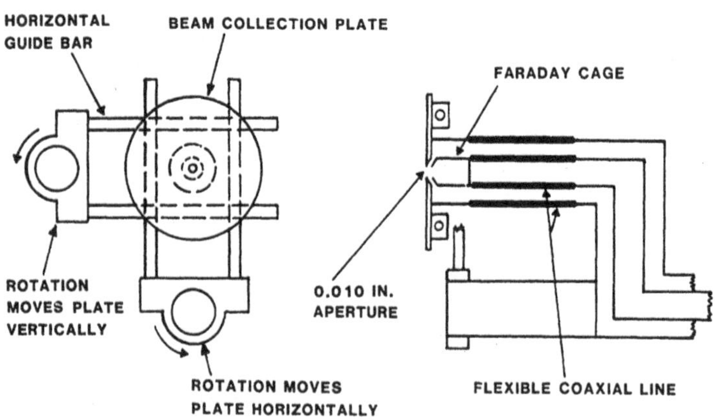

Figure 7.12 Beam Scanning Device Used for Measurements of Beam Current Density

(From A.S Gilmour, Jr. *Proc. High Power Microwave Tube Symposium,* The Hexagon, 1962.)

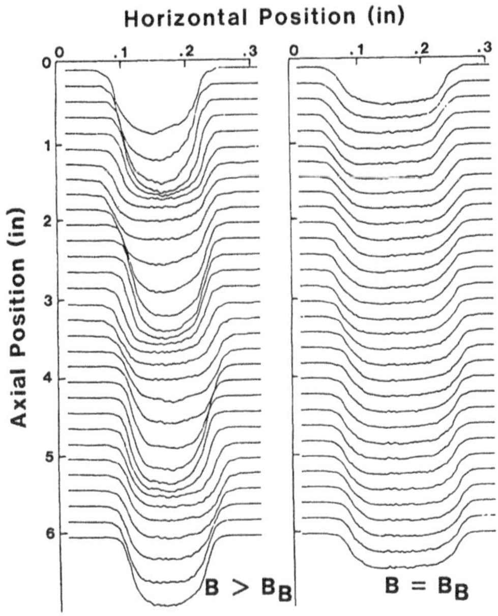

Figure 7.13 Sample Data Obtained Using Beam Scanning Device of Fig. 7.12

(From A.S Gilmour, Jr. *Proc. High Power Microwave Tube Symposium,* The Hexagon, 1962.)

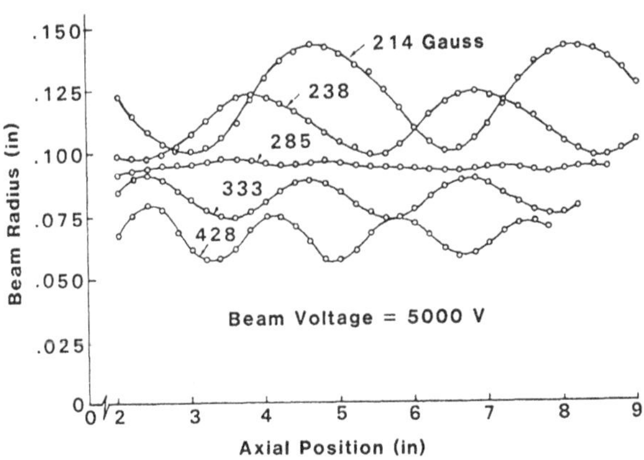

Figure 7.14 Measured Beam Radius *versus* Axial Position and Magnetic Flux Density

(From A.S Gilmour, Jr. *Proc. High Power Microwave Tube Symposium,* The Hexagon, 1962.)

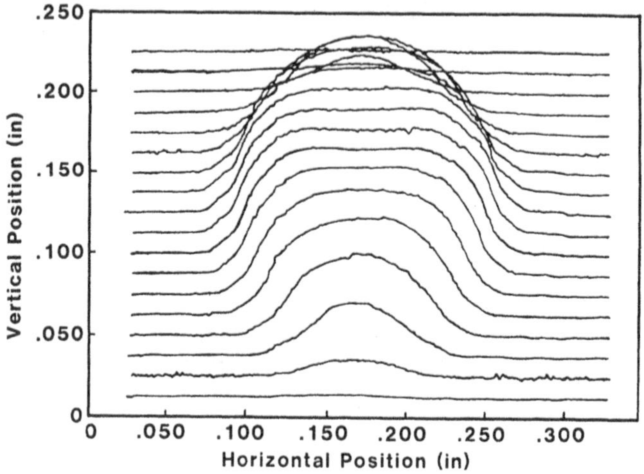

Figure 7.15 Electron Beam Cross Section Obtained under Brillouin Flow Conditions for No Current through Cathode Heater

(From A.S. Gilmour, Jr., *Proc. IRE*, Vol. 49, 1961.)

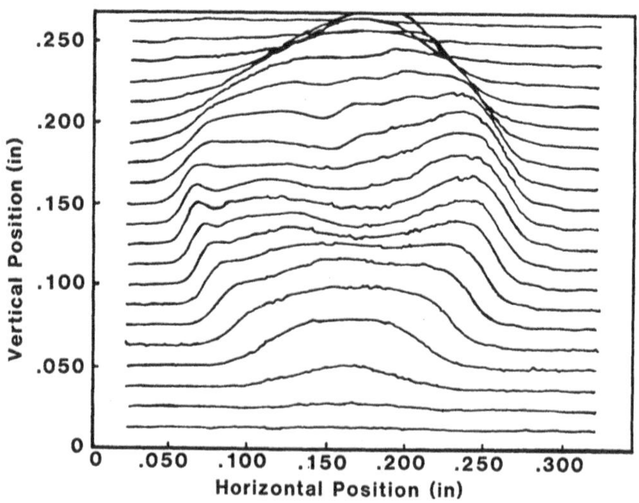

Figure 7.16 Electron Beam Cross Section Obtained under Brillouin Flow Conditions with the Cathode Heater Current at a Maximum Value

(From A.S. Gilmour, Jr., *Proc. IRE*, Vol. 49, 1961.)

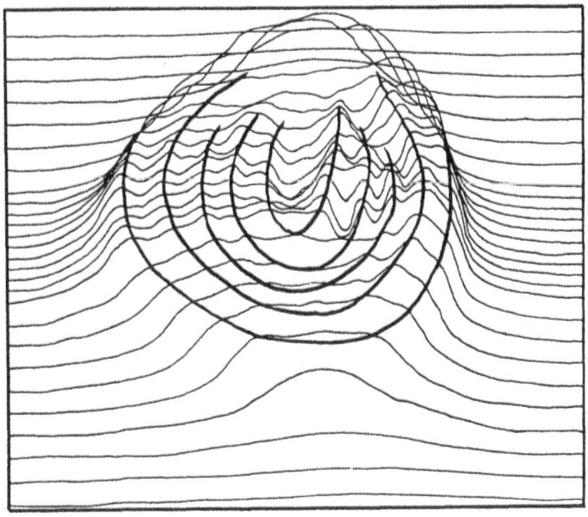

Figure 7.17 Image of Spiral-Wound Cathode Heater in Electron Beam Cross Section

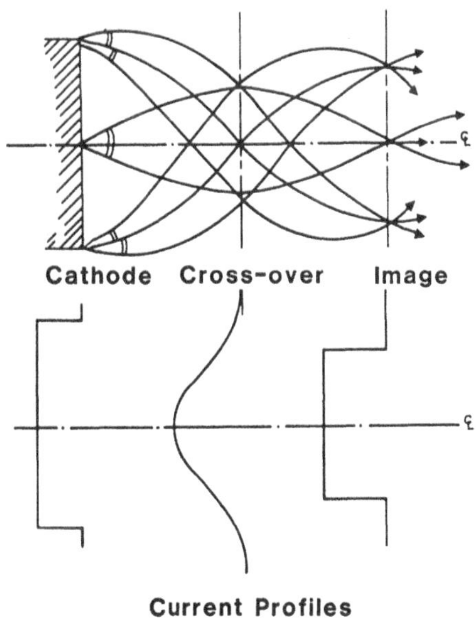

Figure 7.18 Theoretical Electron Trajectories in Beam Resulting from Perturbations at Cathode

7.2 CONFINED FLOW FOCUSING

In practice, Brillouin focusing is rarely used because of the high sensitivity of the beam to perturbations. Instead, as shown in Fig. 7.19, the magnetic circuit providing the focusing field for a tube is purposely designed so that magnetic flux lines pass through the cathode. This makes possible the use of a much larger magnetic field than the Brillouin value for focusing the beam. In addition, this is accomplished without causing scalloping of the beam.

As shown in Fig. 7.20, the use of focusing fields that are a factor of two to three above the Brillouin value reduces beam-diameter perturbations by an order of magnitude from those which occur at the Brillouin field level.

It is helpful in understanding the effect of flux through the cathode in confined flow to assume that the electron trajectories follow the magnetic flux lines in the electron gun. With this assumption, the magnetic field has no affect on the electrons in the gun.

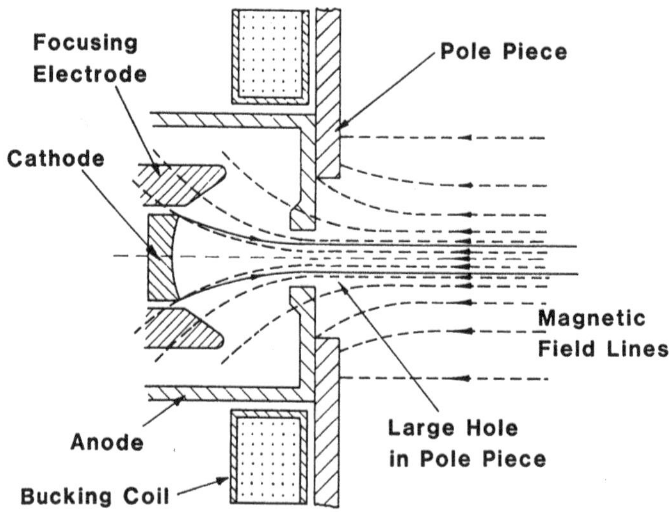

Figure 7.19 Magnetic Circuit Configuration for Launching Smooth Flow at Field Levels Above the Brillouin Value

(From *Power Travelling-Wave Tubes* by J.F. Gittins, Published 1965 by American Elsevier Publishing, Inc.)

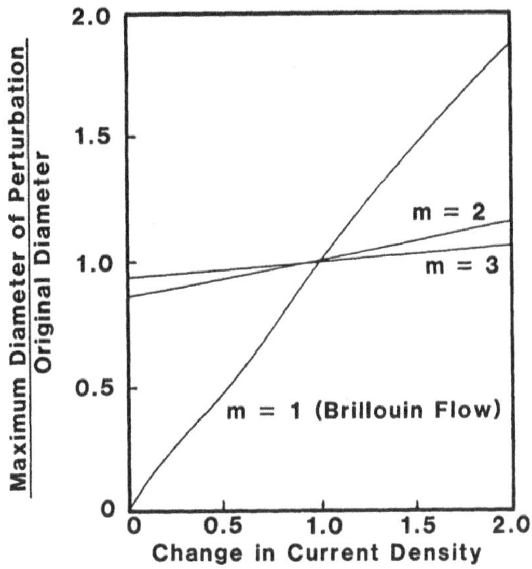

Figure 7.20 Perturbations in Smooth Flow at Fields Above Brillouin Caused by Changes in Current Density

(From *Power Travelling-Wave Tubes* by J.F. Gittins, Published 1965 by American Elsevier Publishing, Inc.)

It should be pointed out that electrons do, in fact, cross flux lines in the gun (as was indicated in Fig. 7.19). The resulting behavior of the beam is essentially the same as it is if it is assumed that the electrons do not cross flux lines.

As the electron beam leaves the gun and enters the main part of the focusing field, the *increase* in flux density encountered must be sufficient to produce a magnetic force that counterbalances the space-charge force of the beam. In fact, for smooth flow, the increase in flux density should be the Brillouin value. That is,

$$B_B = B_z - B_{z0} \tag{7.3}$$

where B_z is the flux density in the main part of the focusing field and B_{z0} is the flux density in the beam as it leaves the gun (see Fig. 7.21).

Because the electron trajectories coincide with flux lines in the gun, the factor by which the flux lines are compressed must be the same as the compression ratio for the electron beam. Thus,

$$B_{z0} = B_c \frac{A_c}{A_b} \tag{7.4}$$

where B_c is the flux density at the cathode, A_c is the area of the cathode, and A_b is the area of the beam.

Finally, for smooth flow (no scalloping),

$$B_B = B_z - B_c \frac{A_c}{A_b} \tag{7.5}$$

This relation is a form of a theorem known as Busch's theorem, which relates the change in angular rotation of a beam to the change in flux density experienced by the beam.

Since the area compression ratio of an electron gun, A_c/A_b, can be large (a value of 50 is not uncommon), only a small flux density at the cathode surface is required to permit B_z to be a factor of two or three larger than the Brillouin value.

When a perturbation in charge density occurs in a confined beam (such as in electron bunching as a result of interaction with an RF field), the electrons are forced to move radially, just as they did in the Brillouin flow case considered early in this chapter. The significant difference between confined flow and Brillouin flow is that the radially moving electrons encounter a flux density of B_z, which can be (see Eq. (7.5)) much larger than B_B. As a result, the radial motion of the electrons can be easily limited by an order of magnitude, as was shown in Fig. 7.20. The relative advantages and disadvantages of Brillouin flow, confined flow and an intermediate technique sometimes called space-charge balanced flow are given in Fig. 7.22.

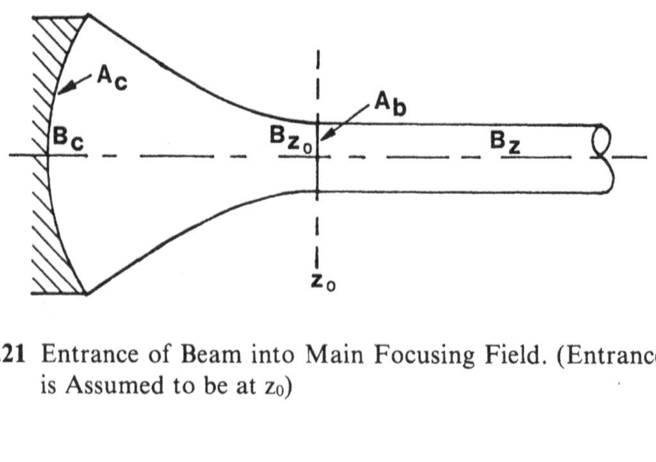

Figure 7.21 Entrance of Beam into Main Focusing Field. (Entrance Plane is Assumed to be at z_0)

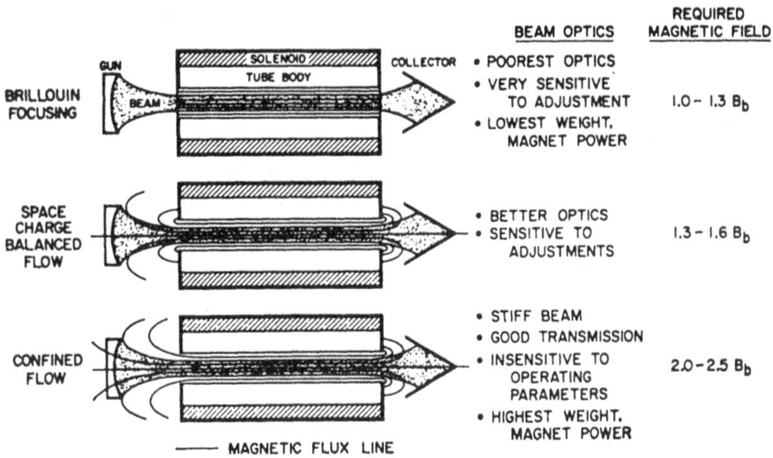

Figure 7.22 Comparison of Focusing Methods

(From Armand Staprans, *et al.*, *Proc. IEEE*, March 1973.)

7.3 PERIODIC PERMANENT MAGNET FOCUSING

When minimum size and weight are essential, periodic permanent magnets are frequently used for focusing electron beams. The weight reduction compared to a solenoid or uniform-field permanent magnet can be one to two orders of magnitude. Permanent magnet materials and the configurations of periodic permanent magnet structures were discussed in Chapter 4.

An electron beam focused by periodic permanent magnets is illustrated in Fig. 7.23. The magnet configuration shown would not be of practical use, but it is convenient for illustrating the principles of periodic permanent magnet focusing.

The basic principle of periodic-magnet focusing is straightforward. As an electron beam enters the field of one of the magnet sections, the forces on the electrons act in exactly the same way as was described in the section on Brillouin focusing. That is, the beam starts to rotate and the interaction of the resulting rotational motion with the axial field produces a radial force that compresses the beam.

As the beam leaves the magnet section, rotation stops and the focusing forces vanish. As a result, the beam expands under the influence of space-charge forces.

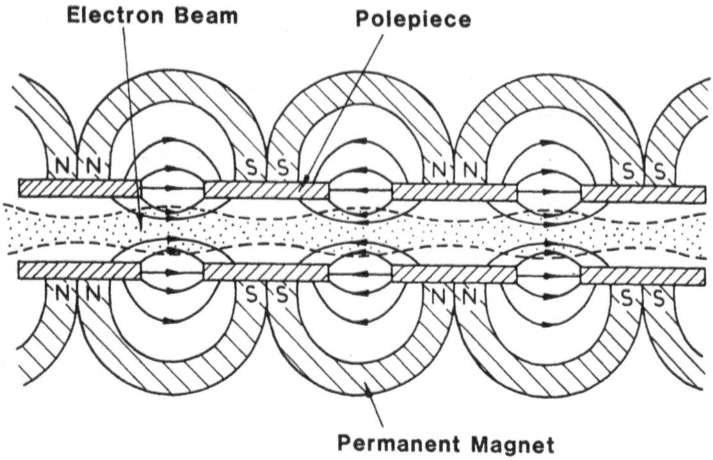

Figure 7.23 Periodic Focusing of an Electron Beam with Permanent Magnets

(From *Power Travelling-Wave Tubes* by J.F. Gittins, Published 1965 by American Elsevier Publishing, Inc.)

The beam then enters another magnet section with the field in the opposite direction from the previous one. As a result, the beam rotates in the opposite direction, but it is focused just as it was in the previous section.

The overall result, as the beam traverses the alternating field of the periodic magnet structure, is that its rotational motion oscillates back and forth, producing alternating periods of magnetic focusing and beam expansion. The appearance of the beam, as shown in Fig. 7.23, is like that of a scalloping Brillouin beam.

Periodic permanent magnet focusing was originally used in low-power helix TWTs. It is now applied to moderately high powered helix TWTs as well as to coupled cavity TWTs and some klystrons.

Because of the undulating shape of the beam, transmission through a tube is not as good as for uniform field focusing. As a result, as will be shown in the chapter on traveling wave tubes (Ch. 10), average power handling capability for a given tube structure is about half an order of magnitude lower than for uniform field focusing.

The analysis of periodic focusing systems is carried out with the assumption that the beam encounters a series of converging lenses (the permanent magnet fields) and expands under the influence of space-charge forces between the lenses, as illustrated in Fig. 7.24.

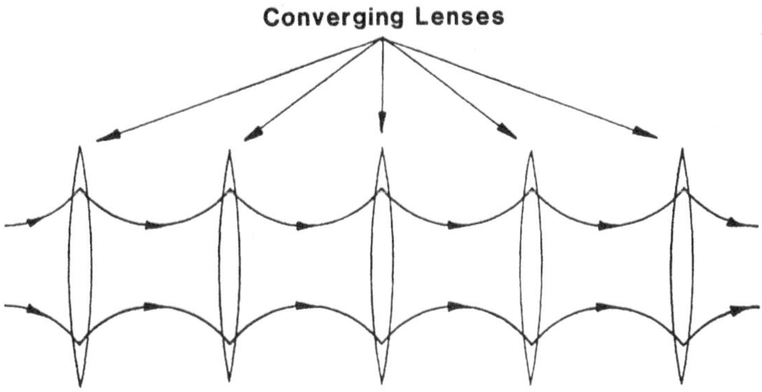

Figure 7.24 Periodic Focusing by a Series of Converging Lenses

(From *Power Travelling-Wave Tubes* by J.F. Gittins, Published 1965 by American Elsevier Publishing, Inc.)

It can be shown theoretically that the root mean square value of the magnetic field required for periodic focusing is the Brillouin field. Periodic focusing systems are not stable at all values of magnetic flux, nor, for a given flux level, at all values of beam current. In the unstable regions, the lenses cannot maintain control and the beam diverges. In addition, even in the regions of stability, for currents other than the design value, excessive interception is often a problem because of large fluctuations in the beam diameter. In particular, the beam is not well focused at voltages below the design value and this can create interception problems when the beam is turned on or off.

7.4 ION EFFECTS IN ELECTRON BEAMS

Gas in a microwave tube is ionized by the electron beam and may degrade performance if the ion density becomes excessive. Linear beam tubes (klystrons and TWTs) are far more susceptible to ion effects than crossed-field tubes because potential wells exist in which ions are trapped. Ion density increases with time so that ion effects are more severe during operation under long pulse or CW conditions. Ions affect tube performance in at least the following three ways:

1. Ions neutralize electron space charge potentials within a beam. Beam diameter becomes smaller ("ion focusing" is said to occur), and so power output is reduced. Also, interception current decreases with

time. In extreme cases, the electron beam remains focused in the collector and melts a hole through the collector wall and destroys the tube.

2. Because the electron space charge potential is reduced by ions, the electron beam velocity is increased. This causes a reduction in the phase shift through a tube.
   The change in beam potential caused by ions may be on the order of 1%. Since beam velocity is proportional to the square root of potential, the change in velocity may be on the order of 1/2%. For a tube phase length of 10,000°, the decrease in phase would be about 50° and would vary with tube operating conditions.
3. Ions may produce electron beam oscillations or noise, which appears as modulation on the power output of a tube ("ion sidebands" are said to occur). The frequency of the oscillations is dependent on electron beam density and ion mass. Frequencies of MHz to tens of MHz are common.

7.4.1 Effect on Emission

Figure 7.25 contains a graphic illustration of one of the effects of ions in an electron beam. An electron beam cross section is shown with a significant peak in electron density in the center of the beam [7.5]. The peak in density results from positive ions in the center of the beam near the electron gun being accelerated toward the cathode. In the region near the center of the cathode the ions cause sufficient neutralization of the electron space-charge cloud so that additional electrons leave the cathode.

The potential distribution in the drift tube containing an electron beam is shown in Fig. 7.26 along with an illustration of ions being trapped in the potential well caused by the electron beam. Of course, the presence of ions in the well reduces the potential of the well and causes the electron beam to be accelerated to a higher velocity. This results in the decrease in phase length of a tube as noted earlier.

7.4.2 Oscillations

When the ion concentration becomes sufficient, oscillations may occur as indicated by Fig. 7.27. This figure shows the amplitude of the current to a probe in a pulsed electron beam in the presence of gas (hydrogen at a pressure of about 10^{-6} Torr).

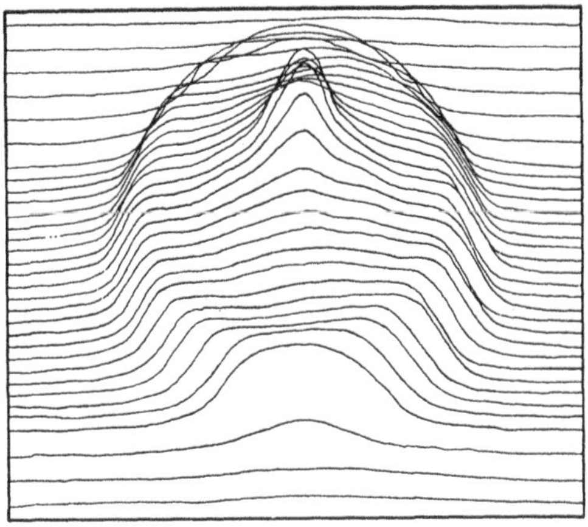

Figure 7.25 Electron Beam Cross Section Showing Effects of Ions

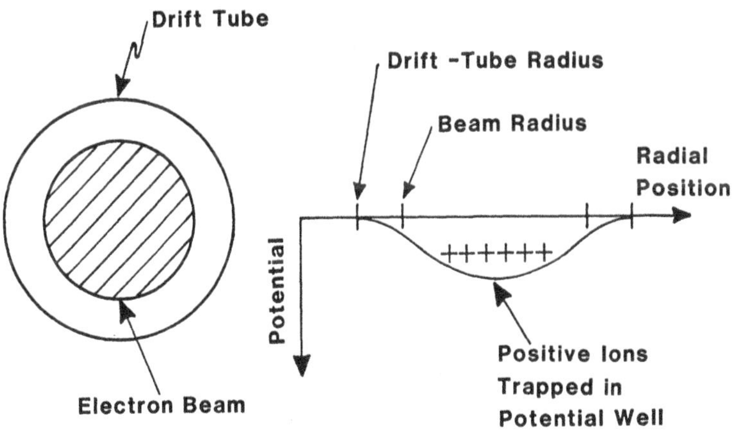

Figure 7.26 Potential Distribution and Ion Concentration in Drift Tube Containing an Electron Beam

At the beginning of the pulse, several cycles of a relatively high frequency of oscillation occur (~ 8 MHz). After about 2 μs, a lower frequency of oscillation (~ 4.5 MHz) occurs. Then, the higher oscillation frequency returns. Finally, at the end of the pulse, the oscillations become very noisy.

Detailed experimental analyses showed that two distinct modes of ion induced beam oscillations occur [7.7]. Measurements were made of the

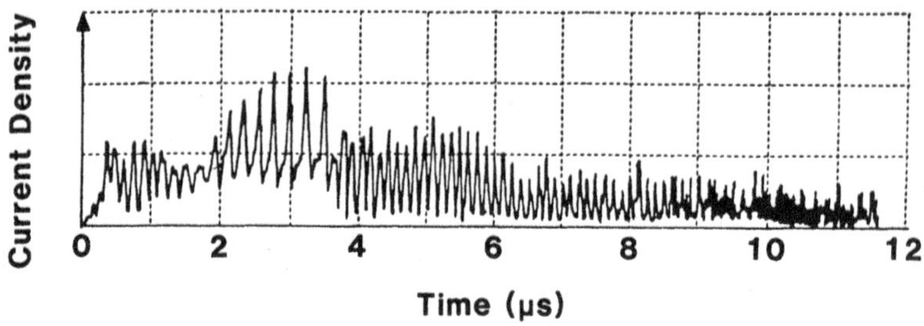

Figure 7.27 Probe Current in Electron Beam Analyzer Showing Two Modes of Ion-Induced Beam Oscillations

(From A.S Gilmour, Jr. RADC-TDR-63-465, vol. 1. December 1963.)

position and shape of the electron beam at various times during a pulse of the type shown in Fig. 7.27.

During high-frequency portions of the pulse, a radial oscillation of the electron beam was found. That is, the beam diameter changed periodically from a relatively large value to a smaller value as shown in Fig. 7.28. This figure shows probe current as a function of radial probe position in the beam at times of maximum and minimum beam diameter. The time increments were equally spaced at the half period of the oscillation frequency.

The radial mode of oscillation may be explained with the aid of Fig. 7.29. At some point in time, the positive ions are nearly uniformly distributed throughout the potential well of the electron beam. The ions are attracted and flow to the center of the beam (bottom of the well). The resulting concentration of ions produces a repulsive force (associated with a positive peak in the potential well not shown in Fig. 7.29), which causes the ions to reverse direction and flow radially outward. As the ions flow outward, the potential well deepens, causing a strong enough electrostatic force to pull them inward again. The radial electrostatic forces in the electron beam depend on the ion concentration and so the electron beam oscillates radially in synchronism with the ions.

The oscillation frequency is dependent on the shape of the potential well (determined by electron distribution and concentration) and on ion mass, and thus is not difficult to calculate. The calculated oscillation frequency agrees well with the observed frequency.

During the low frequency oscillations indicated in Fig. 7.27, a transverse oscillation of the electron beam was found. That is, the peaks and valleys in current in the two to three microsecond time frame of Fig. 7.27 were found

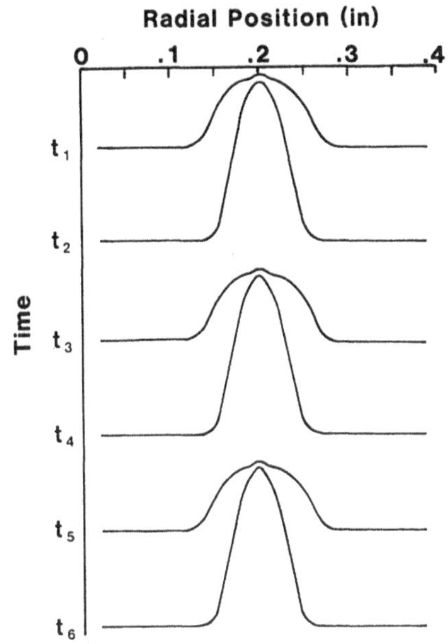

Figure 7.28 Probe Current as a Function of Radial Position for Radial Mode of Oscillation

(From A.S Gilmour, Jr. RADC-TDR-63-465, vol.1. December 1963.)

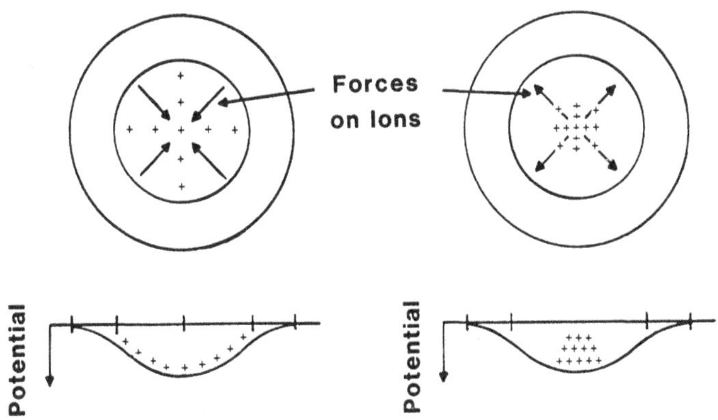

Figure 7.29 Ion Distribution in Radial Mode of Oscillation

to correspond to alternating transverse positions of the electron beam shown in Fig. 7.30. Subsequently, more detailed measurements showed that the beam was actually following a rotating helical path. In fact, the center position of the electron beam during the initiation of transverse oscillations was as shown in Fig. 7.31.

Based on Fig. 7.30, it would appear that the ions were simply oscillating back and forth in the potential well as indicated by Fig. 7.32. In order to explain the rotating motion of the beam, it is necessary to assume an angular motion of the ions superimposed on the radial motion.

The transverse oscillation frequency, like the radial oscillation frequency, is a function of ion mass as well as electron beam density and density profile. Oscillation frequencies from the megahertz range to several tens of megahertz are common.

The presence of ion oscillations is indicated by ion sidebands on the output frequency of a microwave tube. The oscillation frequency is a function of the square root of beam voltage (see Fig. 7.33). This provides a means for checking to determine if gas is the cause of spurious tube output signals.

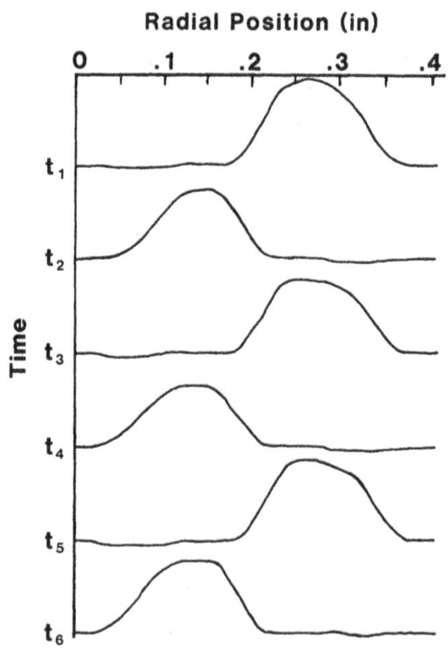

Figure 7.30 Probe Current as a Function of Radial Position for Transverse Mode of Oscillation

(From A.S Gilmour, Jr. RADC-TDR-63-465, vol. 1. December 1963.)

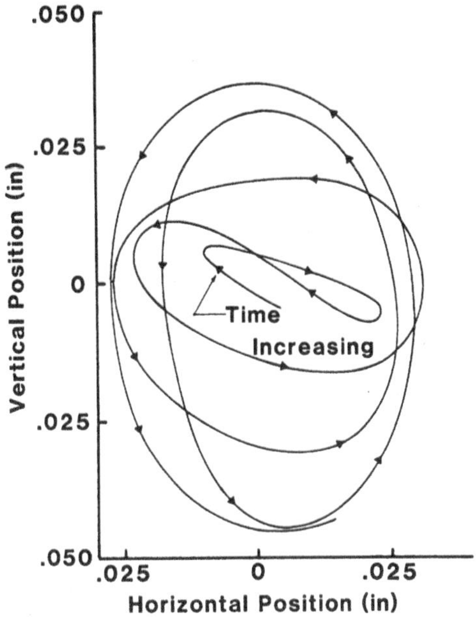

Figure 7.31 Center Position of Beam During Initiation of Transverse Oscillation

(From A.S Gilmour, Jr. RADC-TDR-63-465, vol.1. December 1963.)

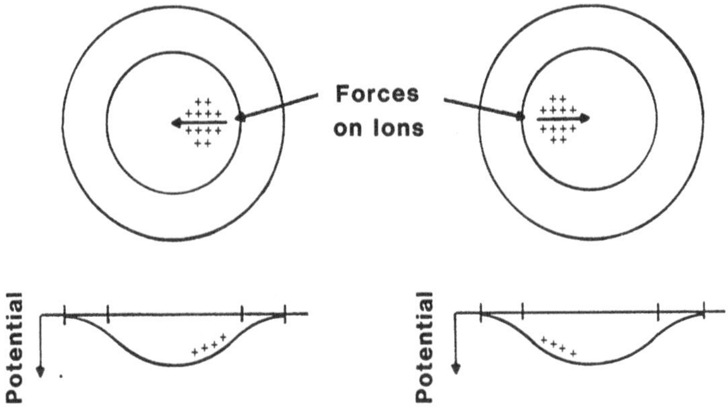

Figure 7.32 Ion Distribution in Transverse Mode of Operation

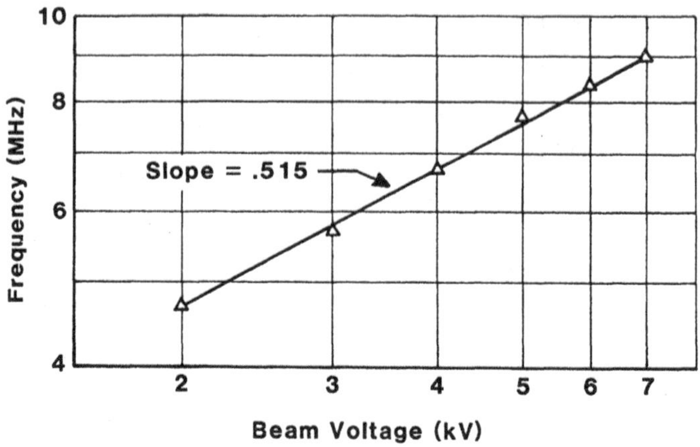

Figure 7.33 Measured Ion-Induced Beam-Oscillation Frequency as a Function of Beam Voltage. (The slope of the measured result is very close to the theoretical slope of 0.5)

(From A.S. Gilmour, Jr. RADC-TDR-63-465, vol. 1. December 1965.)

REFERENCES

[7.1] Recommended for further reading:

M. Chodorow and C. Susskind, *Fundamentals of Microwave Electronics.* New York: McGraw-Hill, 1964.

A. S. Gilmour, Jr., "Basic principles and some aspects of the electromagnetic compatibility of microwave tubes," *Inter-System Electromagnetic Compatibility.* New York: Holt, Reinhart and Winston, 1972.

W. J. Kleen, *Electronics of Microwave Tubes.* New York: Academic Press, 1958.

[7.2] A. S. Gilmour, Jr., G. C. Dalman, and L. F. Eastman, "The current and velocity distributions in a velocity-modulated Brillouin-focused electron beam," *Proc. 4th International Congress on Microwave Tubes,* Eindhoven, The Netherlands: Centrex, 1963, pp. 675–688.

[7.3] A. S. Gilmour, Jr. and D. Hallock, "A demountable beam analyzer for studying magnetically confined electron beams," *Advances in Electron Tube Techniques,* vol. 2, New York: Pergammon Press, 1963, pp. 153–157.

[7.4] A. S. Gilmour, Jr., "Effect of filament magnetic field on the electron beam from a Pierce gun," *Proc. IRE,* vol. 49, 1961, p. 976.

[7.5] A. S. Gilmour, Jr., unpublished.

[7.6] Gabriel Hermann, "Optical theory of thermal velocity effects in cylindrical electron beams," *J. Appl. Phys.,* vol. 29., no. 2, 1958.

[7.7] A. S. Gilmour, Jr., "Transverse and radial oscillations induced in electron beams by ions," *Proc. 5th International Congress on Microwave Tubes.* Paris: Dunod, 1965, pp. 453–456.

Chapter 8
Gridded Tubes

Gridded tubes are used at low microwave frequencies (in the hundreds of megahertz). Above about one GHz, microwave tubes (klystrons, TWTs, or crossed-field devices) are normally used. The gridded tubes that are discussed in this chapter are triodes, tetrodes, and klystrodes.

The klystrode, while basically an old device, has recently received significant attention and promises to be an extremely important device in the low microwave region [8.1]. Also, the klystrode is an excellent point of departure for discussions of klystrons and other microwave tubes because it contains some of the resonant circuit technologies of those devices coupled with the grid modulation technologies of triodes and tetrodes.

8.1 TRIODES AND TETRODES

A triode consists of a thermionic cathode, a control grid, and an anode as shown in Fig. 8.1. The function of the grid is to regulate the potential profile in the region adjacent to the cathode, as indicated in Fig. 8.2, and thereby control the current from the cathode.

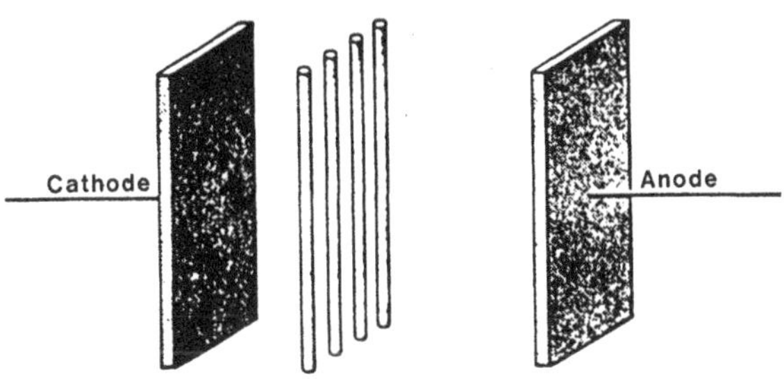

Figure 8.1 Basic Configuration of a Triode

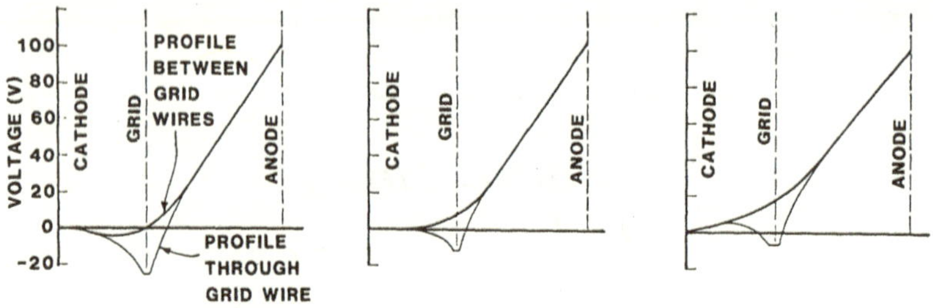

Figure 8.2 Potential Profiles for a Planar Triode without Space Charge (for Three Different Control Grid Voltages)

As Fig. 8.2 indicates, the control grid is normally negative with respect to the cathode, so grid interception of cathode current is essentially zero. It is the region between grid elements in which the potential profile is sufficiently positive to permit electrons to leave the cathode.

In a tetrode, another grid called a screen grid is placed between the control grid and the anode. The primary function of the screen grid is essentially to eliminate current from being induced in the control grid as the result of changes in anode potential. Shown in Fig. 8.3 are typical potential distributions in a tetrode.

As frequencies approach the microwave region (generally considered to be about 300 MHz and above), the frequency of the signal applied to the control grid becomes high enough so that its period is short compared with the time (transit time) that it takes for the electrons to move from the cathode to the grid. As a result, electrons may oscillate back and forth many times in the space between the cathode and grid before either passing the grid and being accelerated to the anode or else being driven back into the cathode.

To alleviate this problem, the control grid is moved closer to the cathode. Ultimately, the spacing between the cathode and the control grid becomes a few thousandths of an inch, the diameter of the grid wires becomes a fraction of a thousandth of an inch, and the diameter of the cathode and control grid (planar configuration) become less than half an inch. An example is a 3 GHz planar triode in which cathode, grid, and anode are circular and planar. The grid-wire diameter is 0.4 mil (10 μm), the grid-to-cathode spacing is about 3 to 4 mils (80 to 100 μm), and the diameter (of the cathode, grid, and anode) is less than 0.4 inch (1 cm). The average power rating is somewhat under 100 W. Figure 8.4 contains a cross-sectional view of a high-frequency planar triode.

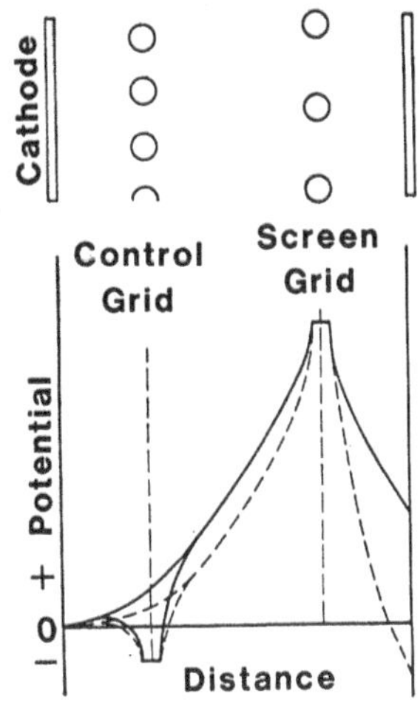

Figure 8.3 Potential Profiles in a Tetrode for Two Plate (Anode) Voltages

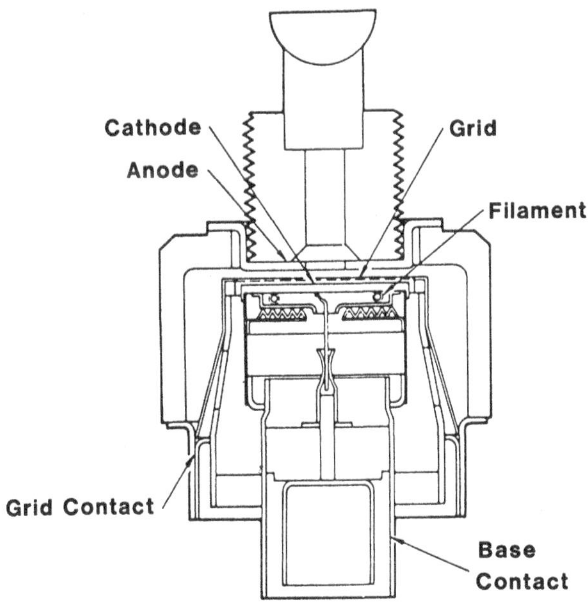

Figure 8.4 Internal Configuration of a Planar Triode

(Courtesy Varian Eimac)

In addition to the transit-time problem, lead inductances and capacitances between tube electrodes become extremely troublesome at high frequencies. As will be seen in examples given in subsequent paragraphs, these problems are partially alleviated by using microwave transmission line techniques to adapt the tube to the circuit.

In spite of the transit-time problems which necessitate close grid-to-cathode spacings, triodes and tetrodes capable of peak power levels of several megawatts and average power levels of several hundred kilowatts are available for operation at hundreds of megahertz [8.2]. In these devices, the RF circuit may be external to the tube, in which case the tube has a large ceramic insulator through which energy is coupled to the external circuit. The internal parts of such a tube are illustrated in Fig. 8.5.

In other cases, much of the RF circuit is incorporated into the design of the tube. An example is the coaxitron illustrated in Fig. 8.6. The cathode, grid, and anode part of this extremely complex structure is deceptively simple, as illustrated in Fig. 8.7. It consists of a cylindrical array of 48 independent linear triodes that are operated in parallel. The directly heated cathode of each is a long ribbon shaped filament that is spring-supported so that it retains its shape and position as temperature is varied. A copper grid block is used to suspend grid wires in close and carefully controlled proximity to the cathode.

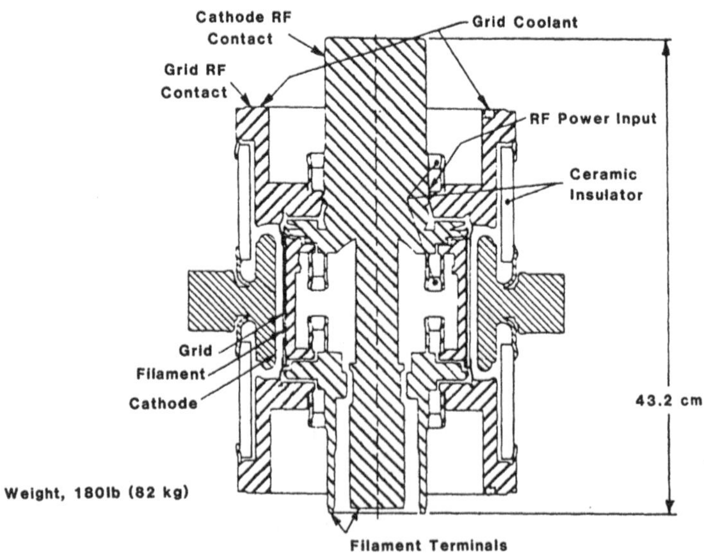

Figure 8.5 Cross-Sectional View of High Power Triode

(From T. E. Yingst *et al., Proc. IEEE,* March 1973.)

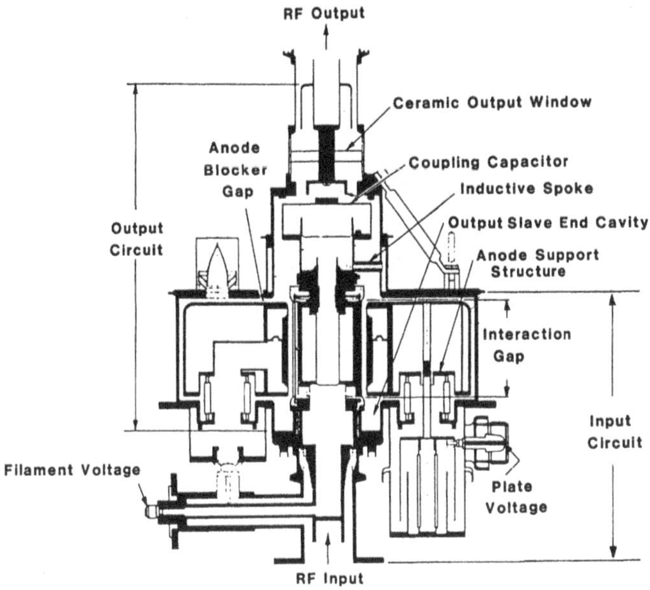

Figure 8.6 High-Power Coaxitron Tube

(From T. E. Yingst *et al., Proc. IEEE,* March 1973.)

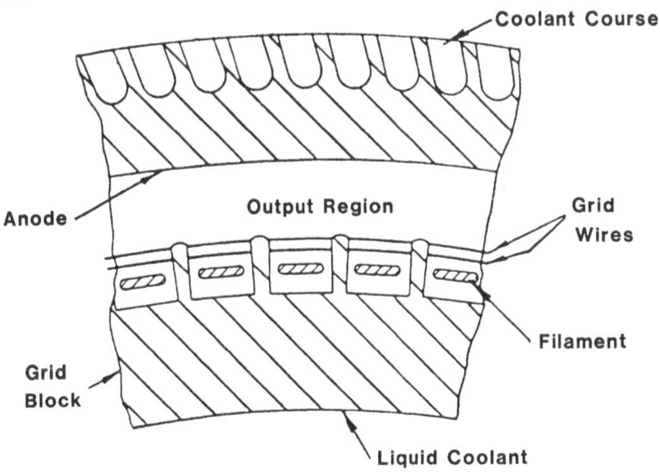

Figure 8.7 End View of Cathode (Directly Heated) Grid and Anode Structure of Coaxitron Tube

(From T. E. Yingst *et al., Proc. IEEE,* March 1973.)

A plot showing peak coaxitron power output *versus* frequency for various load phase adjustments is given in Fig. 8.8. Other coaxitron designs cover the frequency band from 200 to 1300 MHz at average powers up to 300 kW and peak powers exceeding 15 MW.

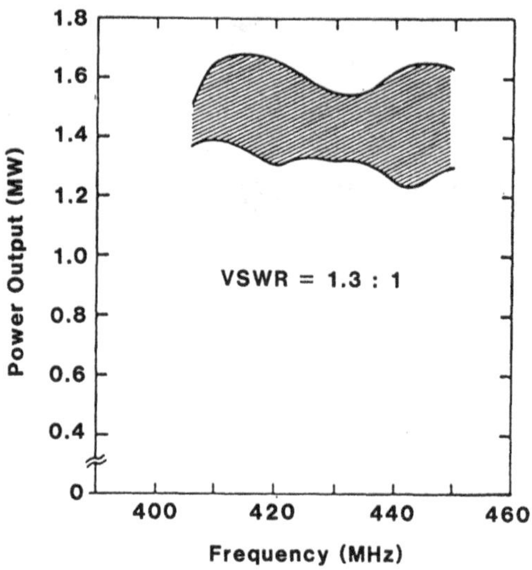

Figure 8.8 Peak Power Output for a Coaxitron with Output Load VSWR Phase Adjusted from Maximum to Minimum Power Output

(From T. E. Yingst *et al.*, *Proc. IEEE,* March 1973.)

8.3 KLYSTRODES

As frequency is increased and dimensions are reduced, a major limitation to power output of conventional gridded tubes is the power that must be dissipated by the anode.

In 1939, Haeff perceived that RF power could be removed from a bunched electron beam (modulated by a grid next to a cathode) by passing it through a resonant cavity. This and similar devices were developed to some extent, but were dropped shortly after World War II because of low gain and bandwidth.

Recently, Preist and Shrader resurrected the Haeff tube [8.1]. The device, known as a klystrode, has five elements as shown in Fig. 8.9. These are the cathode, planar grid, anode, tail pipe, and collector. The tube may be immersed in a uniform magnetic field for beam control. Superficially, the tube is similar to a tetrode, except that the output circuitry is a resonant cavity.

In operation, an alternating RF voltage is applied between the cathode and grid by one of several possible circuit arrangements. Two of these are illustrated in Fig. 8.10. In Fig. 8.10(a) the signal is capacitively coupled to the grid. In Fig. 8.10(b) the signal is inductively coupled with a coupling loop into an RF cavity containing the grid structure. The resulting density-modulated (bunched) electron beam is accelerated toward the apertured anode at high potential.

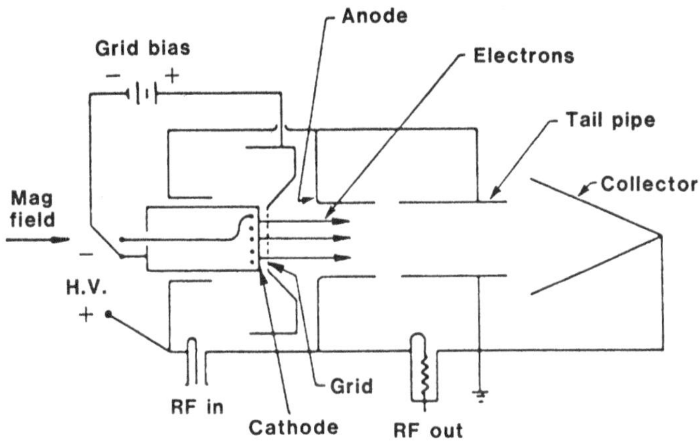

Figure 8.9 Klystrode Schematic

(From Donald H. Preist and Merrald B. Schrader, *Proc. IEEE,* November 1982.)

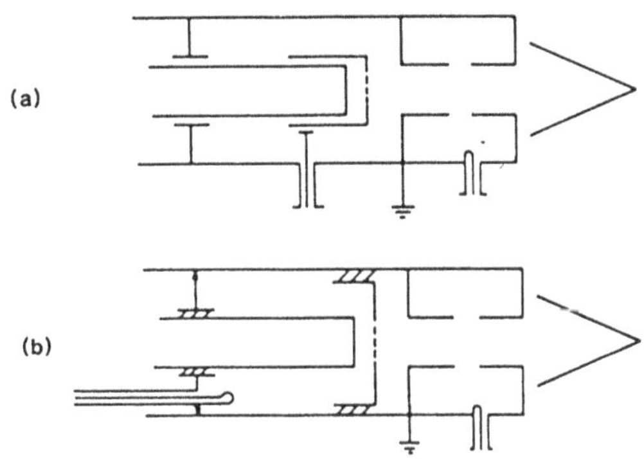

Figure 8.10 Two Possible Input-Circuit Arrangements

(From Donald H. Preist and Merrald B. Schrader, *Proc. IEEE,* November 1982.)

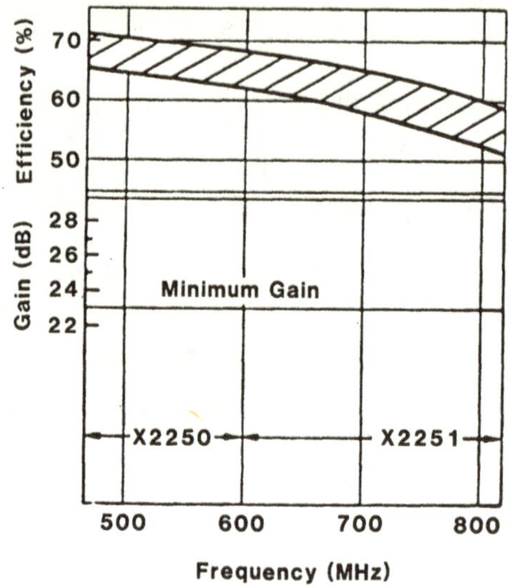

Figure 8.11 Klystrode Model X2250 and X2251 Efficiency and Power Gain *versus* Frequency

(Courtesy Varian Eimac)

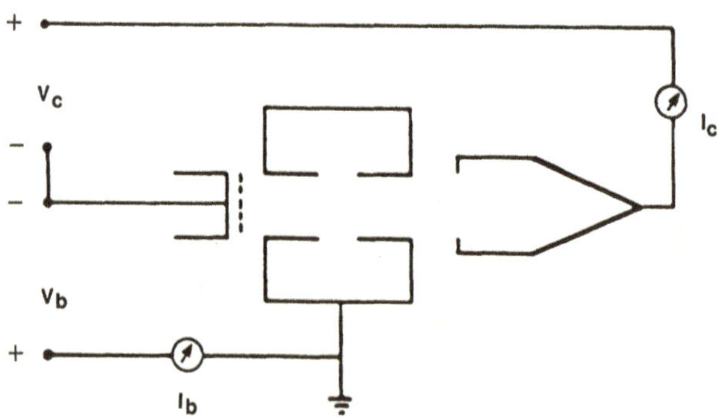

Figure 8.12 Power Supply Connections for Depressed Collector Operation

(From Donald H. Preist and Merrald B. Schrader, *Proc. IEEE,* November 1982.)

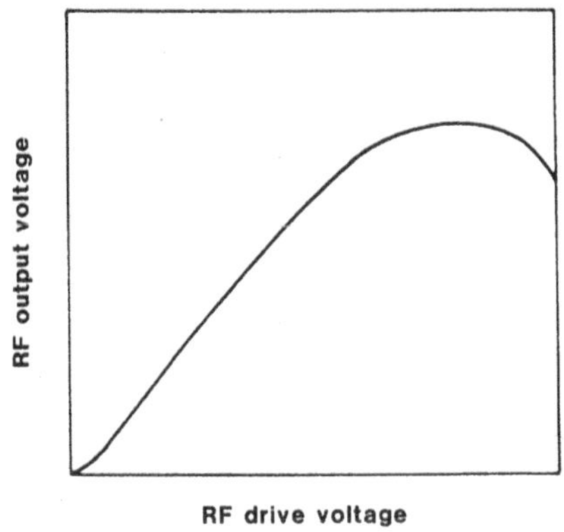

Figure 8.13 Typical Output Voltage *versus* Drive Voltage

(From Donald H. Preist and Merrald B. Schrader, *Proc. IEEE,* November 1982.)

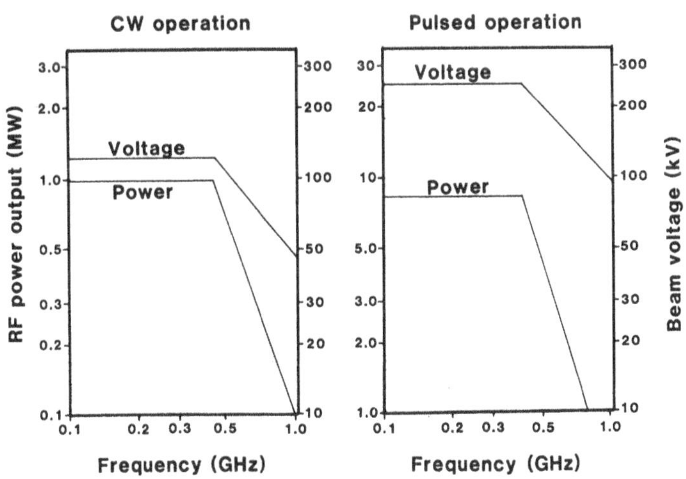

Figure 8.14 Calculated Maximum Klystrode Power Capabilities

(From Donald H. Preist and Merrald B. Schrader, *Proc. IEEE,* November 1982.)

Next, the beam passes through a gap, called the output gap, in the resonant RF cavity and induces an oscillating voltage and current in the cavity. Cavities and the mechanisms by which current is induced in a cavity are discussed in the chapter on klystrons (Ch.9). RF power is coupled from the cavity by an appropriate technique, such as inserting a coupling loop into the RF magnetic field within the cavity. Demonstrated efficiency is high, as shown in Fig. 8.11.

Finally, the beam passes through the tail pipe into the collector. By depressing the potential of the collector, as indicated in Fig. 8.12, some of the dc beam power can be recovered to enhance the efficiency of the device. Efficient multistage depressed collectors are discussed in Chapter 12.

Typical values of gain are in the neighborhood of 25 dB as was shown in Fig. 8.11. The gain characteristic of the klystrode is typical of all linear beam amplifiers. A typical gain curve is shown in Fig. 8.13. At low and medium signal levels, gain is constant. At high levels, gain decreases because conversion efficiency drops.

Preist and Shrader have calculated the maximum power capabilities that can be expected for klystrodes. The results are given in Fig. 8.14. Some of the assumptions used are:

CW Power:

- One MW limit at low frequencies set by size and cost of power supply (not necessarily applicable to all applications)
- Above ~ 450 MHz, control-grid temperature limitation is assumed

Pulse Power:

- A 250 kV limit for beam voltage is used, in line with present practice for klystrons
- A maximum voltage gradient of 100 kV/cm in the electron gun above 350 MHz

An electronic efficiency ~ 60% is observed in practice and was assumed. By using multistage depressed collectors, overall efficiencies in the 70–80% range should be possible.

REFERENCES

[8.1] D. H. Preist and M. B. Shrader, "The klystrode — An unusual transmitting tube with potential for UHF-TV," *Proc. IEEE,* vol. 70, November 1982, pp. 1318–1325.

[8.2] T. E. Yingst, D. R. Carter, J. A. Eshleman, and J. M. Pawlikowski, "High-power gridded tubes — 1972," *Proc. IEEE,* vol. 61, 1973, pp. 357–381.

Chapter 9
Klystrons

The klystron is one of two major classes of devices categorized as linear-beam (O-type) tubes. The other is the traveling wave tube. The family tree of linear-beam tubes given in Fig. 9.1 shows some types of klystrons and traveling wave tubes, and also indicates that there are hybrid devices that combine the klystron and TWT technologies.

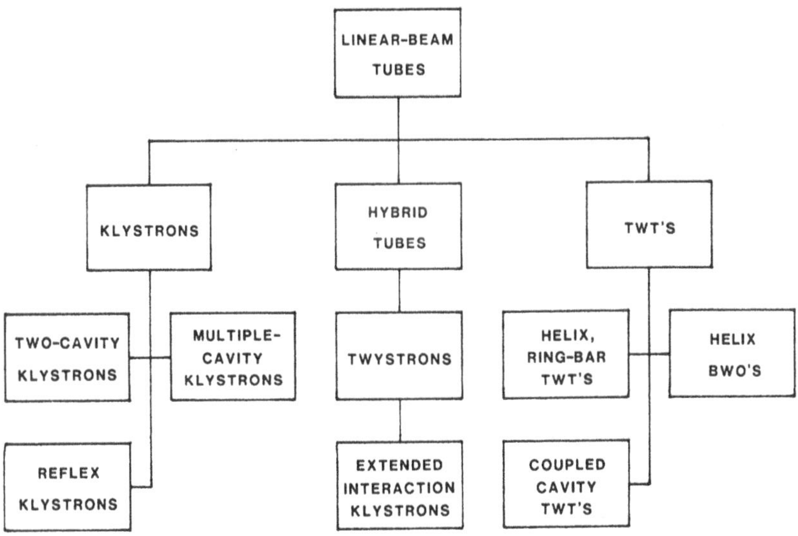

Figure 9.1 Family Tree of Linear-Beam Tubes

(From *Microwave Tube Manual* by Varian Associates, Air Force Publication Number T.0.00-25-251, October 1979.)

Figures 9.2 through 9.5 indicate the differences between the klystron, TWT and hybrid device technologies [9.1]. In the klystron, interactions take place in discrete locations along the beam and there is no coupling of the signal between parts of the interaction structure. The RF signal is carried from cavity to cavity by the electron beam.

In the coupled-cavity version of the TWT, cavities similar in appearance to klystron cavities are used. There are coupling apertures in the walls of the cavities so that the RF energy can propagate along the interaction structure in addition to being carried along by the electron beam. The propagation of the signal along the interaction structure is a primary distinguishing feature between TWTs and klystrons.

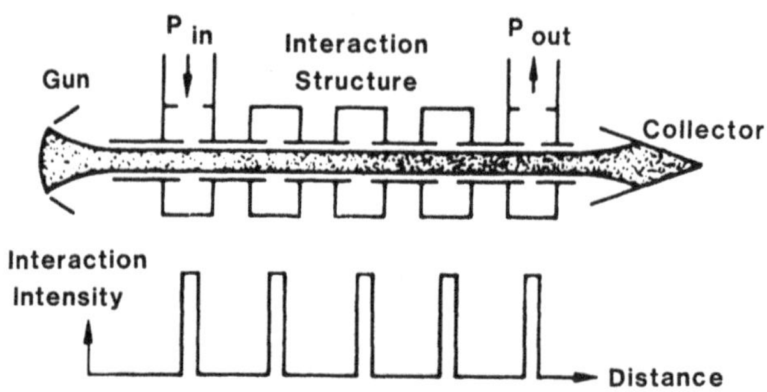

Figure 9.2 Klystron

(From Armand Staprans, *et al.*, *Proc. IEEE*, March 1973.)

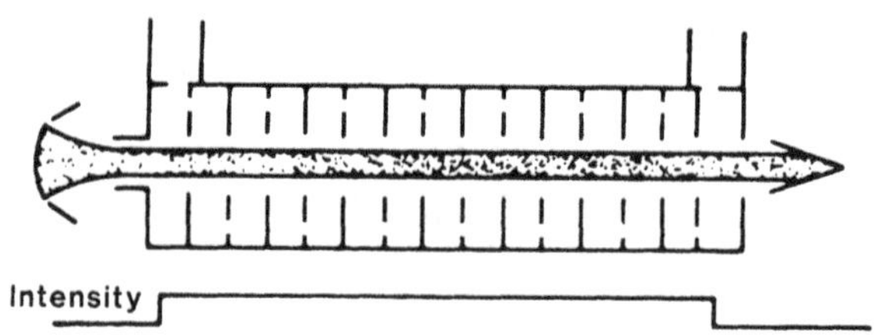

Figure 9.3 Coupled-Cavity TWT

(From Armand Staprans, *et al.*, *Proc. IEEE*, March 1973.)

In the EIK (extended interaction klystron) one or more of the single cavities of the conventional klystron is replaced by a structure containing two or more interaction gaps. Bandwidth is increased over that of the conventional klystron. In addition, at the output end of the tube, the voltage at each interaction gap is reduced, permitting operation at very high power levels.

The twystron consists of a klystron input section and a coupled cavity output section. Twystrons are capable of high power (several megawatts) and large bandwidth (ten to fifteen percent).

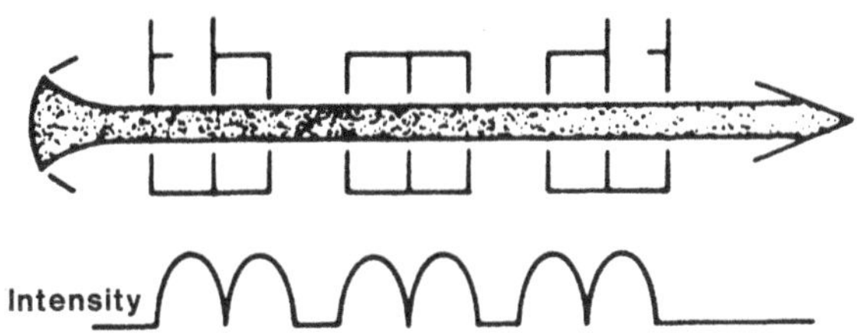

Figure 9.4 Extended Interaction Klystron

(From Armand Staprans, *et al.*, *Proc. IEEE*, March 1973.)

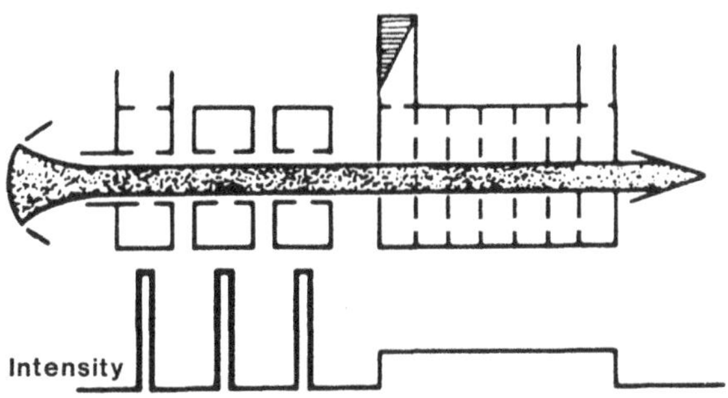

Figure 9.5 Hybrid Twystron Amplifier

(From Armand Staprans, *et al.*, *Proc. IEEE*, March 1973.)

Klystrons will be discussed before the other linear-beam devices (and before crossed-field tubes) because an explanation of the interactions between the beam and circuit in the klystrons is straightforward. It is also a logical extension of the klystrode (inductive output tube) discussion in Chapter 8.

The initial discussion of the interactions in traveling wave tubes in Chapter 10 will be directly based on the klystron discussions of this chapter.

9.1 THE INVENTION OF THE KLYSTRON

Research leading to the development of the klystron had its start in the mid- and late-1930s. A report by R.H. Varian and S.F. Varian in 1939, "A High Frequency Oscillator and Amplifier," described the invention of the klystron [9.2].

The following abstract from the Varians' paper contains a concise and very clear description of the basic operating principals of the klystron. As a side note, this abstract could well serve as a model of clarity to many authors who appear intent on disguising rather than describing their work.

A High Frequency Oscillator and Amplifier

Russell H. Varian and Sigurd F. Varian

Abstract

A dc stream of cathode rays of constant current and speed is sent through a pair of grids between which is an oscillating electric field, parallel to the stream and of such strength as to change the speeds of the cathode rays by appreciable but not too large fractions of their initial speed. After passing these grids the electrons with increased speeds begin to overtake those with decreased speeds ahead of them. This motion groups the electrons into bunches separated by relatively empty spaces. At any point beyond the grids, therefore, the cathode ray current can be resolved into the original dc plus a nonsinusoidal ac. A considerable fraction of its power can then be converted into power of high frequency oscillations by running the stream through a second pair of grids between which is an ac electric field such as to take energy away from the electrons in bunches. These two ac fields are best obtained by making the grids form parts of the surfaces of resonators of the type described in [this Journal].

The Varians called their device a "klystron" after the Greek verb "klyzo," which expresses the breaking of waves on a beach. The significance of this description will become apparent when details of the klystron are examined.

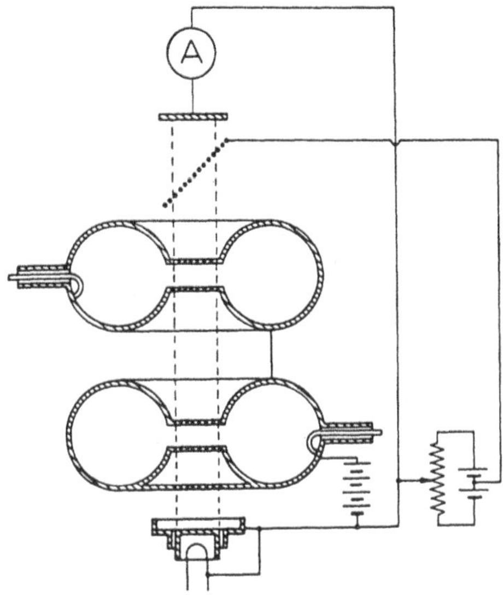

Figure 9.6 Basic Klystron Configuration Shown by the Varian Brothers

(From R. H. Varian and S. F. Varian, *J.A.P.*, May 1939.)

Figure 9.6 contains the basic two-cavity configuration shown by the Varians. The electron beam (stream of cathode rays) from the cathode at the bottom of the figure is velocity modulated by the lower resonator (sometimes called a "rhumbatron" by the Varians and their associates). Electron bunching occurs as the beam drifts toward the upper cavity. The resulting ac current in the beam induces current in the second cavity and results in RF power out of the klystron which is substantially larger than the RF input power.

9.2 Cavity Resonators

Cavity resonators somewhat like those in Fig. 9.6 were discussed briefly in Chapter 2. From a circuit point of view, each resonator can be considered to be the result of modifying a parallel resonant LC circuit so that it will operate at microwave frequencies. This is illustrated in Fig. 9.7. Thus, as frequency is increased, the multiturn-inductor, multiplate-capacitor circuit (a), is eventually replaced by a circuit in which the inductor has a single half-turn and the capacitor has a single pair of closely spaced parallel plates (b).

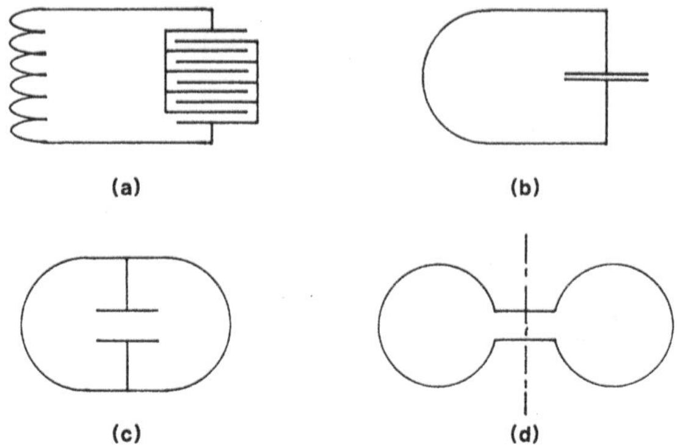

Figure 9.7 Modification of Parallel Resonant Circuit for Operation at Microwave Frequencies

To further increase frequency, the capacitance is reduced by increasing plate spacing and inductance is reduced by adding inductors in parallel(c).

Finally, with the addition of more inductors, the inductive portion of the circuit becomes a hollow toroid (d). The magnetic field is concentrated in the toroidal portion of the circuit and the electric field is concentrated in the central (capacitive) portion. At resonance, energy oscillates back and forth between the toroidal and parallel-plate portions of the cavity, just as it does in a parallel LC circuit. By replacing the parallel-plate portion of the cavity in Fig. 9.7(d) with parallel grids, it is possible to pass a low-power electron beam through the center portion as was indicated in Fig. 9.6.

At high power levels, the cavity grids must be eliminated to prevent them from being destroyed by the electron beam. The natural evolution of Fig. 9.7(d) is Fig. 9.8(a). To simplify fabrication, the cavity usually takes the shape shown in Fig. 9.8(b). This type of cavity is sometimes referred to as a doubly re-entrant cavity. The ends of the tubular electron beam tunnels (sometimes referred to as the drift tube tips) are the re-entrant portions of the cavity. The capacitance of the cavity results from the energy stored in the electric field between the drift tube tips.

The electric field pattern (dashed lines in Fig. 9.8) is not as well defined in the gridless cavity as it is in the gridded cavity. Still, the field exists in the central region of the cavity and is available for interaction with an electron beam.

In Fig. 9.6, small coupling loops attached to the ends of coaxial lines are shown for injecting a signal into a cavity or taking energy from a cavity.

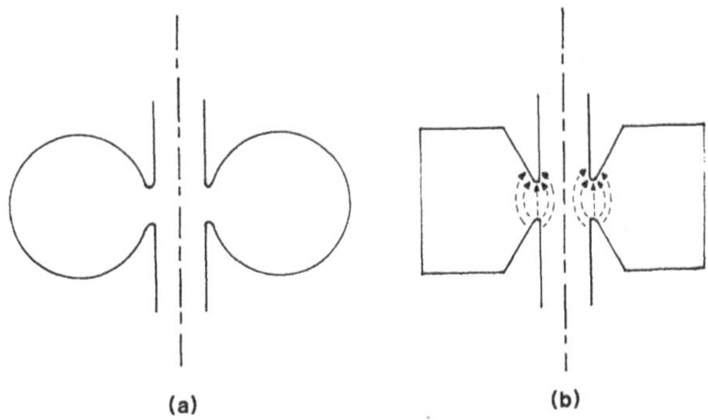

Figure 9.8 Cavities for High-Power Klystrons

The principle of operation for these coupling loops is straightforward. To inject a signal into a cavity, an RF current, I_1, is injected into the coupling loop with the coaxial line as indicated in Fig. 9.9. As a result, some magnetic field lines are established within and around the coupling loop and some circle through the toroidal portion of the cavity. These field lines induce a current flow, I_2, in the walls of the cavity. The resulting flow of charge produces an electric field, E, across the capacitive portion of the cavity.

As the input current, I_1, oscillates at the RF input frequency, the magnetic field, the current I_2 and the electric field, E, are all caused to oscillate.

When energy is extracted from a cavity by a coupling loop, the situation described in the previous paragraph is reversed. A modulated electron beam passing through the capacitive portion of the cavity induces current, I_2, in the walls of the cavity. The way in which current is induced is described in a later section of this chapter.

Associated with the current, I_2, there is an electric field across the gap, E, and a magnetic field within the toroid. A portion of the magnetic field links the coupling loop and induces a current, I_1, in the output coaxial line and in the load attached to the line. The direction of the induced current is such that it opposes the flux producing the current.

Klystron cavities can be modeled as shown in Figs. 9.10 through 9.12. There is, of course, the inductance and capacitance of the cavity, the impedance of the electron beam (usually resistive), and a resistor representing the losses in the cavity.

The coupling loop is represented as the primary (for energy input) or secondary (for energy output) winding of a transformer. Impedance matching is usually required to optimize power transfer. At an input cavity, the

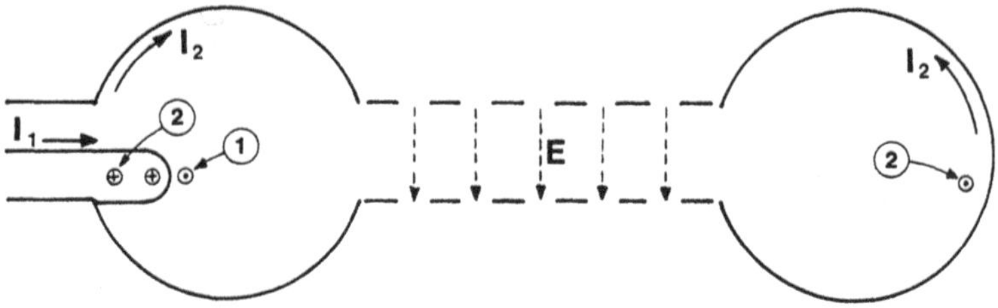

Figure 9.9 Use of Coupling Loop to Inject Signal into Cavity

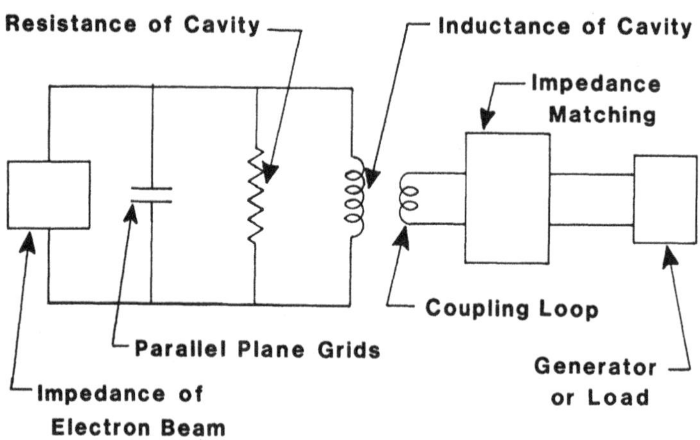

Figure 9.10 Equivalent Circuit of Beam, Cavity, and Coupling Loop Circuit

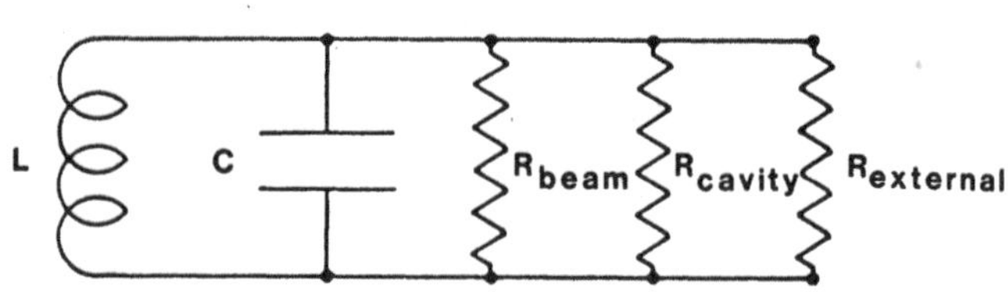

Figure 9.11 Transformed Equivalent Circuit

impedance ratio that must be accommodated by the impedance-matching networks and coupling loop may be on the order of 10,000:50 (10,000 Ω in the cavity and 50 Ω in the coaxial line). When the impedances in the cavity are transformed out through the coupling loop, the circuit of Fig. 9.11 is obtained.

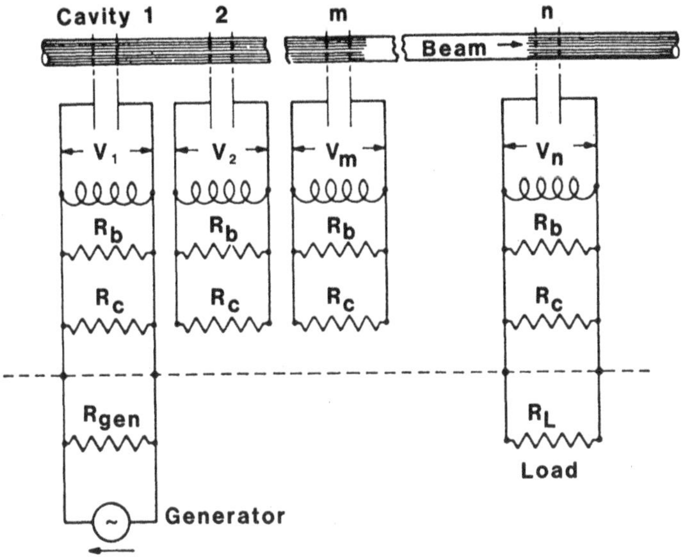

Figure 9.12 Equivalent Circuit for a Multicavity Klystron

(From Armand Staprans, *et al.*, *Proc. IEEE*, March 1973.)

Finally, when several cavities are present in a klystron, the equivalent circuit becomes as shown in Fig. 9.12. Notice that for the input and output cavities, impedances external to the tube (indicated by the dashed line) are represented. For multicavity klystrons there may be several cavities between the input and output cavities that are not connected in any way to circuits outside the tube. This is also indicated in Fig. 9.12.

The function of the intermediate cavities is to enhance the gain and bandwidth of the klystron. This will be discussed in subsequent sections.

9.3 VELOCITY MODULATION [9.3]

In a triode, the electron beam is density modulated by the grid. As was pointed out in Chapter 8, electrons have such a low average velocity in the region between the cathode and grid that at high frequencies the signal applied to the grid may change polarity many times before the electron finally passes the grid (if it ever does).

In the klystron, by comparison, the dc electron beam is accelerated to a high velocity before it is passed through the modulating electric field of the input cavity. In Fig. 9.13, the input cavity is represented by a pair of grids spaced by distance *d*. A sinusoidal voltage is applied to the grids.

When the high-velocity electron beam is passed through the modulating field, some electrons have their velocities increased (when the right-hand grid of Fig. 9.13 is positive with respect to the left-hand grid), some electrons have their velocities decreased when the voltage is reversed, and a few have their velocities unchanged. As a result, the electrons are said to be velocity modulated.

Because the velocity change of the electrons is generally small, the electrons remain very close to their dc beam positions, and so essentially no density modulation occurs within the modulating gap.

After the beam leaves the modulating gap, those electrons that have had their velocity increased gradually overtake the slower electrons. This process is illustrated in Fig. 9.14, which is known as an Applegate diagram. Axial electron positions are plotted as a function of time so that the slopes of the various lines represent electron velocities.

The trajectories of electrons leaving the modulating gap when the voltage is negative (indicated at the bottom of the figure) are represented by lines of reduced slope. Similarly, the trajectories of accelerated electrons are represented by lines of increased slope.

In Fig. 9.14, space-charge forces between electrons are ignored, and so the electron trajectories are shown to cross. In practice, at small and medium drive signal levels, space-charge forces prevent electron trajectories from crossing. That situation is discussed later in this chapter. At large signal levels, electron trajectories do cross, and so the following discussion is relevant to that situation.

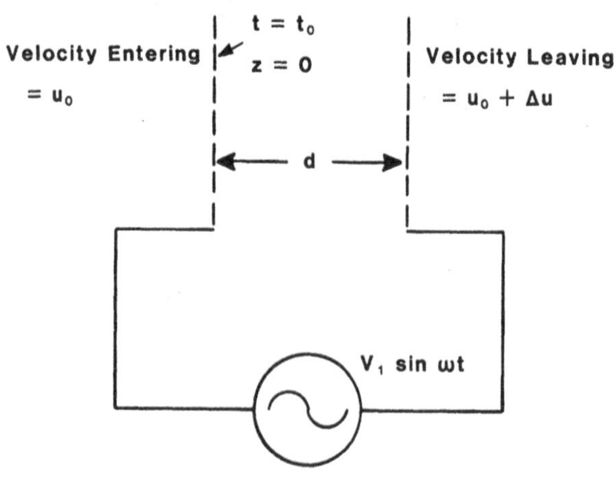

Figure 9.13 Velocity Modulation of Electron Beam

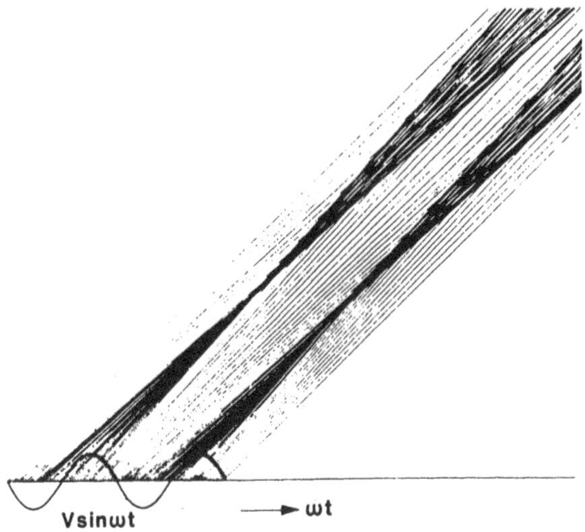

Figure 9.14 Distance-Time Curves for the Electrons in a Velocity-Modulated Beam in a Field-Free Drift Space (Applegate Diagram)

(From *Electronics of Microwave Tubes* by W.J. Kleen, Copyright 1958 by Academic Press. Reproduced with permission.)

As a result of the convergence of electron trajectories illustrated by the Applegate diagram, electron bunching is said to occur. The traveling electron bunches represent an RF current in the electron beam. In the absence of space-charge forces, the calculation of the RF current in the beam is relatively straightforward. The result of the calculation is illustrated in Figure 9.15.

Near the modulation gap, the RF beam current is sinusoidal. As the electron bunches form, sharp peaks develop in the current waveform and the harmonic content becomes very high. The location of the second cavity for maximum efficiency in removing power from the beam at the fundamental frequency should be at the axial position where the fundamental component of the current is largest.

The amplitude of the fundamental current (and of the harmonics) can be determined mathematically by Fourier analysis. The result is shown in Figure 9.16, and is

$$I_1 = 2i_0 J_1 (X) \tag{9.1}$$

where i_0 is the dc beam current, $J_1(X)$ is a first-order Bessel function, and X is called the bunching parameter, which is given by

$$X = \frac{\omega z}{u_0}\frac{\alpha\beta}{2} \tag{9.2}$$

In this expression, α is called the depth of modulation and is the ratio of the peak voltage, V_1, across the modulating gap to the dc beam voltage, V_0. The quantity β, is called the gap coupling coefficient.

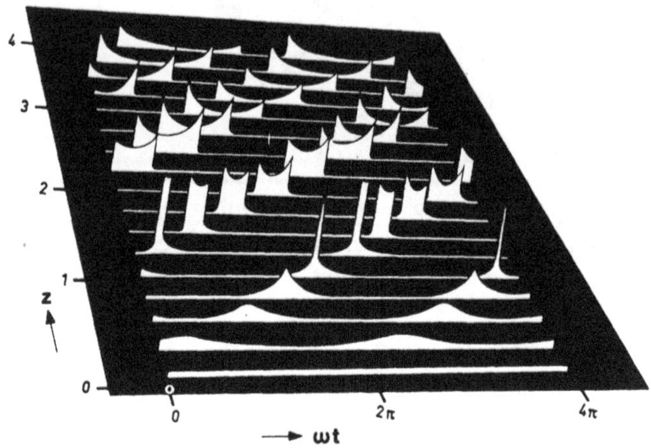

Figure 9.15 RF Currents in a Velocity-Modulated Electron Beam in the Absence of Space-Charge Forces

(From *Electronics of Microwave Tubes* by W.J. Kleen, Copyright 1958 by Academic Press. Reproduced with permission.)

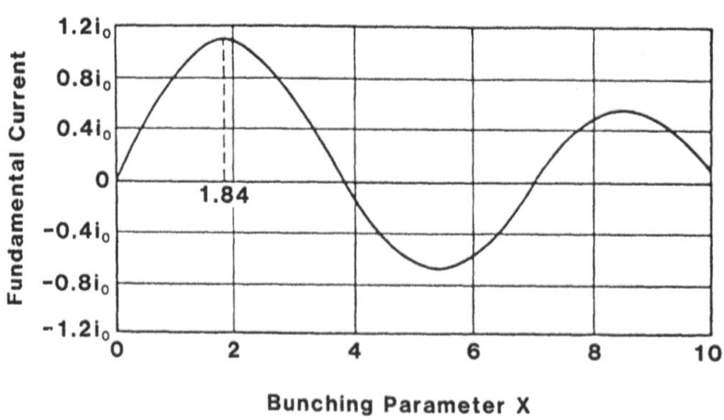

Figure 9.16 Fundamental Component of Current in a Velocity-Modulated Electron Beam in the Absence of Space-Charge Forces

The gap coupling coefficient accounts for the fact that the value of voltage across the gap changes as an electron traverses the gap. For example, for a very wide gap, the voltage could swing through a complete positive half-cycle and a complete negative half-cycle as an electron traverses the gap. The phase change in voltage, which is also the gap transit angle, ϕ_0, for the electrons, would be 2π radians. The net velocity change for the electrons would be zero.

If the gap width were reduced from the 2π radian case, there would be a net velocity change for the electrons, and so β would be greater than zero. Ultimately, as the width of the gap becomes very small, the electrons experience a constant voltage as they pass through the gap and the gap coupling coefficient becomes unity. The expression for the gap coupling coefficient and a plot as a function of transit angle are given in Fig. 9.17

Referring back to Fig. 9.16, the value of the bunching parameter at which the current is maximum is 1.84. This means, from (9.2), that the optimum location, z_{opt}, of the second cavity should be at

$$z_{opt} = 1.84 \frac{u_0}{\omega} \frac{2}{\alpha\beta} \tag{9.3}$$

Notice that this optimum location is a function of operating frequency, modulation voltage (from α) in input gap width (from β).

Fortunately, for the klystron engineer, the effects of frequency, drive voltage, and gap width disappear when space-charge forces are taken into consideration. As a result, in contrast to the situation indicated in (9.3), the location of the second cavity is fixed.

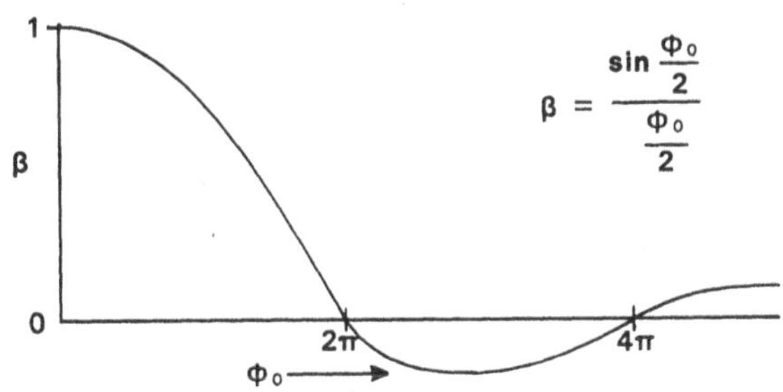

Figure 9.17 Gap Coupling Coefficient as a Function of Gap Transit Angle

9.4 RF CURRENT GENERATION IN A CAVITY

When a bunched electron beam is passed through a cavity, RF current is induced in the cavity as is shown qualitatively in Fig. 9.18. As a bunch of electrons approaches grid 1 (shown in (a)) electrons within grid 1 and the nearby cavity walls flow to the grid-2 side of the cavity. Thus, grid 2 becomes negative with respect to grid 1 and the bunch of electrons is decelerated and gives up energy as it passes between the grids, Fig. 9.18(b). Finally, as the bunch of electrons leaves grid 2, Fig. 9.18(c), electrons within grid 2 and the nearby cavity walls flow back towards grid 1.

When an analysis of the current induced by the beam is carried out, it is found that the RF current induced in the cavity is identical to the RF current in the beam. The voltage developed across the gap in the cavity depends on the cavity impedance. At resonance, the cavity impedance is high, and so the voltage induced by the fundamental component of beam current (assumed to be at the resonant frequency) is high.

Large harmonic currents may be induced in the cavity, but, at the harmonic frequencies, the cavity impedance is low (unless another mode of resonance is excited) so that harmonic cavity voltage is low.

The maximum RF voltage that can be established across a cavity is on the order of the beam voltage. An RF voltage higher than the beam voltage could cause electrons to be reflected, which would reduce the induced current and voltage.

At a peak RF voltage equal to the beam voltage and the peak fundamental current indicated in Fig. 9.16, the conversion efficiency from dc beam power to RF output power can readily be shown to be 58 percent. In practice, conversion efficiencies in excess of 70 percent have been achieved. This is partly because space-charge forces reduce the harmonic content of the RF beam current and partly because a cavity tuned to the second harmonic frequency was used to help optimize electron-bunch formation.

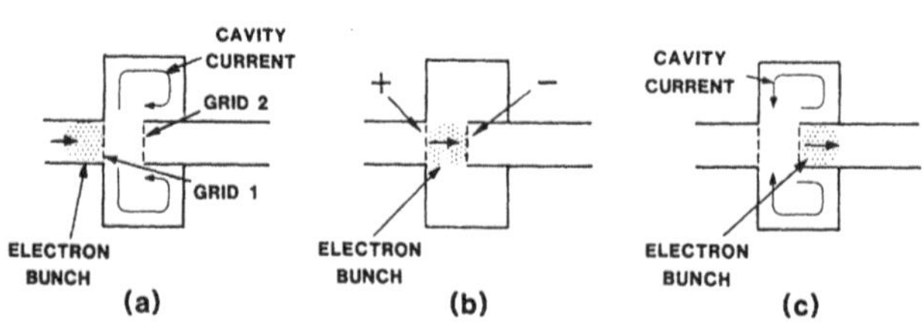

Figure 9.18 Generation of RF Current in a Cavity

(From *Microwave Tube Manual* by Varian Associates, Air Force Publication Number T.0. 00-25-251, October 1979.)

9.5 VELOCITY MODULATION WITH SPACE-CHARGE FORCES [9.3]

When space charge forces were ignored in the bunching process following velocity modulation, only the velocities of the electrons were determined. From their velocities, electron positions could be determined, and the current could be determined from the electron positions.

When space-charge forces are taken into consideration in the bunching process, the analysis becomes significantly more complex. The five quantities shown in Fig. 9.19 must be taken into consideration in order to calculate the desired quantity, which is the current density, J, in the beam. Subscripts zero indicate dc values of the quantities before modulation. Subscripts one indicate the changes in the quantities that result from the modulation process. While the analysis will not be included in the following discussion, some of the quantities listed in Fig. 9.19 will be referred to subsequently and that is the reason for showing them at this point.

Within the cavity gap, the process of velocity modulation is no different than it was when space charge forces were ignored. This is because the electrons barely move from their dc positions, and so no RF space charge forces are generated.

As electrons converge in the bunching process after the beam has been velocity modulated, space-charge forces build up, impeding the rate at which fast electrons overtake slow electrons. In addition, slow electrons have their velocities increased because of the bunch that forms behind them (toward the modulating cavity). The resulting electron trajectories are illustrated in Fig. 9.20, which is an Applegate diagram that has been modified to show the effects of space-charge forces.

At the axial position denoted by $\lambda_p/4$, the fast electrons have been slowed to the dc velocity of the beam and the slow electrons have been accelerated to the dc velocity of the beam. Thus, at that point, all electrons have the same velocity. Also, at $\lambda_p/4$, the electron density and the RF current reach maximum values. For small to medium RF signal levels, the RF current is essentially sinusoidal, so this is the position at which the second cavity of the klystron should be placed.

As the beam moves beyond the $\lambda_p/4$ position, electrons ahead of the center of a bunch are accelerated by the space-charge force and electrons behind the center are decelerated. As a result, the bunch disperses so that at $\lambda_p/2$, there are no variations in electron density. The velocity distribution in the beam at $\lambda_p/2$, is exactly the same as it was at z=0. At $\lambda_p/2$, the slow electrons (at $z=0$) have become fast electrons and the fast electrons have become slow electrons.

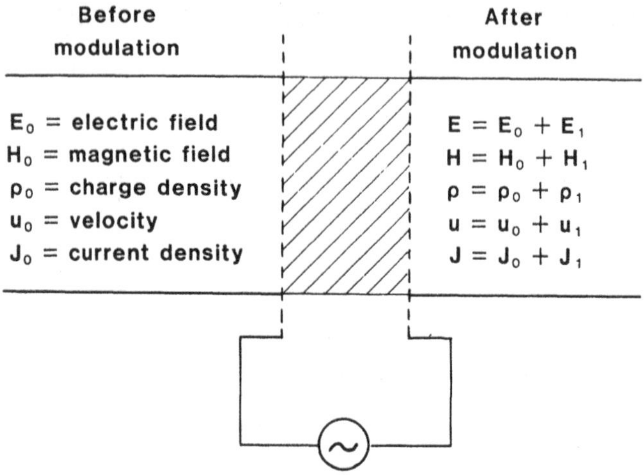

Figure 9.19 Quantities Used in Analyzing Velocity Modulation with Space-Charge Forces

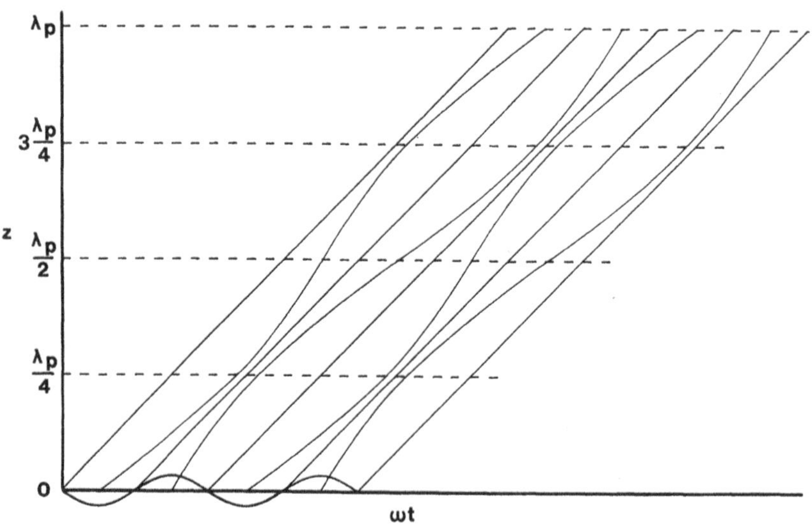

Figure 9.20 Modified Applegate Diagram Indicating Space-Charge Effects

As the beam moves beyond $\lambda_p/2$, the entire process that took place in the $0 \leq z \leq \lambda_p/2$ region is repeated. Finally, at $z = \lambda_p$, the velocities of all electrons have returned to their intitial values (at z=0).

If the beam continued beyond the position λ_p, electrons would continue to oscillate between bunches. The frequency at which they would oscillate is the electron plasma frequency, which is given by

$$\omega_p = \sqrt{\frac{\eta \rho_0}{\epsilon_0}} \tag{9.4}$$

Notice that the plasma frequency is a function of the density of the electron beam only.

The axial position, λ_p, given in Fig. 9.20 is the wavelength of the plasma oscillation. The wavelength is given by

$$\lambda_p = \frac{u_0}{f_p} = \frac{2\pi\, u_0}{\omega_p} \tag{9.5}$$

and so the wavelength is a function of beam velocity and density only. This result is extremely important to the klystron engineer because, unlike the situation when space-charge forces were ignored, the cavity location is not a function of signal level, gap width, or frequency of operation (refer to Eq. (9.3) for cavity position without space-charge forces).

It is interesting to note that space-charge forces are the key to the successful use of thermionic cathodes in microwave tubes. (See discussion of space-charge-limited emission, section 5.1.3, in the chapter on cathodes.) Space-charge forces are also the key to the successful operation of a multicavity klystron, since these forces govern cavity locations.

To summarize, the electron velocity, density, and current vary sinusoidally with distance at the plasma frequency. These variations are illustrated in Fig. 9.21.

9.5.1 Plasma Frequency Reduction

The space-charge forces within electron bunches vary with the size and shape of an electron beam. This is illustrated in Fig. 9.22, where a beam of finite dimensions is compared with an infinitely wide beam. In the infinitely wide beam, the electric fields (and, thus, the space-charge forces) are constrained to act only in the axial direction. In a finite beam, the electric fields are radial as well as axial with the result that the axial component is reduced in comparison to the infinite beam. With reduced axial space-charge forces, the plasma frequency is reduced and the plasma wavelength is increased [9.4, 9.5].

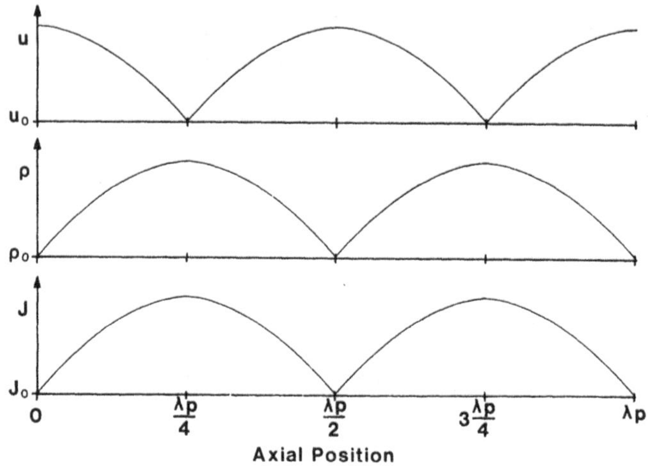

Figure 9.21 Variations of Peak Values of Velocity, Density, and Current with Axial Position

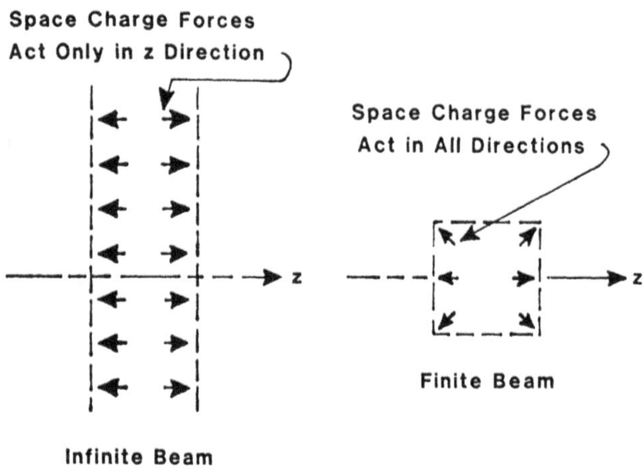

Figure 9.22 Comparison of Space-Charge Forces in an Electron Bunch in an Infinite Beam and in a Finite Beam

It is common to denote the reduced plasma frequency for a finite beam by ω_q and the corresponding wavelength by λ_q. The amount by which the plasma frequency is reduced is a function of the beam radius, b, and the ratio n, of the beam-tunnel radius to the beam radius, as shown in Fig. 9.23.

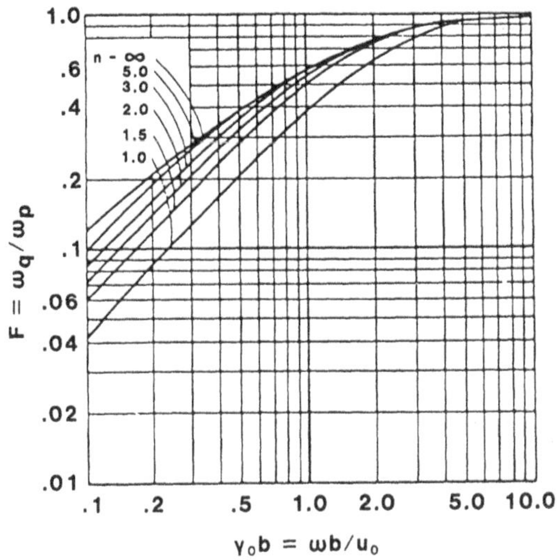

Figure 9.23 Plasma-Frequency Reduction Factor, *F*, for a Confined Beam

(From *Space Charge Waves and Slow Electromagnetic Waves* by A.H.W. Beck, published 1985 by Pergammon Press. Copyright by A.H.W. Beck, Reproduced with permission.)

As an example of the effect of the reduced space-charge forces, if $\gamma_0 b =$ 1.0 and $n = 1.5$, then F = 0.5. Therefore, $\lambda_q = 2\lambda_p$, which means that in a klystron the cavities would be placed twice as far apart as would be indicated by infinite-beam calculations.

9.5.2 Velocity Modulation of a Brillouin Beam

In the preceding discussions (Section 9.5) space-charge forces were taken into consideration, but no radial motions of the electrons were included. For confined flow conditions, when a magnetic focusing field that is large compared with the Brillouin value is used, the confined flow discussion is correct. When flow is not confined, radial motions of the electrons occur, resulting in an RF modulation of the radius of the beam [9.6].

In the Brillouin flow case, as a bunch starts to form, the beam radius increases just enough so that the charge density in the beam remains constant. This situation is illustrated in Fig. 9.24. Each bulge in the beam radius corresponds to a bunch and the resulting RF current is on the surface of the beam. Space-charge forces are generated, just as they are in a confined beam, and the maximum bunching occurs at a position one-quarter of a plasma wavelength from the modulating gap.

Because the modulation appears on the outer radius of the beam, this is where the electric fields occur which produce the space-charge forces.

The RF electric field pattern for a Brillouin beam is compared to that for the confined beam in Fig. 9.25. Notice that the field lines can bow inward as well as outward from the surface bunch on the Brillouin beam. As a result, the axial component of the field and the space charge forces are smaller than for the confined beam. This results in a lower plasma frequency and larger plasma wavelength for the Brillouin beam than for the confined beam.

The plasma frequency reduction factor for a Brillouin beam is given in Fig. 9.26. For a value of $\gamma_0 b$ of 1.0 and n = 1.5, the value of F is 0.37. For identical conditions for the confined beam, F was 0.5. The practical result of this is that the cavities in a klystron with a Brillouin beam must be somewhat farther apart than those in a klystron with a confined beam.

Experiments have been performed to verify that the RF current is on the surface of a Brillouin beam [9.7]. The RF beam analyzer is shown in Fig. 9.27. This was the analyzer used for the dc beam measurements described in the chapter on linear beams (Ch. 7). The analyzer was modified for RF current measurements.

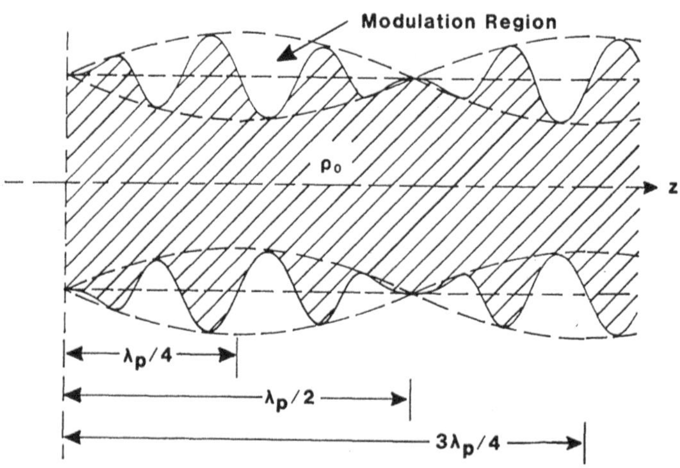

Figure 9.24 Velocity Modulated Brillouin Beam

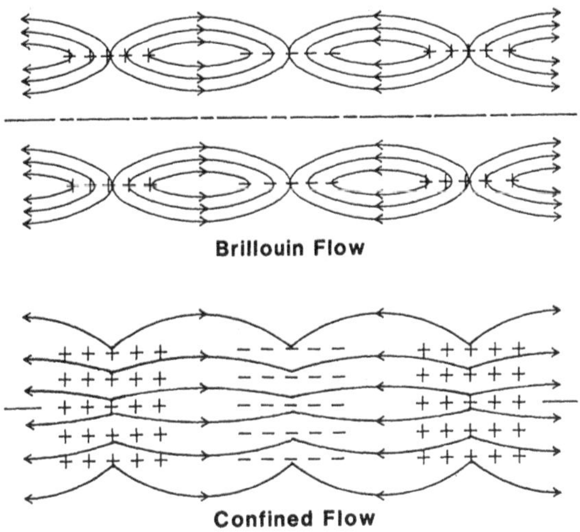

Figure 9.25 Comparison of RF Electric Field Patterns Near a Bunch in a Brillouin Beam and a Confined Beam

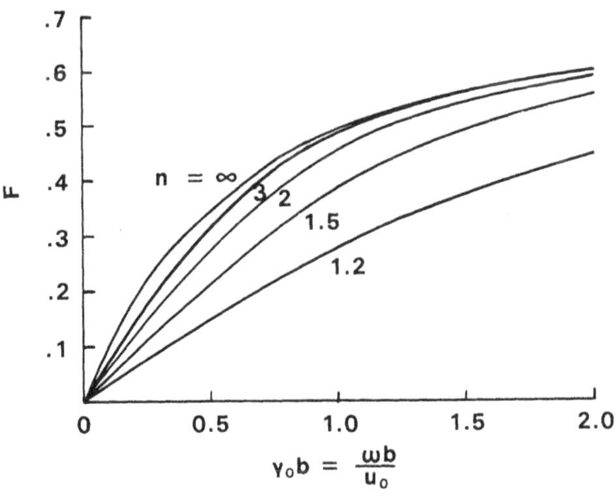

Figure 9.26 Plasma Frequency Reduction Factor for Brillouin Beam

(From *Space Charge Waves and Slow Electromagnetic Waves* by A.H.W. Beck, published 1985 by Pergammon Press. Copyright by A.H.W. Beck, Reproduced with permission.)

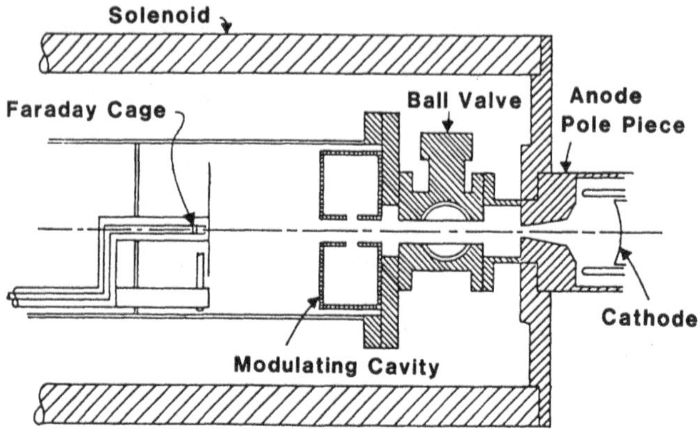

Figure 9.27 RF Beam Analyzer for Measuring Surface Currents in Brillouin Beam

(From A.S Gilmour, Jr. *IEEE Trans. on Electron Devices,* July 1972.)

The beam from a shielded Pierce gun was focused with a solenoid. A vacuum ball valve was used to permit changes to be made in the measurement apparatus without exposing the gun and cathode to air (which could have changed the beam characteristics).

The beam was velocity-modulated at approximately 2000 MHz. A beam-scanning mechanism with a .010-in aperture was used for sampling currents at various transverse and axial positions.

The circuit used for detecting average and RF currents is given in Fig. 9.28. Either the fundamental, second harmonic, or third harmonic current could be selected for measurement with a bandpass filter.

Plots of the average and fundamental components of current as functions of the transverse and axial position in the beam are shown in Fig. 9.29. The depth of modulation, α, was about .05. Because of the small RF drive signal, the average current distribution was very similar to that observed in a dc beam.

The RF current was entirely on the surface of the beam. The predicted space-charge wave behavior of the electrons is clearly illustrated by four half-plasma wavelengths of variation of the beam current.

The fundamental current distributions at higher RF drive levels are shown in Fig. 9.30 [9.8]. The RF drive powers of 10 W, 50 W, and 200 W corresponded to approximate values of α of 0.1, 0.25, and 0.5, respectively. Notice that at the medium drive level the space-charge wave behavior became severely distorted. At the highest drive level, the space-charge wave behavior disappeared completely.

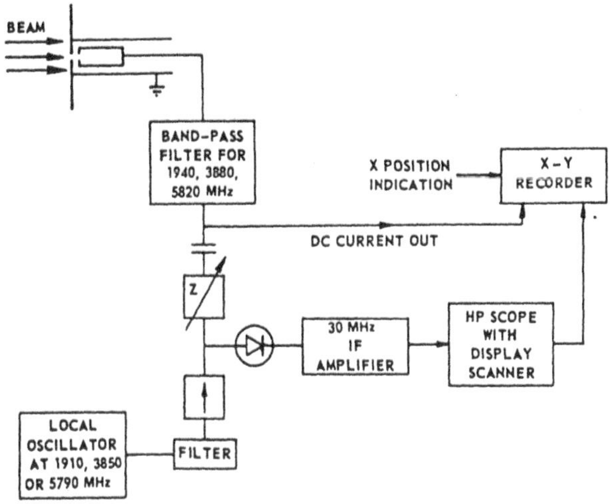

Figure 9.28 Circuit Used for Detecting Average and RF Currents in Velocity Modulated Brillouin Beam

(From A.S Gilmour, Jr. *IEEE Trans. on Electron Devices,* July 1972.)

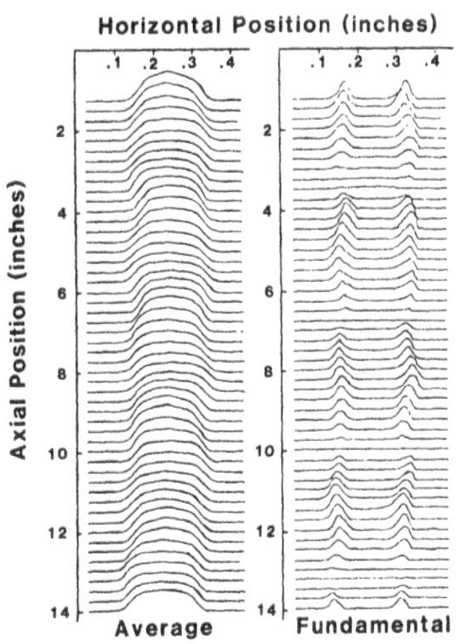

Figure 9.29 Average and Fundamental Current Distributions for $\alpha \cong .05$

(From A.S Gilmour, Jr. *Proc. High Power Microwave Tube Symposium,* The Hexagon, 1962.)

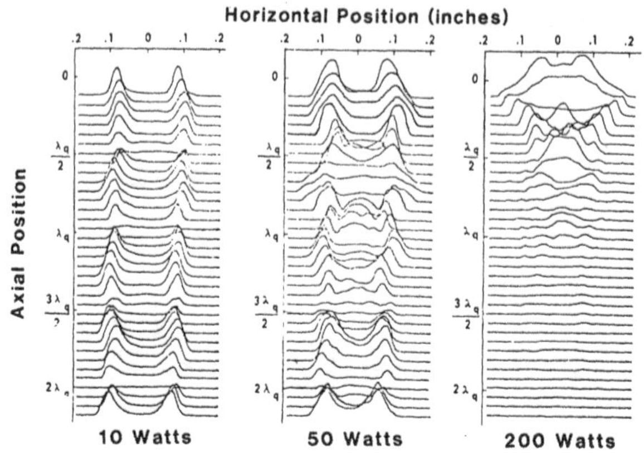

Figure 9.30 Distribution of Fundamental RF Current for Various Modulation Powers (for 10 W, $\alpha \cong 0.1$; for 50 W, $\alpha \cong 0.25$; for 200 W, $\alpha \cong 0.5$)

(From A. S. Gilmour, Jr., *IEEE Trans. on Electron Devices*, July 1972.)

Electron energy measurements were made to help explain the beam behavior shown in Fig. 9.30. First, the peak values of the electron energies were determined at relatively low values of α using the circuit shown in Fig. 9.31.

The Faraday cage used for collecting the electrons passing through the .010-in aperture in the beam-scanning mechanism was biased to a negative voltage (4800 V) comparable to the dc beam voltage. In addition, a 10,000 MΩ resistor was placed in series with the bias batteries. With this large value of resistance, only a fraction of a microampere of current to the Faraday cage caused an additional bias of several hundred to several thousands of volts. A fraction of a microampere was less than 0.1% of the normal peak Faraday cage current and consisted, therefore, of only those electrons having the maximum energy.

Shown in Fig. 9.32 are the peak energies (expressed in terms of the bias voltage on the Faraday cage) up to an α of approximately 0.1. At small values of α, the variation of energy was a series of half-sinusoids, as expected. When α exceeded 0.05, the energy profiles became distored. At values of α much above 0.1, the profiles became so severely distored as to be essentially meaningless. This is better illustrated by Fig. 9.33, which shows the energy distributions of the electrons at various axial position and for three drive power levels. As a reminder, the RF drive powers of 10 W, 50 W, and 200 W corresponded to approximate values of α of 0.1, 0.25, and 0.5 respectively.

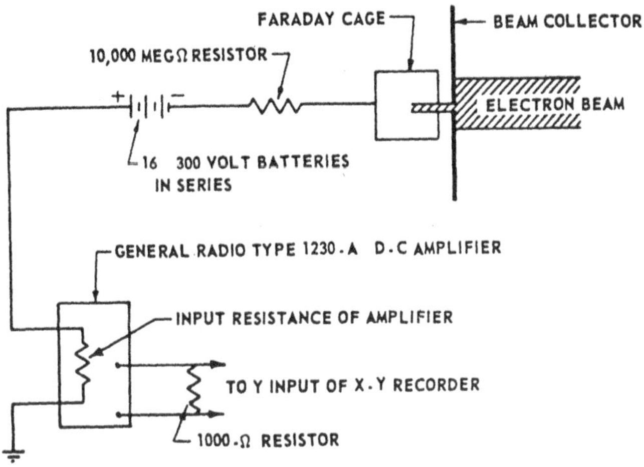

Figure 9.31 Circuit Used to Provide Peak Beam Energy Indication to X-Y Recorder

(From A.S Gilmour, Jr. *IEEE Trans. on Electron Devices,* July 1972.)

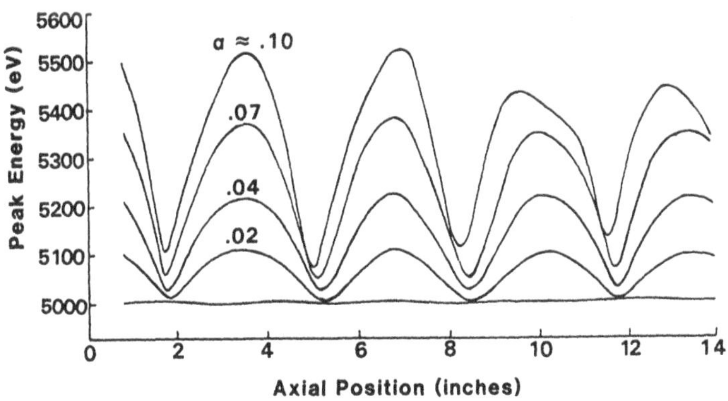

Figure 9.32 Peak Beam Energy Expressed in Terms of Faraday Cage Bias in Volts as a Function of Axial Position and RF Drive Level for Brillouin Flow Conditions

(From A.S Gilmour, Jr. *Proc. High Power Microwave Tube Symposium,* The Hexagon, 1962.)

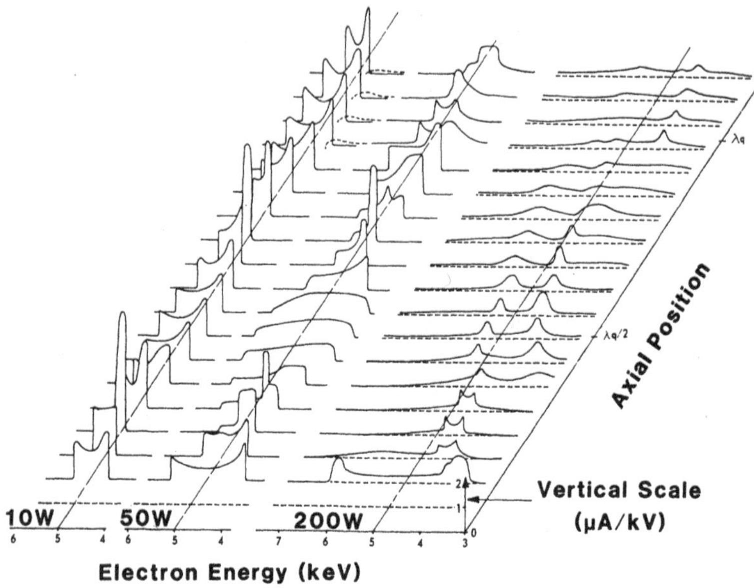

Figure 9.33 Energy Distribution of Current as a Function of Axial Position for Drive Powers of (a) 10 W ($\alpha \sim .1$) (b) 50 W ($\alpha \sim .25$), (c) 200 W ($\alpha \sim .5$)

(From A. S. Gilmour, Jr., *IEEE Trans. on Electron Devices*, July 1972.)

Data for Fig. 9.33 were obtained by sweeping the Faraday cage voltage and measuring cage current as a function of voltage. The resulting curve was then differentiated to determine the energy distribution of the current. In Fig. 9.33, energy, expressed in terms on the Faraday cage potential in kilovolts, is the horizontal axis. The vertical scale is in microamperes of current to the cage per kilovolt of potential applied to the cage. The third coordinate is axial position in the beam.

The 10 W ($\alpha \cong 0.1$) curves show that the electrons left the modulating cavity (at axial positions equal to zero) with an estimated energy distribution from 4.5 to 5.5 keV. At $\lambda_q/4$, where the first bunch occurred, the horizontal spread of the curve narrowed to ~ 5 keV ± 100 eV, so that all of the electrons essentially had the dc velocity of the beam as is indicated by the peak in current. As the beam drifted beyond $\lambda_q/4$, the energy distribution broadened to a maximum at $\lambda_q/2$ and then returned to a minimum value at 3 $\lambda_q/4$, where another bunch was formed. The distributions shown in Fig. 9.33(a) are very nearly as would be expected for space-charge wave behavior of a beam.

In Fig. 9.33(b) (50 W and $\alpha \cong 0.25$), the electrons left the modulating cavity with an energy distribution of over 2 keV. At $\lambda_q/4$, the energy distribution does not become narrow as it did at 10 W. Instead, it appears that high energy electrons move through the bunch which was being formed. This is even more evident in Fig. 9.33(c), where there are no major peaks in current, indicating that the electrons have a broad energy distribution at all axial positions. Thus, electron velocities were apparently large enough so that they were not significantly affected by space-charge forces.

9.6 KLYSTRON AMPLIFIERS

In the previous sections, the mechanisms of velocity modulation of electrons and subsequent RF current growth in an electron beam have been discussed. A simple two-cavity klystron amplifier has the second cavity placed at the axial location of the first current peak. A gain in the neighborhood of 20 dB is obtained.

In principle, the gain of the klystron amplifier can be increased indefinitely by adding additional cavities along the beam. In practice, with a six-cavity amplifier, a stable gain in excess of 110 dB has been observed. Normally, the gain is limited to 50 to 60 dB. At higher levels, the shot noise of the beam is increasingly amplified, and so the signal-to-noise ratio of the output is reduced. In addition, to obtain stability, there must be no feedback from output to input, either because of leakage of transmission line connections, RF modulation of collector leads, and feedback to the electron gun, or reflected electrons within the beam.

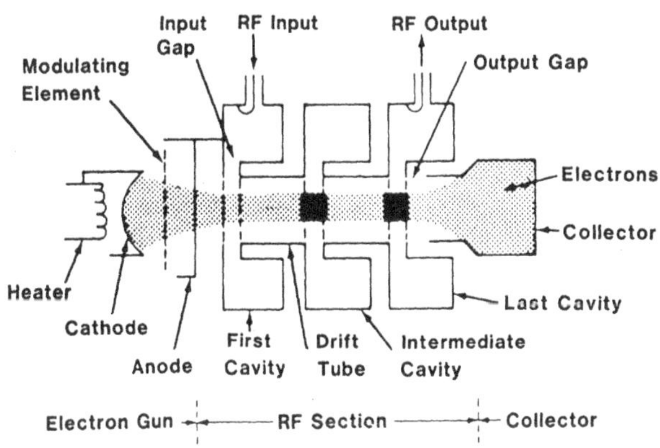

Figure 9.34 Elements of a Three-Cavity Klystron Amplifier

(From *Microwave Tube Manual* by Varian Associates, Air Force Publication Number T.0.00-25-251, October 1979.)

The elements of the three-cavity klystron amplifier are shown in Fig. 9.34. Other than the collector, which will be discussed in a subsequent chapter, the only element that has not yet been discussed is the intermediate cavity.

The function of the intermediate cavity is to increase the gain, and usually the bandwidth, of the amplifier. The location of the intermediate cavity is one quarter of the plasma wavelength from the input cavity. Current is induced in the intermediate cavity as was discussed in Section 9.4. This current and the impedance of the cavity produce a voltage across the gap that enhances the modulation of the beam to produce further current growth as indicated by Fig. 9.35. The amplitude of the change in RF current in the beam at $\lambda_q/4$, depends on factors such as gap length and impedance (which, in turn, depends on cavity frequency in relation to operating frequency).

The location of the third cavity is shown to be less than $\lambda_q/4$ from the second cavity in Fig. 9.35. The reason is the result of the nonlinear beam behavior described in the previous section. The velocities of the electrons are so high that space-charge forces have little effect during the bunching process. As a result, that optimum bunch (containing maximum fundamental current) occurs before the beam has drifted a quarter of the plasma wavelength.

In klystrons with more than three cavities, the next-to-final cavity is called the penultimate cavity. The distance between the penultimate and the final cavity is said to be foreshortened. That is, it is less than a quarter of a plasma wavelength for the same reason as was given in the previous paragraph for the three-cavity klystron.

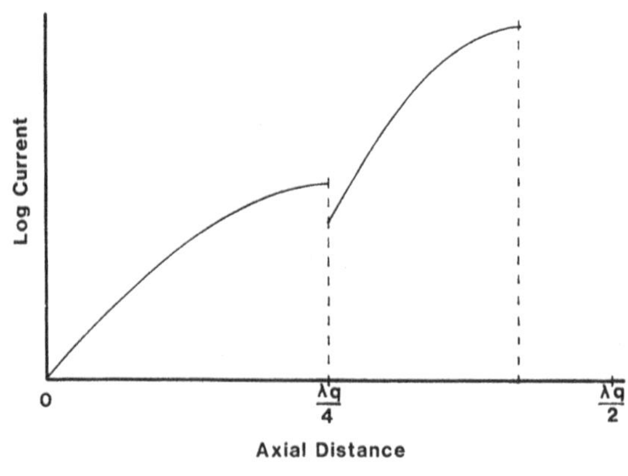

Figure 9.35 RF Current Growth in Beam in a Three-Cavity Klystron

9.6.1 Tuning

Klystrons may be synchronously tuned, efficiency tuned, or broadband tuned. With synchronous tuning, all cavities are tuned to the same frequency and gain is maximized.

To high-efficiency tune a klystron, it is first synchronously tuned and then the penultimate cavity is tuned upward in frequency to maximize power output. By tuning upward in frequency, the penultimate cavity is inductive at the operating frequency. As a result, the voltage across the cavity gap lags the current induced in the cavity by the electron beam. Thus, the voltage is phased with respect to the electron bunches so that electrons between bunches are swept into bunches. That is, as a bunch approaches the cavity, the gap voltage slows down electrons ahead of the bunch. As the bunch leaves the cavity, the gap voltage accelerates electrons immediately behind the bunch. As a result, bunching is enhanced so that the RF current induced in the output cavity is increased.

Typical tuning characteristics of a penultimate cavity are shown in Fig. 9.36. With a klystron operating at, or near, full output power, it is important not to tune the penultimate cavity down in frequency into the shaded region indicated in Fig. 9.36. This is because there is no load on the penultimate cavity, and so the voltage induced could be high enough to cause breakdown across the cavity gap.

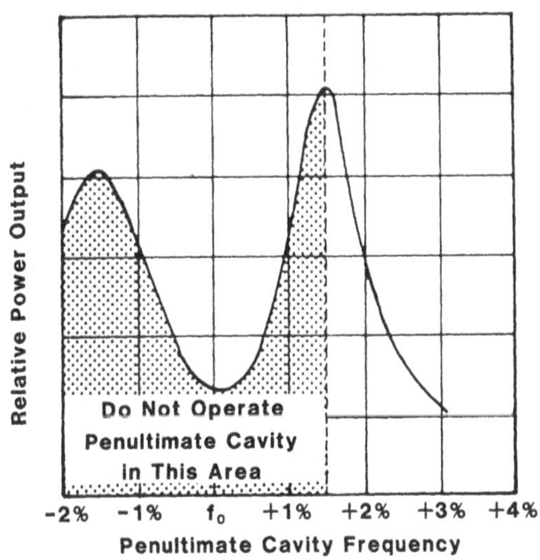

Figure 9.36 Penultimate Cavity Tuning Characteristics

(From *Microwave Tube Manual* by Varian Associates, Air Force Publication Number T.0.00-25-251, October 1979.)

To broadband tune a klystron, it is first synchronously tuned at a low power level. Then, as indicated in Fig. 9.37, the intermediate cavities are detuned downward and upward to achieve the desired bandwidth and gain flatness. The bandwidth of the output cavity determines the ultimate bandwidth of the tube. For broad bandwidth, a high perveance beam is normally used to achieve low beam impedance, which, in turn, results in a low value of beam-loading impedance in the output cavity. Also, broader bandwidth may be achieved using an extended interaction (multiple-coupled-cavity) output circuit.

Figure 9.38 contains typical curves comparing output power as a function of drive frequency for an efficiency-tuned and a broadband-tuned klystron. Finally, Fig. 9.39 shows the actual output characteristics for the VA-884C and VA-884D klystrons when they are broadband tuned.

The center frequency, f_0, of operation of a klystron can be varied while maintaining the desired broadband characteristics by simultaneously tuning all cavities. This is done mechanically with a channel tuner mechanism like that illustrated in Fig. 9.40. With this mechanism, which is sometimes manually operated and sometimes motor driven, a gear-coupled drive is used to simultaneously change the position of the tuning plunger in all cavities. An example of the resulting shift in operating frequency is given in Fig. 9.41.

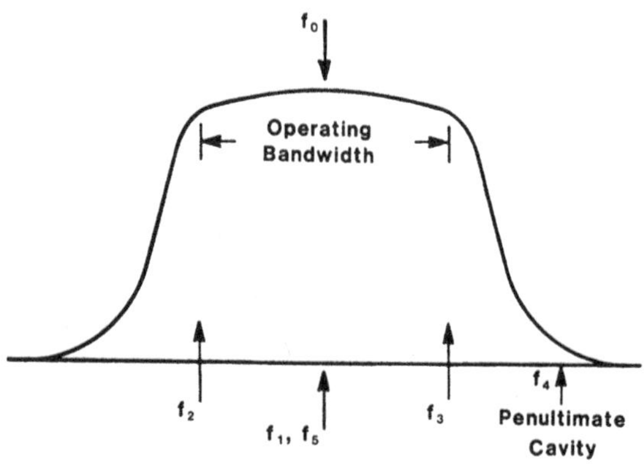

Figure 9.37 Typical Cavity Resonant Frequencies for a Broadband Tuned Five-Cavity Klystron

(From Armand Staprans, *et al.*, Proc. IEEE, March 1973.)

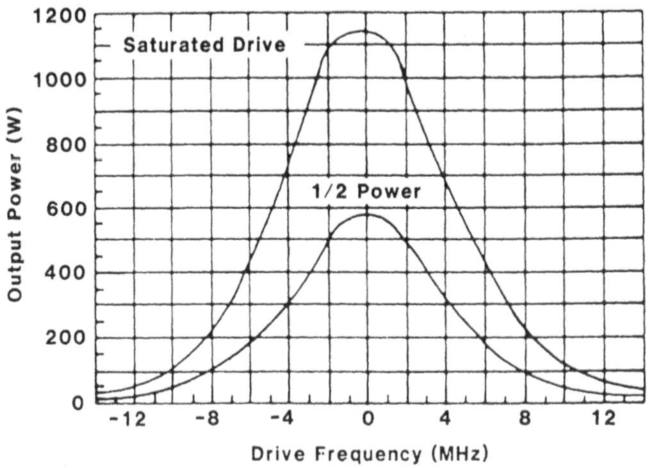

(a)

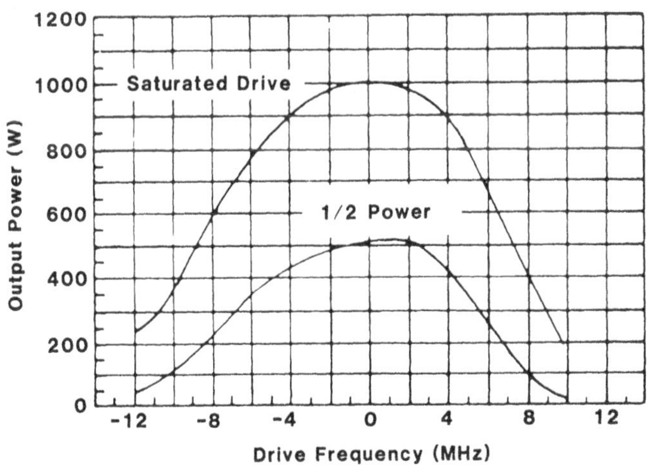

(b)

Figure 9.38 Output Power *versus* Drive Frequency for (a) Efficiency-Tuned and (b) Broadband-Tuned Klystron

(From *Microwave Tube Manual* by Varian Associates, Air Force Publication Number T.0.00-25-251, October 1979.)

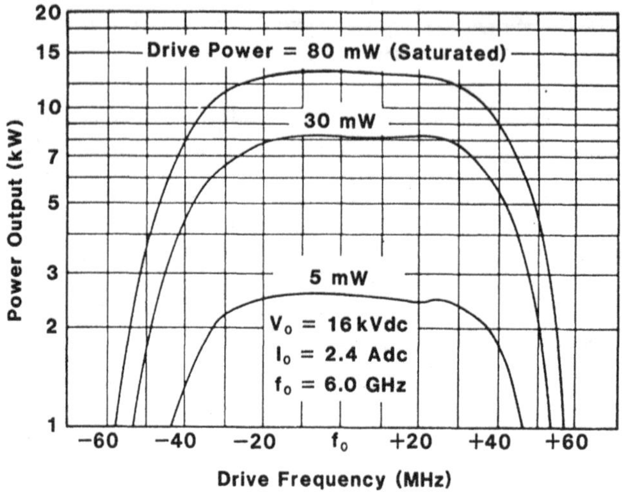

Figure 9.39 Output Characteristics for the VA-884C and VA-884D Klystrons (Saturated Output at P_d (drive power) = 80 mW)

(From Armand Staprans, *et al.*, *Proc. IEEE*, March 1973.)

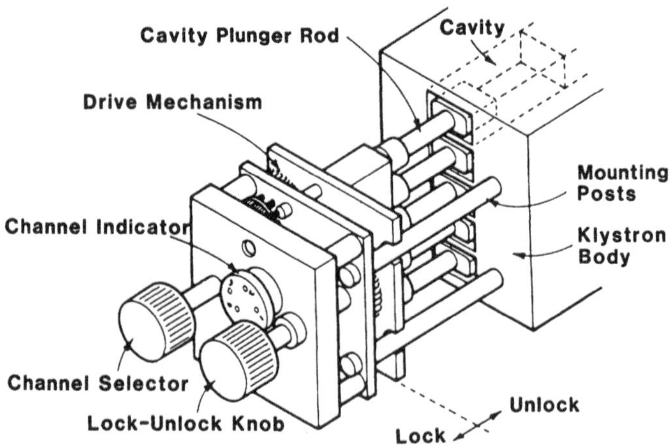

Figure 9.40 Klystron Channel Tuner Mechanism

(From Armand Staprans, *et al.*, *Proc. IEEE*, March 1973.)

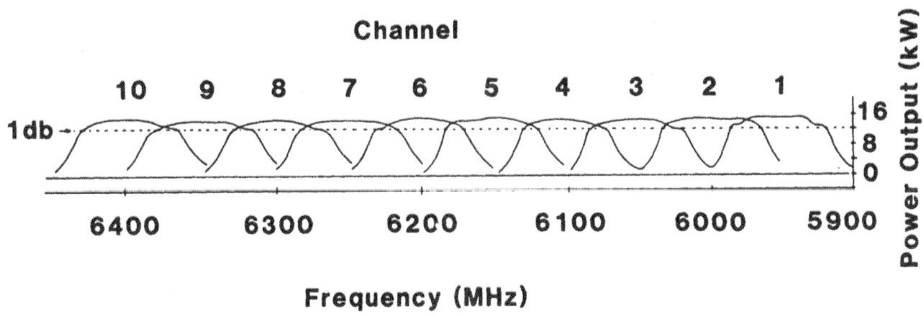

Figure 9.41 Channel Tuning Characteristics of a VA-884D Klystron

(From Armand Staprans, *et al.*, *Proc. IEEE,* March 1973.)

9.6.2 Gain Characteristics

Typical gain characteristics for a klystron are shown in Fig. 9.42. As is typical for all linear-beam tubes (but not for crossed-field amplifiers), output power varies linearly with drive power at low power levels. At high drive levels, there is a point where optimum bunching occurs in the output cavity and output power is maximized. Beyond that drive level, output power decreases.

When a klystron is broadband tuned, there is significant decrease in gain compared with synchronous or efficiency tuning. This is clear from Fig. 9.42, which shows an increase in drive power of nearly two orders of magnitude required for broadband-tuned operation as compared with efficiency-tuned operation.

The output power of a klystron (as well as for other microwave tubes) varies dramatically with beam voltage as shown in Fig. 9.43. This is as expected for at least two main reasons. In the first place, beam power, P_0, various as the 5/2 power of beam voltage, V_0, because

$$P_0 = I_0 V_0$$

but $\quad I_0 = pV_0^{3/2}$

so $\quad P_0 = pV_0^{5/2}$

where, of course, I_0 is beam current and p is perveance. Thus, an increase in beam voltage of a factor of 1.5 (as shown in Fig. 9.43) is accompanied by an increase in beam power of a factor of 2.8. For constant tube efficiency, RF power output would go up by the same amount. Efficiency is not constant, however, but decreases with decreasing voltage because of changes in beam bunches and diameter at voltages below the design voltage.

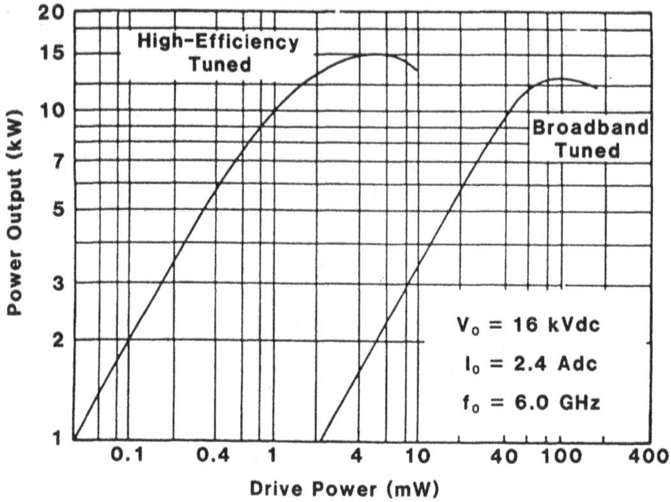

Figure 9.42 Gain Characteristics for the VA-884C and VA-884D Klystrons

(From Armand Staprans, *et al.*, *Proc. IEEE*, March 1973.)

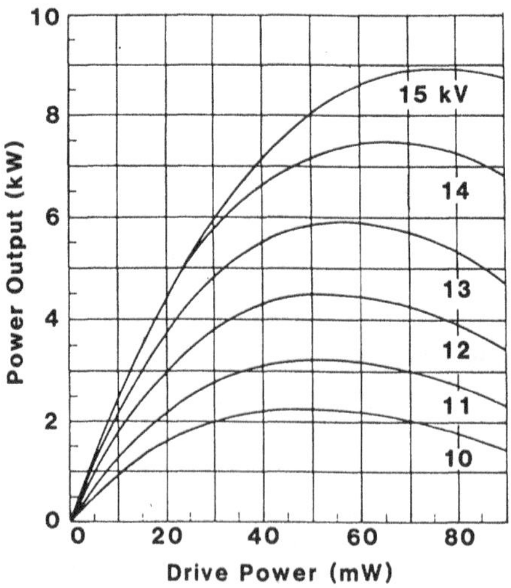

Figure 9.43 Output Power as a Function of Drive Power and Beam Voltage

(From *Microwave Tube Manual* by Varian Associates, Air Force Publication Number T.0.00-25-251, October 1979.)

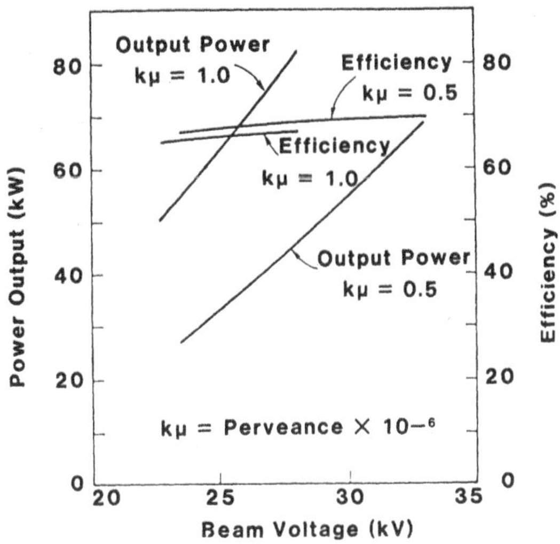

Figure 9.44 Performance of X-3074B High-Efficiency Klystron

(From Armand Staprans, *et al.*, *Proc. IEEE*, March 1973.)

Klystrons are capable of very high efficiencies as indicated by Fig. 9.44. Through the use of multistage depressed collectors (to be discussed in Ch. 12), it is predicted that efficiencies in the 80 to 85% range should be possible.

Average power levels as high as about one megawatt have been achieved in experimental klystrons. Peak powers of several tens of megawatts are available in klystrons for a number of applications.

The typical bandwidth of a multiple-cavity klystron is on the order of one percent. A bandwidth as high as 10% is possible in the megawatt power range by using the coupled-cavity output circuit technique described for extended interaction klystrons in Chapter 11.

9.7 REFLEX KLYSTRONS

A reflex klystron is a single-cavity klystron oscillator. As shown in Fig. 9.45, an electron beam passes through a cavity gap (normally, grid type) into a region where the negative voltage on a repeller (reflector) electrode causes the electrons to reverse direction.

Assuming that a signal exists across the cavity grids, then the electron beam is velocity modulated as was described in previous sections. As the electron beam drifts into the repeller region, bunching occurs as indicated in Fig. 9.46. If the bunched electrons return to the modulating gap at the correct time, they induce a current that reinforces the current in the cavity. Thus, the electron beam provides the feedback loop that causes the device to oscillate.

In Fig. 9.46, the phase at which the electron bunch returns to cause oscillations is 2-3/4 cycles after the beam initially passed through the bunching cavity. By changing the repeller voltage, the phasing of the return of the electrons to the modulating cavity can be changed. As a result, the electron bunch can be caused to return to the modulating gap 2-3/4, 3-3/4, 4-3/4 cycles, *et cetera*, after initially passing through it. As indicated in Fig. 9.47, oscillations can occur in any one of several modes that correspond to the phase angle through which the signal changes as the electrons drift through the repeller region.

As is also indicated in Fig. 9.47, the oscillation frequency in any mode of operation is a function of the time during which the electrons return in that mode. For example, if the repeller voltage is above the mode-center value, then the electron transit time is less than the mode-center value, and the frequency is above the mode-center value. Thus, the RF current induced in the cavity is at a frequency above the resonant frequency of the cavity (mode-center frequency), and so the frequency of operation is above the resonant frequency. Power output is reduced because operation off-resonance reduces the cavity impedance and gap voltage.

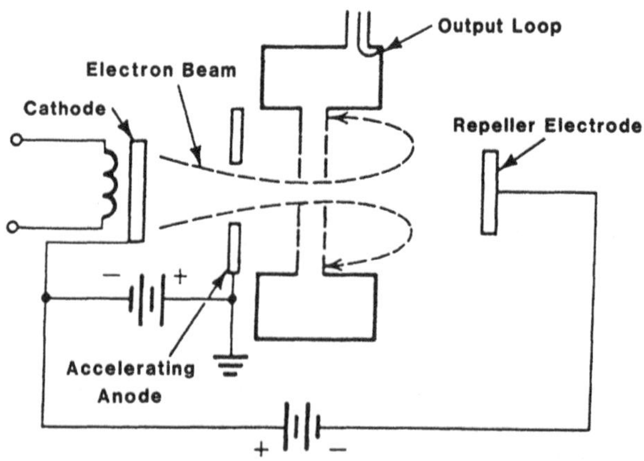

Figure 9.45 Reflex Klystron Principles

(From *Microwave Devices* by J. T. Coleman, Copyright 1982 by Reston Publishing Co. Reproduced with permission.)

Reflex klystrons are low power (usually less than a watt) devices that have historically been used as local oscillators. They are low noise devices that are not difficult to phase lock. As is shown in Fig. 9.48, reflex klystrons are capable of operating well into the millimeter wave region and they remain as the signal source of choice for many millimeter applications.

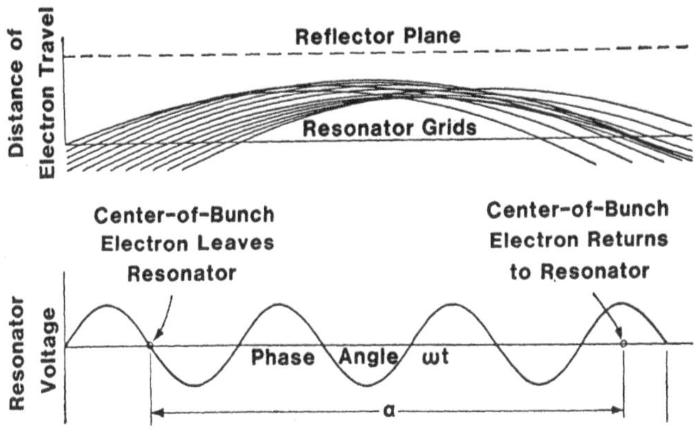

Figure 9.46 Electron Bunching in a Reflex Klystron

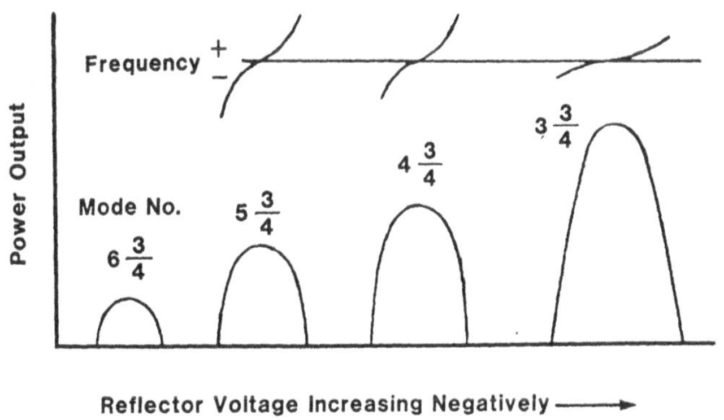

Figure 9.47 Power Output *versus* Reflector Voltage for a Reflex Klystron

(From *Microwave Tube Manual* by Varian Associates, Air Force Publication Number T.0.00-25-251, October 1979.)

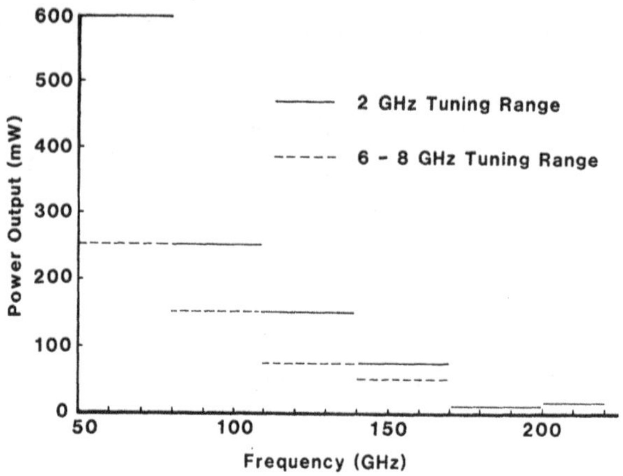

Figure 9.48 Power Output Capability of Reflex Klystrons in the Millimeter Wave Region

(From Varian of Canada Data Sheet 2561-65M, 12/70.)

REFERENCES

[9.1] A. Staprans, E.W. McCune, and J.A. Ruetz, "High-power linear-beam tubes," *Proc. IEEE*, vol. 61, March 1973, pp. 299-330

[9.2] R.H. Varian and S.F. Varian, "A high frequency oscillator and amplifier," *Journal of Applied Physics*, vol. 10, May 1939, pp. 321-327

[9.3] Recommended for further reading:

W.J. Kleen, *Electronics of Microwave Tubes*. New York: Academic Press, 1958.

M. Chodorow and C. Susskind, *Fundamentals of Microwave Electronics*. New York: McGraw-Hill, 1964.

A.S. Gilmour, Jr., "Basic principles and some aspects of the electromagnetic compatibility of microwave tubes," *Inter-System Electromagnetic Compatibility*. New York: Holt, Reinhart and Winston, 1972.

[9.4] G.M. Branch and T.G. Mihran, "Plasma frequency reduction factors in electrons beams," *IRE Trans. Electron Devices*, vol. ED-2, Apr. 1955, pp. 3-11.

[9.5] A.H.W. Beck, *Space Charge Waves and Slow Electromagnetic Waves*. New York: Pergammon Press, 1958.

[9.6] W.W. Rigrod and J.A. Lewis, "Wave propagation along a magnetically focused cylindrical electron beam," *Bell System Technical Journal*, 33, March 1954.

[9.7] A.S. Gilmour, Jr., G.C. Dalman, and L.F. Eastman, "The current and velocity distributions in a velocity-modulated Brillouin-focused electron beam," *Proc. 4th International Congress on Microwave Tubes*, Eindhoven: Centrex, 1963, pp. 675-688.

[9.8] A.S. Gilmour, Jr., "Radial and axial current and velocity distribution in large-signal velocity-modulated electron beams," *IEEE Trans. on Electron Devices*, vol. ED-19, no. 7, July 1972, pp. 886-890.

Chapter 10
Traveling Wave Tubes

As was pointed out at the beginning of the chapter of klystrons (Ch.9), the traveling wave tube (TWT) is one of two major categories of devices known as linear-beam or 0-type tubes. The other is, of course, the klystron.

While many different RF circuits have been devised for use in TWTs, two predominate at present:

1. Helix, for broadband applications;
2. Coupled cavity, for high-power applications.

While other circuits are mentioned in this chapter, the majority of the discussions center on helix and coupled-cavity TWTs.

10.1 THE INVENTION OF THE TRAVELING WAVE TUBE

The traveling wave tube was invented in England in 1942 by Rudolf Kompfner [10.1], who was trained as an architect in Vienna. Figure 10.1 is from Kompfner's notebook and shows his first sketch of a traveling wave

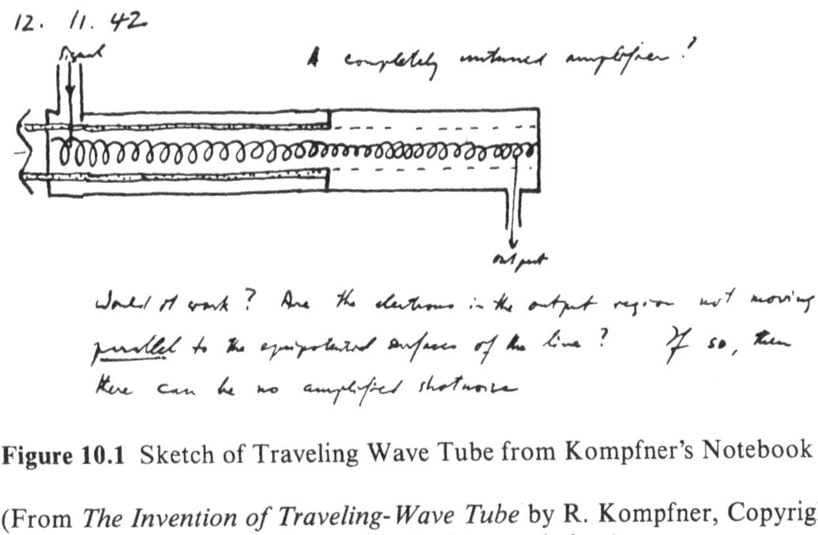

Figure 10.1 Sketch of Traveling Wave Tube from Kompfner's Notebook

(From *The Invention of Traveling-Wave Tube* by R. Kompfner, Copyright 1963 by Francisco Press, Reproduced with permission.)

tube. Notice the notations in Kompfner's sketch. "A completely untuned amplifier! Would it work?" Subsequent figures from Kompfner's notebook show his conception of the RF field lines (Fig. 10.2), which was not correct, and a coworker's version (Fig. 10.3 by Dr. R.R. Nimmo), which was correct. In the following section, this field pattern and its interaction with an electron beam are discussed.

10.2 TRAVELING WAVE INTERACTION [10.2]

The main elements of a helix TWT are shown in Fig. 10.4. Conceptually, the electron gun, the electron beam, and the collector are the same as those for a klystron. However, the RF circuit (helix delay line structure) is completely different from the klystron circuit.

In the helix circuit, the RF wave travels at approximately the same speed as the electron beam and the interaction with the beam is continuous rather than discrete as it is in the case of the klystron.

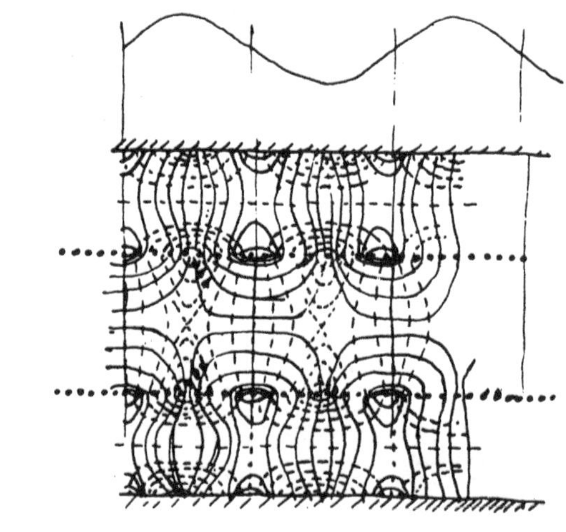

Figure 10.2 Kompfner's Conception of RF Field Lines in Traveling Wave Tube

(From *The Invention of Traveling-Wave Tube* by R. Kompfner, Copyright 1963 by Francisco Press, Reproduced with permission.)

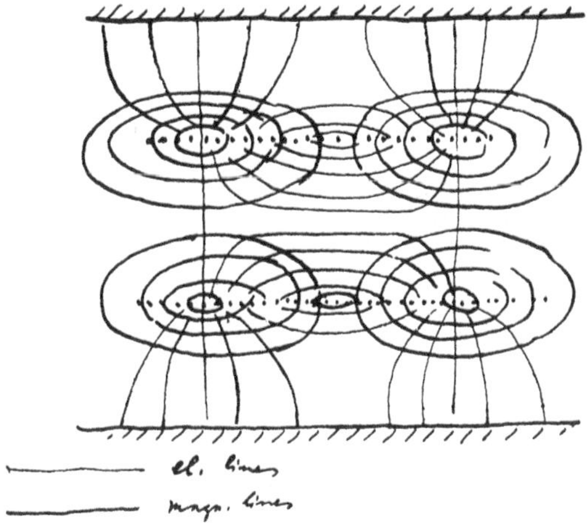

Figure 10.3 Correct Field Lines Sketched in Kompfner's Notebook by R.R. Nimmo

(From *The Invention of Traveling-Wave Tube* by R. Kompfner, Copyright 1963 by Francisco Press, Reproduced with permission.)

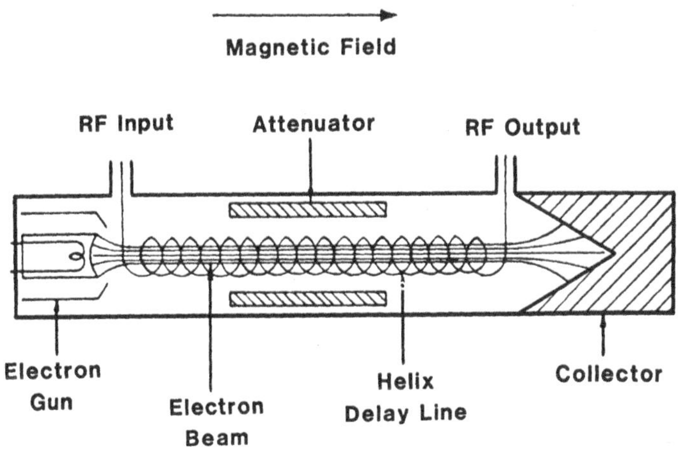

Figure 10.4 Basic Helix TWT

(From John T. Mendel, *Proc. IEEE,* March 1973.)

It is helpful in explaining the RF field pattern for a helix to first consider a single-wire transmission line above a ground plane as shown in Fig. 10.5. Charges on the center conductor and the RF electric field pattern are shown at an instant in time. The RF magnetic field pattern is not shown and will not be considered because the RF magnetic forces on the electron beam in a microwave tube are insignificant compared with the forces produced by the electric field. If the RF generator is on the left end of the coaxial line in Fig. 10.5, and if a matched load is on the right end, then the charge and field patterns move to the right with constant amplitude. The velocity of propagation is that of light and is independent of frequency, so the transmission line is completely nondispersive.

Next, assume that the single-wire transmission line is formed into a helix as shown in Fig. 10.6. An RF signal applied to the left end of the helix will travel at a velocity near that of light in a helical path along the conductor. The average velocity in the axial direction will be the velocity in the helical direction reduced by the pitch of the helix. The polarity of the signal will alternate every half-wavelength along the helical conductor. In Fig. 10.6, two complete turns are required for each half-wavelength.

Electric field lines extend from regions of positive charge to regions of negative charge with a configuration similar to that shown for the transmission line in Fig. 10.5. Other than the reduced velocity, the primary difference between the field pattern for the helix and that for a single-wire transmission line is that there is an electric field inside the helix with large axial components.

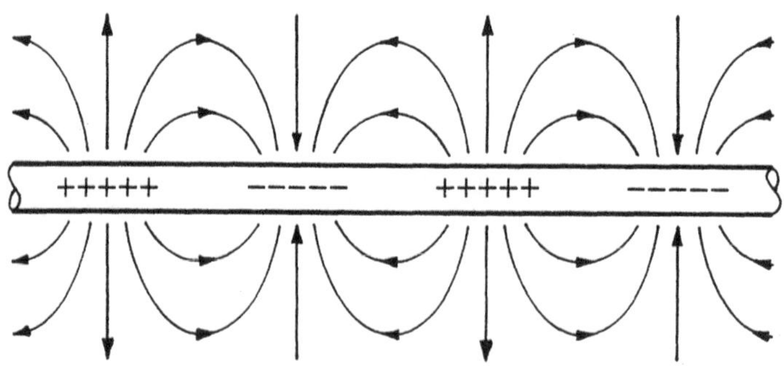

Figure 10.5 Sketch of RF Charge and Electric Field Pattern for a Single-Wire Transmission Line above a Ground Plane

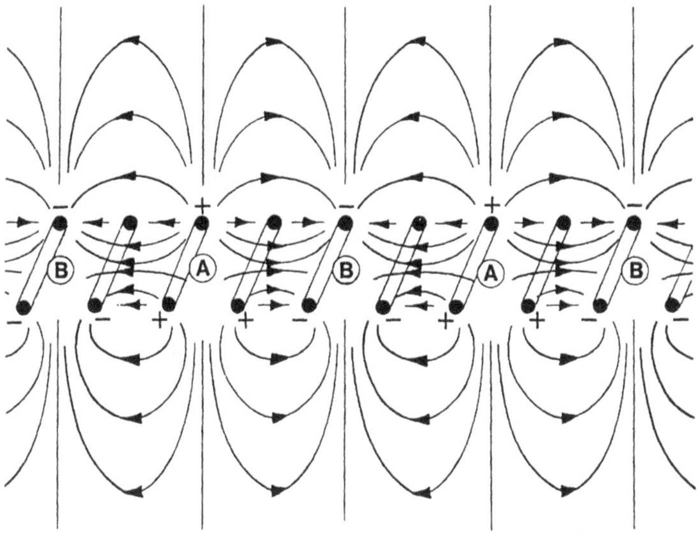

Figure 10.6 RF Charge and Electric Field Pattern for a Helix

When an electron beam is injected along the axis of the helix, the axial electric field components accelerate some electrons and decelerate others. In Fig. 10.6, the forces on the electrons would be toward the region denoted by (A) and away from the regions denoted by (B). The field pattern would vary sinusoidally in the axial direction as indicated in Fig. 10.7

If the axial velocity of the electric field and the beam are the same, then electrons will experience a continual force toward region (A) as the beam travels through the helix. As a result, an electron bunch will start to form in region (A).

The fields produced by the bunching electrons in the beam will cause electrons on the helix to move away from the region adjacent to (A) and toward the (B) regions. As a result, the field produced by the helix is enhanced, and this, in turn, increases the bunching of the beam.

To produce growth of the field in the circuit, energy is extracted from the electrons. Thus, the velocity of the circuit field must be slightly less than the electron velocity so that the electron bunches move into the decelerating field region. The mutual interactions of the beam and circuit result in an exponential growth of the signal level as is indicated by Fig. 10.8

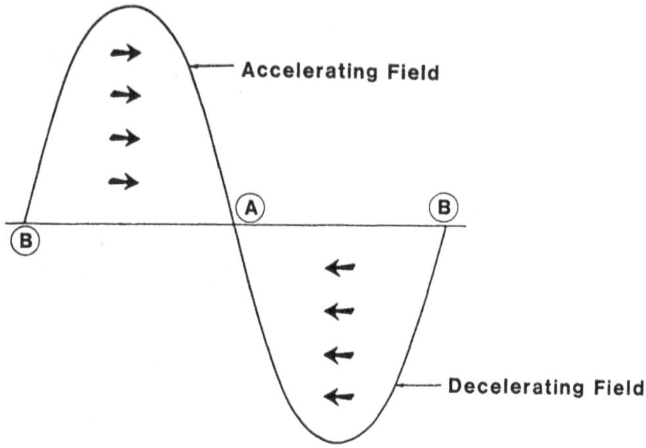

Figure 10.7 Axial Field that Provides Velocity Modulation and Bunching of Electron Beam

As the bunching process continues, space-charge forces increase and it becomes increasingly difficult for the circuit field to enhance the bunching further. In addition, as energy is extracted from the beam to cause the circuit wave to grow, the beam slows down. This causes the bunches to move out of phase with the growing signal. Finally, a point is reached where the bunches start to disperse and the amplitude of the circuit wave starts to decrease. For maximum efficiency, the signal should be removed from the tube at this point.

10.3 ANALYSIS OF TRAVELING WAVE INTERACTION

There are many approaches to the analysis of the interactionof an electron beam with a traveling wave on a circuit. The classical small-signal theory given by J.R. Pierce is a logical extension of the qualitative description given in the preceding section.

While it is not appropriate to give the details of the mathematics of the Pierce theory, it is appropriate and helpful in understanding TWT operation to examine the approach used by Pierce and some of the results of this theory. Many of the parameters defined by Pierce are now part of the accepted vocabulary of the TWT industry.

In the Pierce theory, the RF circuit of the TWT is modeled as a transmission line as shown in Fig. 10.9. This model serves as a starting point, but then the analysis is generalized so that the exact form of the model is not important.

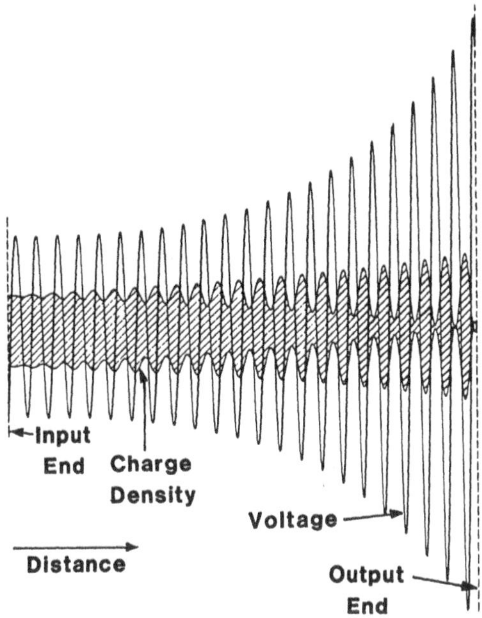

Figure 10.8 Voltage and Charge Density Build-up in a TWT

(From *Microwave Theory and Techniques* by Reich, Ordung, Krauss and Skolnik, Copyright 1953 by D. Van Nostrand, Inc. Reproduced with permission.)

The electron beam with current i is assumed to pass close to the circuit and induce current density, J, into the circuit. The voltage on the circuit, in turn (for example, between points A and B), produces velocity modulation of the beam. The equation for the induced current is combined with the equation for the signal on the circuit to determine the propagation characteristics (amplitude and velocity) of the waves on the circuit.

The resulting equation is a quartic, which, in general, has four roots corresponding to four different propagation characteristics for the beam-circuit combination. These propagation characteristis are functions of many parameters including the impedances of the circuit and the beam, the beam velocity, the circuit loss, and space-charge effects. It is instructive to examine first the special synchronous case when the electron beam velocity equals the cold circuit phase velocity (circuit phase velocity in the absence of the electron beam), ignoring loss and space-charge effects. Then, modifications can be made to the synchronous solution to account for beam velocity variations, losses, and space charge.

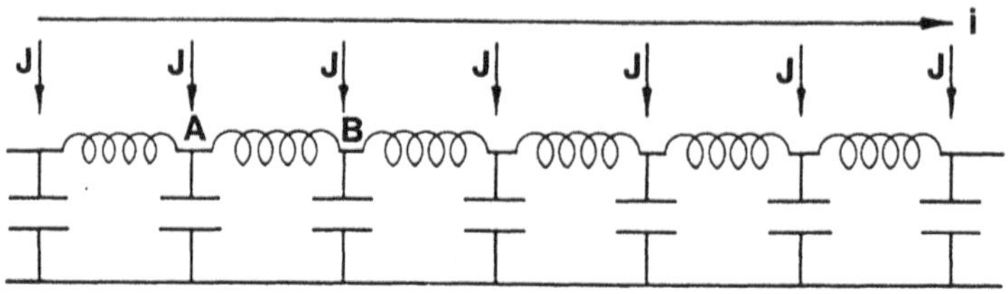

Figure 10.9 Model of Transmission Line Used by Pierce in Small-Signal Theory

(From *Traveling Wave Tubes* by J.R. Pierce, Copyright 1950 by D. Van Nostrand Company, Inc. Reproduced with permission.)

10.3.1 Synchronous Case

The synchronous case is when the electron beam velocity is the same as the velocity of the signal on the circuit in the absence of the beam. By making some astute approximations based on the fact that the velocity of the signal on the circuit should remain close to the beam velocity if gain is to be realized, it is possible to determine the propagation characteristics of the four waves represented by the four roots to the quartic equation mentioned in the previous section.

The resulting propagation constants for the four waves are listed below:

$$-\Gamma(1) = -j\beta_e - j\frac{\beta_e C}{2} + \beta_e C\frac{\sqrt{3}}{2} \tag{10.1}$$

$$-\Gamma(2) = -j\beta_e - j\frac{\beta_e C}{2} - \beta_e C\frac{\sqrt{3}}{2} \tag{10.2}$$

$$-\Gamma(3) = -j\beta_e + j\beta_e C \tag{10.3}$$

$$-\Gamma(4) = +j\beta_e - j\beta_e\frac{C^3}{4} \tag{10.4}$$

In these equations, β_e is ω / u_0 and is the propagation constant for the electron beam. The quantity C is Pierce's gain parameter defined by

$$C^3 = \frac{K}{4R_0} \tag{10.5}$$

where K is the circuit impedance and R_0 is the beam impedance. The circuit impedance relates to the electric field, which is available for interaction with the electron beam. The circuit impedance is not the same impedance as is generally employed in ordinary microwave circuit investigations. A strong electric field results from a high circuit impedance and produces an intense interaction with the beam.

The value of C is small (usually in the 0.01 to 0.1 range). While the value of beam impedance is simply V_0/I_0 (ratio of dc beam voltage to beam current), the value of circuit impedance is usually difficult to calculate and differs substantially from circuit to circuit.

Each propagation constant in (10.1) through (10.3) contains the propagation constant of the electron beam, $-j\beta_e$, along with one or two small quantities (which are small because they are multiplied by C). As a result, each wave is a forward wave (traveling in the same direction as the electron beam) propagating at a velocity close to the electron beam velocity.

In (10.1) and (10.2), the quantity $-j(\beta_e C/2)$ indicates that the waves are traveling slower than the electron beam. In (10.3), the quantity $+j\beta_e C$ indicates that the wave is traveling faster than the electron beam.

The fourth wave contains $+j\beta_e$, which indicates that it travels in the direction opposite to that of the electron beam. The quantity C^3 in (10.4) is so small that the speed of the fourth wave is only slightly greater than the beam speed.

In (10.1) and (10.2), the real part of the propagation constant indicates that the wave is either a growing wave (+) or decaying wave (-).

Finally, in summary, the propagation constants $-\Gamma(1)$ through $-\Gamma(4)$ represent the four waves illustrated in Fig. 10.10.

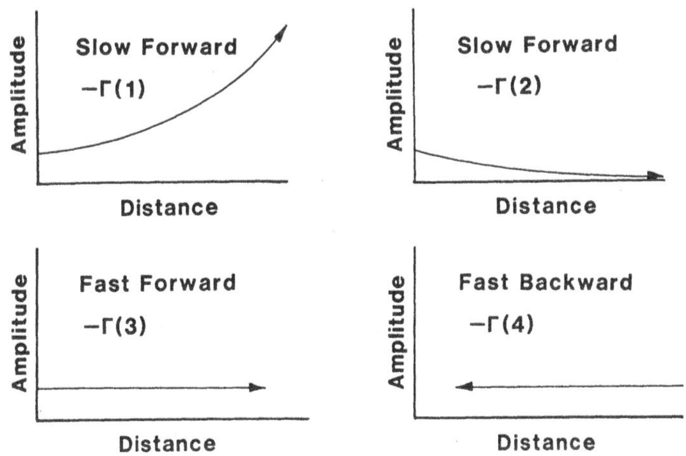

Figure 10.10 Waves for Synchronous Case without Losses or Space Charge

The wave of significance for traveling wave amplification is, of course, the first wave. From the real parts of the propagation constants, it is possible to calculate the gain, G, for the lossless synchronous case without space charge. The result is shown in Fig. 10.11. The quantity C is, as usual, Pierce's gain parameter and N is the distance along the RF structure in wavelengths.

The dashed line in Fig. 10.11 represents the increasing wave, and is generally given by

$$G = A + BCN \tag{10.6}$$

The value of B is calculated directly from the real part of the propagation constant for the growing wave. The value of A represents the portion of the total applied voltage that drives the increasing wave. The remainder of the applied voltage is divided among the other waves.

For the synchronous case without losses or space charge, the value of B is 47.3 The value of A is -9.54 dB and results from the equal division of the input signal among the three forward waves.

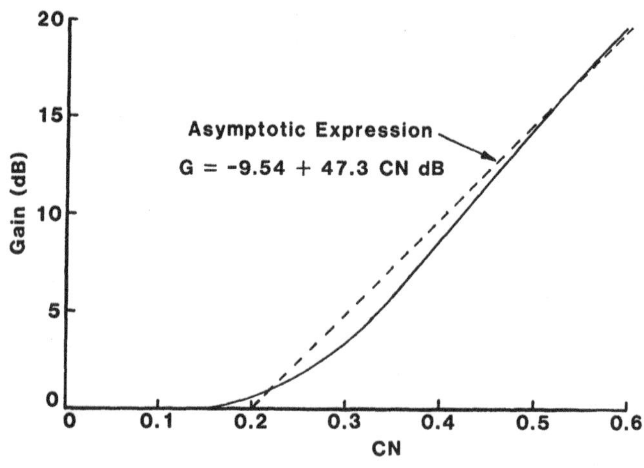

Figure 10.11 Power Gain for Synchronous Case without Losses or Space Charge

(From *Traveling Wave Tubes* by J.R. Pierce, Copyright 1950 by D. Van Nostrand Company, Inc. Reproduced with permission.)

10.3.2 Velocities Other Than Synchronous

When the electron beam travels at a velocity other than the velocity of the wave on the circuit in the absence of the beam, solutions to the quartic equation for the propagation constants are obtained numerically.

At this point, it is helpful to introducc Piercc's velocity parameter, b. This parameter is used in Eq. 10.7 to relate the electron velocity, u_0, to the phase velocity, v_1, of the wave in the absense of electrons:

$$u_0 = (1 + Cb)\, v_1 \tag{10.7}$$

For $b > 0$, the electrons go faster than the undisturbed wave and for $b < 0$, the electrons go slower than the undisturbed wave. For $b = 0$, the synchronous case considered in the previous section occurs.

Equation (10.7) gives the propagation constant for the wave in the absence of electrons:

$$-\Gamma_1 = -j\,\frac{\omega}{v_1} = j\beta_e(1 + Cb) \tag{10.8}$$

In addition to the velocity parameter, Pierce defines an incremental propagation constant, δ, by the expression

$$-\Gamma = -j\beta_e + \beta_e C\,\delta \tag{10.9}$$

In general, δ is a complex number and may be expressed by

$$\delta = x + jy \tag{10.10}$$

If $x > 0$, the wave increases in amplitude as it travels, and if $x < 0$, the wave decreases as it travels. If $y > 0$, the wave travels faster than the electrons and if $y < 0$, the wave travels more slowly than the electrons.

As an example, for the growing wave in the synchronous case (Eq. (10.1)), $x = \sqrt{3}/2$. The value of y is $-1/2$, so the wave travels more slowly than the electrons.

Values of x and y can be computed for the three forward waves as a function of the velocity parameter, b. The growing wave is of primary interest and the components of the incremental propagation constant for that wave (x_1 and y_1) are given in Fig. 10.12 as a function of b for the lossless case without space charge.

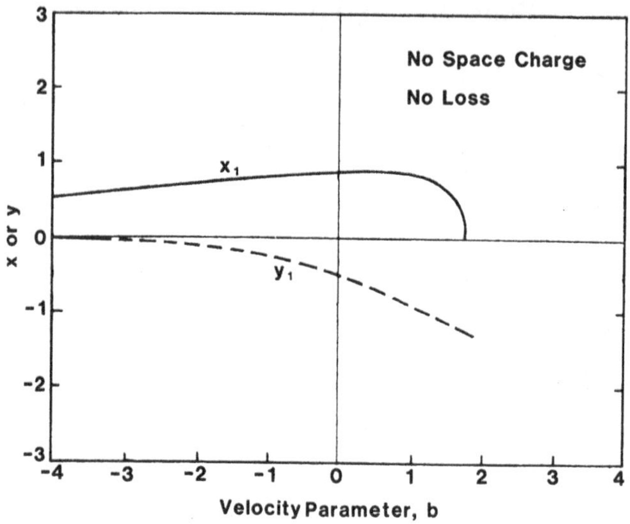

Figure 10.12 Components of Incremental Propagation Constant for Growing Wave

(From *Traveling Wave Tubes* by J.R. Pierce, Copyright 1950 by D. Van Nostrand Company, Inc. Reproduced with permission.)

There are at least two important points to be noted in Fig. 10.12:

1. The value of x_1 is positive for $b < 2$, so that gain occurs over a wide range of electron velocities.
2. The value of y_1 is always negative, so that the growing wave always travels slower than the electron beam, even when the beam itself travels slower than the undisturbed circuit wave ($b < 0$). In fact, the electrons must travel faster than the increasing wave in order to give energy to it.

10.3.3 Effect of Loss

In addition to the losses that occur naturally in RF circuits, additional losses may be introduced to attenuate the backward wave. The backward wave normally originates from reflections either inside or outside the tube and can serve as a feedback mechanism to cause oscillations or large variations in gain (because of beating effects with the input signal).

The effect of loss may be taken into account by introducing Pierce's circuit attenuation parameter, d, in the propagation constant in the absence of electrons as follows:

$$-\Gamma_1 = -j\beta_e (1 + Cb) - \beta_e Cd \tag{10.11}$$

The value of d may be calculated from

$$d = 0.0184\, L/C \tag{10.12}$$

where L is the circuit loss in dB/wavelength.

When the analysis is carried out, it is found that the real parts of the incremental propagation constants are reduced. In particular, the reduction in x_1 for the growing wave leads directly to a reduction of B in the gain equation (Eq. (10.6)). Figure 10.13 illustrates the effect of d on B. The reduction in B resulting from loss can be easily compensated for by increasing the length of the circuit so that the gain, BCN, remains at the desired level. As a side note, depending on the loss distribution, the additional circuit length may further attenuate the backward wave.

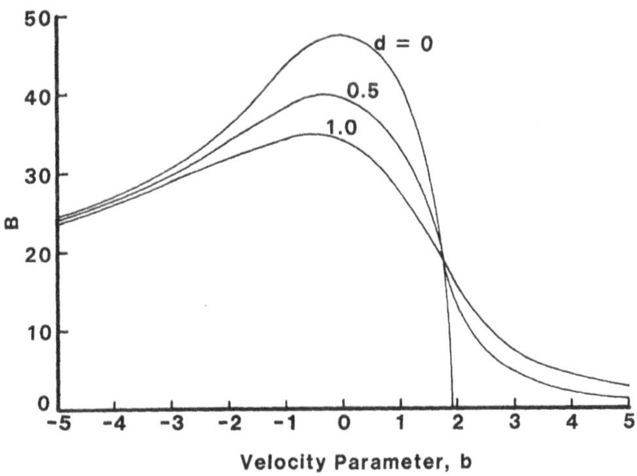

Figure 10.13 Effect of Loss on *B* Factor in Gain of Increasing Wave

(From *Traveling Wave Tubes* by J.R. Pierce, Copyright 1950 by D. Van Nostrand Company, Inc. Reproduced with permission.)

10.3.4 Effects of Space Charge

As was described in Chapter 9, space-charge forces reduce the intensity of bunching in a velocity modulated beam and lead to plasma oscillations for beams that drift long distances.

The effect of space charge may be accounted for by adding the electric field caused by the space charge to the electric field established by the circuit. A modified equation is obtained for the incremental propagation constants. This equation contains a quantity QC, which is known as Pierce's space-charge parameter, and is given by

$$QC = 1/4 \; \frac{\omega_p^{\,2}}{\omega^2 C^2} \tag{10.13}$$

where ω_p is the angular plasma frequency described in the discussion on klystrons (Ch.9). For a finite beam, ω_p should be replaced by the appropriate reduced plasma frequency, ω_q.

The other Pierce parameters (C, b, and d) have had straightforward and logical descriptions. Unfortunately, the physical meaning of the space-charge parameter is not so easy to understand. The square of the plasma frequency is proportional to the density of the electron beam. Also, the ratio of the plasma frequency to the RF frequency is a measure of the space-charge forces tending to disperse the electron beam bunching. Thus, QC is a measure of the seriousness of space-charge effects in a traveling wave tube. These effects will become more apparent after the incremental propagation constants have been examined.

The incremental propagation constants can be determined numerically as before (Fig. 10.12). Results obtained for the growing wave and for QC values of 0.25, 0.5, and 1.0 are shown in Fig. 10.14. In comparing Fig. 10.14 with Fig. 10.12, several effects of space charge are apparent. The most important is that the range of beam velocities (indicated by b) over which gain occurs (x_1 positive) is significantly restricted. As QC is increased, the beam velocity range becomes increasingly smaller and shifts further to the right.

The physical explanation for the restricted ranges of y_1 and b may be found by examining the space-charge-wave behavior of the beam. In Chapter 9, electron plasma oscillations were described in connection with the bunching of an electron beam in a klystron. The way in which these oscillations occur is illustrated in Fig. 10.15. Electron bunches are shown relative to an electron beam at various times. In Fig. 10.15(a), the space-charge forces in the bunches at time t_1 cause the bunches to disperse so that at a quarter of a plasma period later (time t_2) the beam density is uniform. Another quarter of a plasma period later, at time t_3, fast electrons come together with slow electrons to reconfigure bunches. Subsequently, the process is repeated over and over so that oscillations occur.

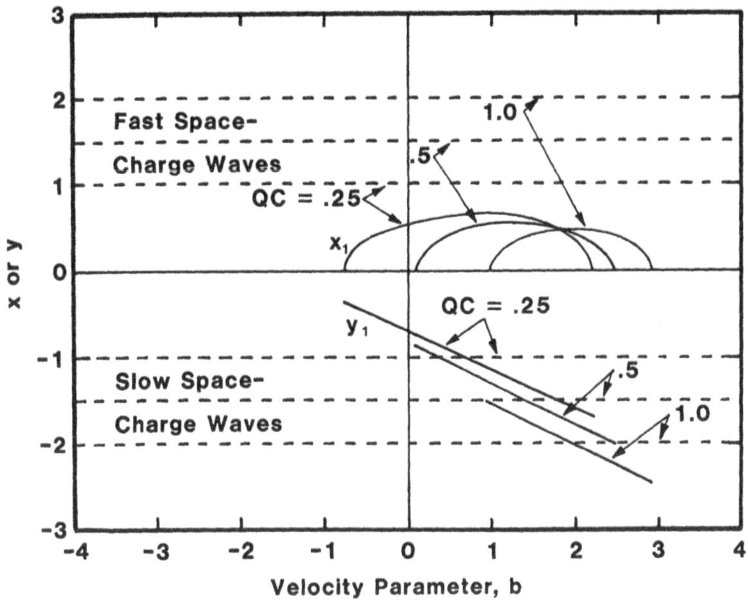

Figure 10.14 Effect of Space Charge on Incremental Propagation Constants for Growing Wave and No Loss

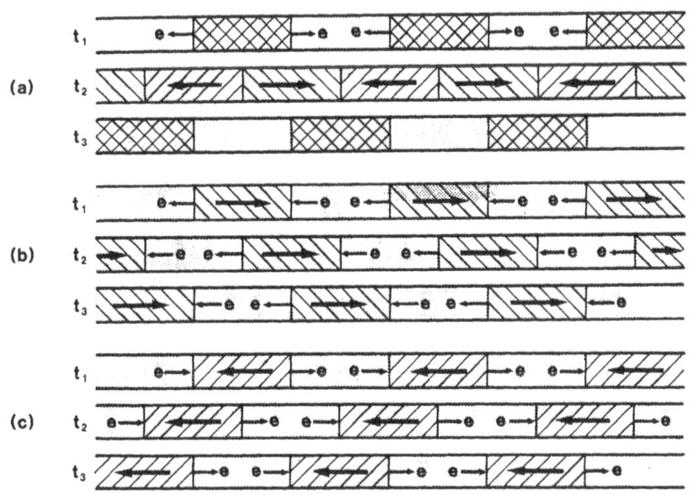

Figure 10.15 Fast (b) and Slow (c) Space-Charge-Wave Components of Electron Plasma Oscillations (a)

The oscillation illustrated in Fig. 10.15(a) can be thought of as the superposition of bunches traveling faster than the electron beam (Fig. 10.15(b)) and bunches traveling slower than the electron beam (Fig. 10.15(c)). The situation that occurs when bunches travel faster than the beam is referred to as a fast space-charge wave. Electrons ahead of a bunch are gathered up by the speeding bunch and electrons at the back of the bunch are expelled by space-charge forces.

Similarly, a slow space-charge wave is said to exist when electron bunches travel more slowly than the electron beam. Electrons behind a bunch catch up with the bunch and are slowed by space-charge forces. Electrons in the front of a bunch are expelled by space-charge forces.

Fast space-charge waves and slow space-charge waves can be independently excited on an electron beam. Figure 10.14 shows that it is the bunching associated with the slow space-charge wave that produces gain in a TWT. The phase velocity of the signal on the circuit must be close to the velocity of the bunches in the slow space-charge wave for gain to occur.

In Fig. 10.14, it is also shown that the maximum gain decreases as the beam density (and QC) are increased. This is to be expected because, as the beam density increases, space-charge forces increase and the intensity of the bunching process decreases, which, in turn, means that the RF current in the beam decreases.

10.3.5 Large-Signal Effects

As the signal level grows in a TWT, the approximations made in small-signal theories become less and less valid. Two relatively independent phenomena occur which lead to saturation. These are illustrated in Fig. 10.16, which shows the results of a computer analysis of signal growth and saturation in a TWT [10.3].

At the left side of the figure, the signal level is small and grows exponentially as expected from the small-signal theory. The electron velocity is greater than the phase velocity of the undisturbed circuit ($b > 0$), so the electron bunches form in the region of maximum decelerating field. As the signal level grows, the first indication of nonlinear behavior is in the plot of electron density where the variation with distance becomes nonsinusoidal. As intense electron bunches are formed, the average electron velocity decreases as beam energy is converted to RF energy in the growing circuit voltage. As the beam slows down, the centers of the bunches move backward with respect to the maximum decelerating field.

As the beam moves through the region of maximum efficiency and maximum circuit voltage, the electron bunches move from the decelerating field into an accelerating field. The average electron velocity starts increasing, indicating that energy is being taken from the circuit wave, thus, leading to the decrease in circuit voltage and efficiency.

In addition to thc change in beam velocity, saturation is also caused by the breakup of the electron bunches in the large-signal region. This results from space-charge forces and from the appreciable velocity spread of the electrons within the high-electron-density regions (evident from the electron-velocity plot in Fig. 10.16). To the user of a TWT, the significance of the

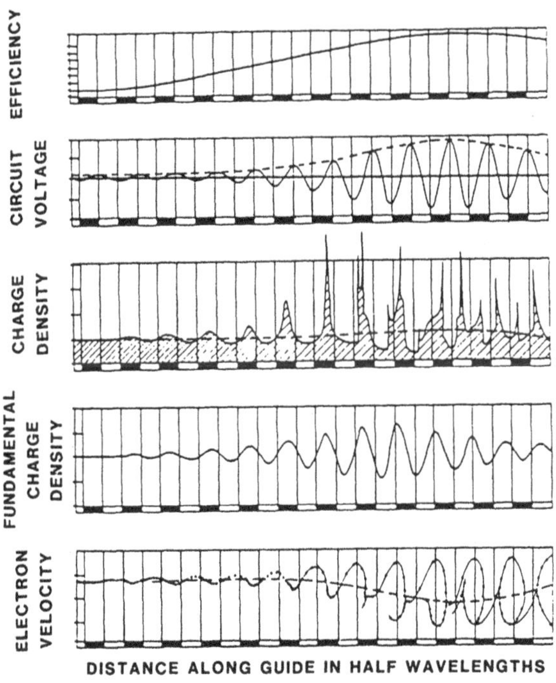

Figure 10.16 Computer Analysis of Interaction in Typical TWT

(From *Power Travelling Wave Tubes* by J.F. Gittins, Published 1965 by American Elsevier, Inc.)

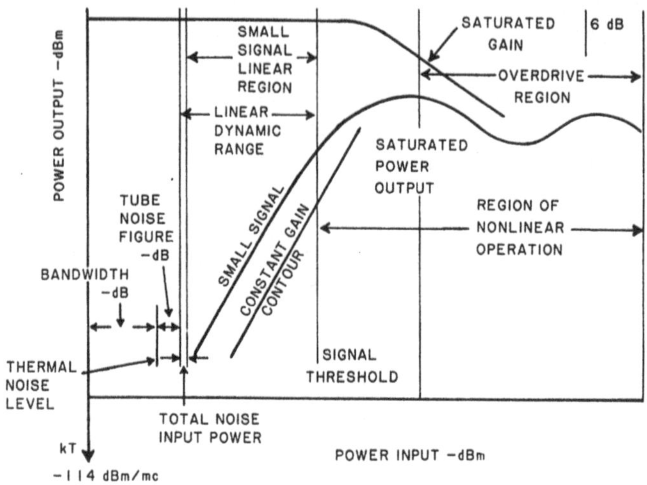

Figure 10.17 Dynamic Characteristics of a TWT

(From Huges Aircraft Co., *TWT and TWTA Handbook*)

gain and saturation characteristics described in the preceding sections is perhaps best shown by a plot like that in Fig. 10.17 of power out *versus* power in. At small-signal levels, changes in output power are proportional to changes in input power and gain is constant. As saturation is approached, gain decreases and operation becomes nonlinear.

At input power levels above saturation, output power decreases and then eventually increases again. This is because the point of optimum bunching moves backward along the helix as input power is increased. The bunches slip out of phase with the circuit wave, and then, at very high drive levels, move back into phase.

At very low signal levels, noise generated in the tube becomes comparable to signal level and this sets the minimum value of input signal that can be usefully amplified.

10.4 ATTENUATORS AND SEVERS

In the analysis of traveling wave amplification, four waves are found to exist. One of these travels in the backward direction from the collector end of the tube to the gun end. This wave can carry power reflected from the load or output region backward through a TWT to the input. If there is a mismatch at the input, a portion of the signal will be reflected to provide a feedback signal. The results can be oscillations or severe variations in gain as a function of frequency.

If, as is indicated in Fig. 10.18, the reflection coefficient at the output terminal is ρ_o and the reflection coefficient at the input terminal is ρ_i, then oscillations may occur when

$$G - L - \rho_o - \rho_i > 0$$

where G is the gain of the tube and L is the loss of the circuit.

Although the tube may be very well matched to the signal source and to the load, there will normally be reflections from the input and output terminals because of the difficultly in making impedance transitions from the RF structure. As a result, it is difficult to make ρ_o and ρ_i less than 0.1. Usually, the cold loss of the circuit cannot be made much more than 6 dB so that the limiting gain is on the order of 26 dB.

To prevent large variations in gain with frequency, the actual gain cannot be as high as 26 dB, and, in fact, is limited to about 20 dB. To permit increased gain, attenuators and severs are used in TWTs.

A common form of attenuation is shown in Fig. 10.19. Portions of the dielectric support rods in a helix TWT are coated with a lossy material such as a carbon film. The thickness of the film is tapered to zero at the ends of the attenuation region to obtain a good impedance match over a wide range of frequencies.

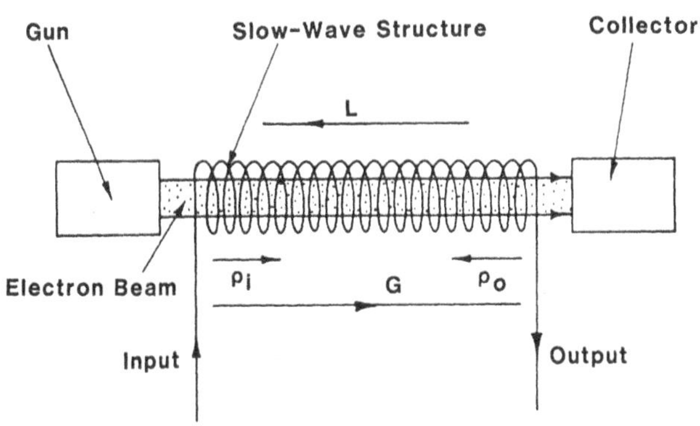

Figure 10.18 Illustration of Quantities Used to Determine Oscillation Condition

(From *Power Travelling Wave Tubes* by J.F. Gittins, Published 1965 by American Elsevier, Inc.)

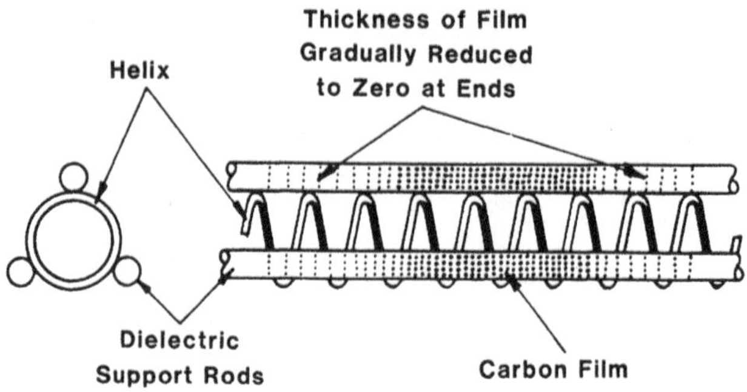

Figure 10.19 Carbon Film Attenuator for Use with a Helix

(From *Power Travelling Wave Tubes* by J.F. Gittins, Published 1965 by American Elsevier, Inc.)

The forward waves as well as the backward wave are attenuated by the carbon film. The loss in gain in the forward direction may be at least partially compensated for by an increase in length of the interaction region.

Unfortunately, if a good impedance match is to be realized, the attenuator must extend for a considerable length along the RF structure. On the other hand, the amplitude of the circuit wave is small in the region of significant attenuation. As a result, the electric fields of the circuit lose control of the bunching process and the velocity spread in the beam increases. The increase in beam velocity spread in the attenuation region results in less effective bunching after the attenuation and a reduction in efficiency. As a result, at high power and when efficiency is important, the carbon film technique for producing attenuation is not attractive.

In high-power tubes where efficiency is important, the most common technique for suppressing the backward wave is through the use of one or more severs. An illustration of a TWT with a severed helix is given in Fig. 10.20. The sever prevents reflections from the load and the output terminal from reaching the input terminal. Although the forward growing wave on the input section of the helix is lost at the sever, current and velocity modulation remain on the electron beam to carry the signal across the sever region for further amplification in the output section.

As the beam drifts across the sever region, bunching degrades and affects the final efficiency of the tube. As a result, the sever region should be as short as possible. Also, the amount of degradation in bunching in the sever region depends on the intensity of bunching that occurs in the input section of the slow-wave structure. Thus, for maximum efficiency, the gain of the input section should be relatively low and gain of the output section should be as high as possible.

Just as there is a limit of about 20 dB to the practical gain of a tube without a sever, there is a similar limit to the gain of each section of a tube with severs. Referring again to Fig. 10.20, the reflection coefficients at the output, ρ_o, at the sever, ρ_s, and at the input, ρ_i, along with the circuit attenuations L_o and L_i, must be taken into consideration in determining practical gain levels. Because it is usually possible to keep the reflection coefficients at the sever terminations less than those for the input and output, the gain per section of a severed tube can be higher than that for a single-section tube. On the other hand, there is some signal loss in the sever region so that an overall gain from a two-section tube of about 40 dB is practical. With additional severs, higher gains are possible.

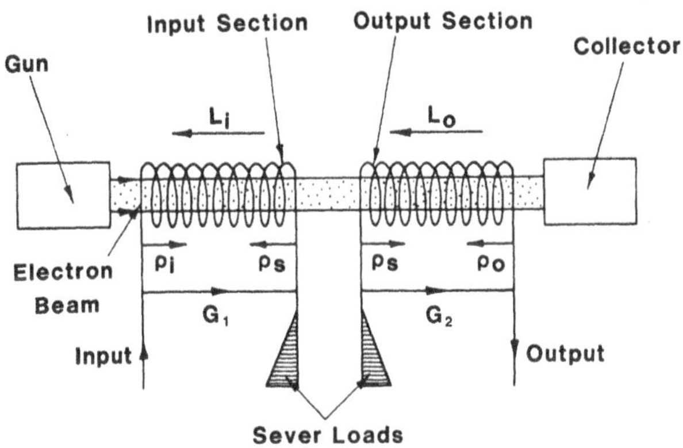

Figure 10.20 TWT Containing a Two-Section Severed Helix

(From *Power Travelling Wave Tubes* by J.F. Gittins, Published 1965 by American Elsevier, Inc.)

10.5 THE DISPERSION CURVE

In previous sections, the significance of the phase velocity of a circuit and its relation to the electron beam velocity have been discussed. The dispersion curve is used to represent the velocity characteristics of a circuit.

Normally, the axes for a dispersion plot are angular frequency, ω, of the RF signal and propagation constant, β. Typical ω - β plots are given in Fig. 10.21.

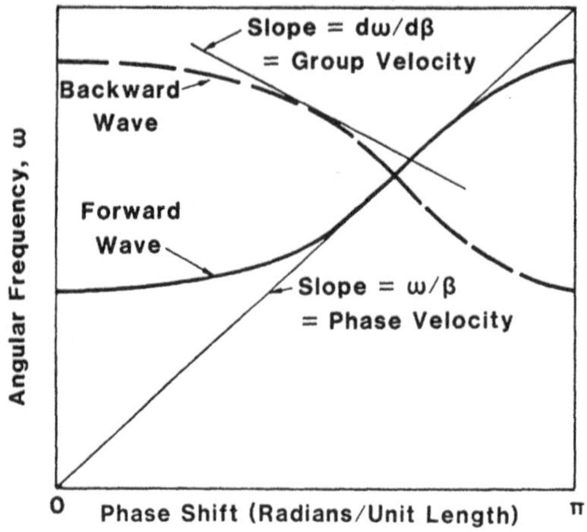

Figure 10.21 Typical ω-β Diagram Showing Dispersion Curves for a Forward-Wave and for a Backward-Wave Circuit.

Two velocities are of significance in ω - β plots. The phase velocity, v_p is simply the ratio of ω to β (because β is ω / v_p). Thus, the slope of a line through the origin to any point on a dispersion curve represents the phase velocity at that point. The velocity of an electron beam can also be represented by a line through the origin. In the regions where the electron-beam velocity line intersects a dispersion characteristic, the beam velocity is synchronous with the phase velocity of the circuit. As a result, these regions of intersection indicate the frequency range over which amplification can be expected to occur.

The group velocity of an RF signal on a circuit is the velocity with which the wave energy propagates along the circuit. The group velocity is represented by the slope of the dispersion characteristic. The group velocity and the phase velocity can have different magnitudes and different directions.

The concepts of phase velocity and group velocity are often confusing. The simple example illustrated in Fig. 10.22 may serve to clarify these concepts. Waves are shown breaking at an angle θ on a beach. The group velocity, v_g, of the waves is the velocity with which the waves propagate across the surface of the water. The phase velocity is the velocity with which the crest of a wave strikes the beach. A person walking along the beach and desiring to keep up with the wave crest would have to walk at the phase velocity.

Similarly, for an electron beam to maintain synchronism with the wave on a circuit, it must travel at or near the phase velocity.

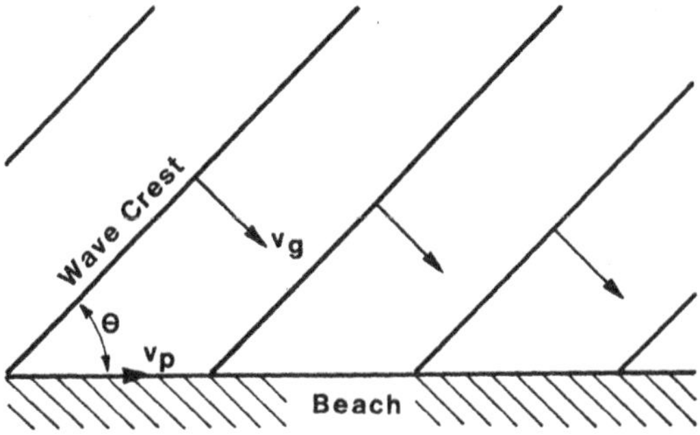

Figure 10.22 Illustration of Difference Between Phase Velocity and Group Velocity

The significance of the information presented in a dispersion curve plotted on ω-β coordinates is not always straightforward. Thus, more convenient coordinates are sometimes used. An example for a coupled-cavity structure is illustrated in Fig. 10.23. Here, the phase shift per section of the RF structure is related to the frequency of operation. A constant velocity line intersects the circuit characteristic at about 3.2. GHz. At this frequency the phase shift per section of the circuit is about 80 degrees. This phase shift and the period at 3.2 GHz define a time interval (80/360 times the RF period at 3.2 GHz), which, with the distance between circuit sections, specifies the velocity an electron beam must have in order to be synchronous with the circuit wave. Dispersion curves for various RF structures (helices, coupled cavity circuits, *et cetera*) will be shown and discussed later in this chapter.

In Fig. 10.21, a dispersion curve is shown for which the group velocity was negative. A circuit with a negative group velocity is a backward wave circuit. The concept of backward-wave interaction can be illustrated by again examining waves breaking on a beach. In this case, as shown in Fig. 10.24, a sea wall with equally spaced openings is located along the shoreline and the waves are moving to the left. A person walking along the beach would be able to see the waves only when passing an opening. The person could walk from right to left and pass the openings at the same time that a particular wave crest did. This situation is like that in Fig. 10.22, except that the directions are reversed.

Alternatively, the person could walk from left to right (from 1 to 2, for example) and pass the openings in synchronism with successive wave crests moving from right to left (from 3 to 2, for example). This situation represents a backward-wave interaction because the group velocity is to the left and

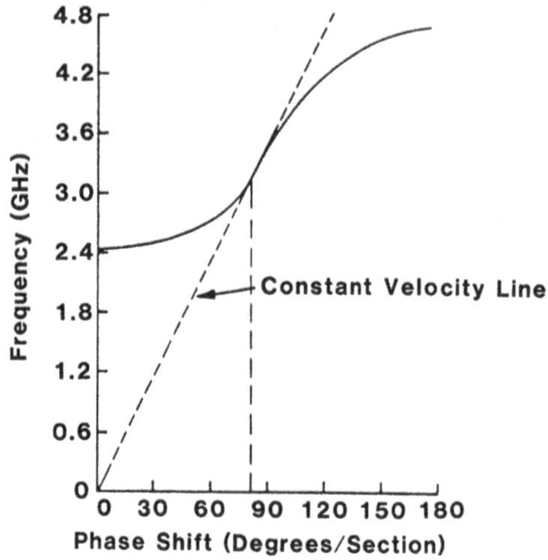

Figure 10.23 Dispersion Curve for Coupled-Cavity Structure

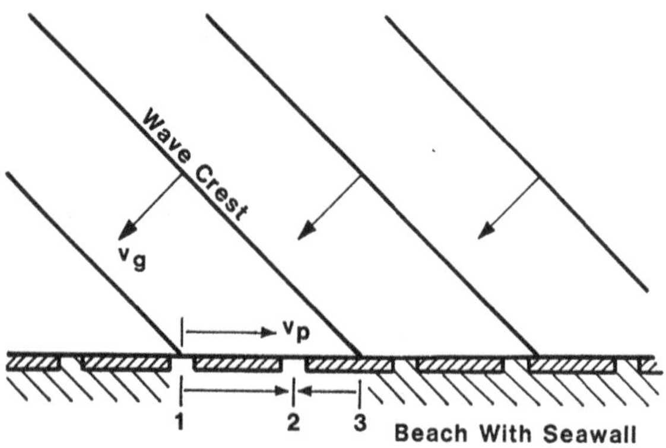

Figure 10.24 Illustration of Backward-Wave Interaction

the phase velocity is to the right. In devices using backward-wave circuits, the electron beam interacts with the phase velocity of the circuit in the forward direction, but, in analogy to Fig. 10.24, the power flow on the circuit is in the backward direction.

Backward wave devices can be amplifiers or oscillators. In either case, the RF power produced comes out at the cathode end of the device.

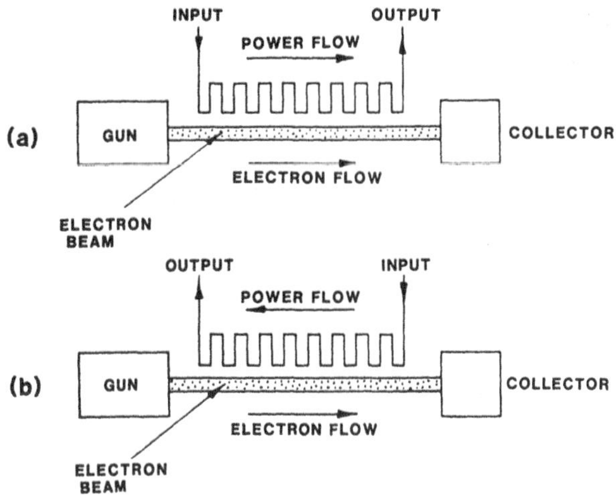

Figure 10.25 Diagram Showing Directions of Electron and Power Flow for (a) Normal Traveling Wave Tube (Forward-Wave Amplifier) and (b) Backward-Wave Amplifier

(From *Power Travelling Wave Tubes* by J.F. Gittins, Published 1965 by American Elsevier, Inc.)

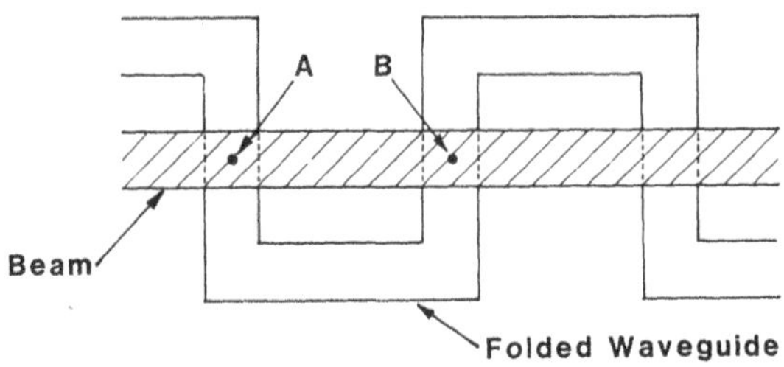

Figure 10.26 Illustration of a Circuit that Can Produce Forward-Wave or Backward-Wave Interaction

The concept of a backward group velocity with a forward phase velocity for backward-wave interaction with an electron beam can be better understood with the example illustrated in Fig. 10.26. A waveguide is shown which is folded so that an electron beam can be passed through it several times. The electric field in regions A and B is in the direction of travel of the electron beam so that interaction with the beam can occur.

Consider first the case of forward-wave amplication (as in the conventional TWT). If the RF signal phase shift is 90° as electrons travel from A to B, then for amplification the electrical length of the waveguide must be 270° to compensate for the 180° field reversal that occurs when the wave goes around the bends in the waveguide.

For backward-wave amplification, the electrical length of the waveguide from B to A must be 90°. This 90° plus the 180° field reversal resulting from the bends in the waveguide produces a total phase shift from B to A of 270°. As electrons drift from A to B, an additional 90° phase shift occurs. Thus, the electrons arrive at B in phase and reinforce the wave in the waveguide.

This stroboscopic-like effect in which electrons are periodically exposed to the field is typical of all backward-wave interactions. It is analogous to the phenomenon observed on television or in the movies when the framing speed of the camera is such that rotating objects, such as the wheels on cars, appear to turn backward.

Shown in Fig. 10.27 is a dispersion curve for backward-wave amplification. This dispersion curve shows operation over a band of frequencies from 5.4 to 5.9 GHz. Synchronous voltage lines (the synchronous voltage is the beam voltage corresponding to the synchronous electron velocity) are shown at two points in the band.

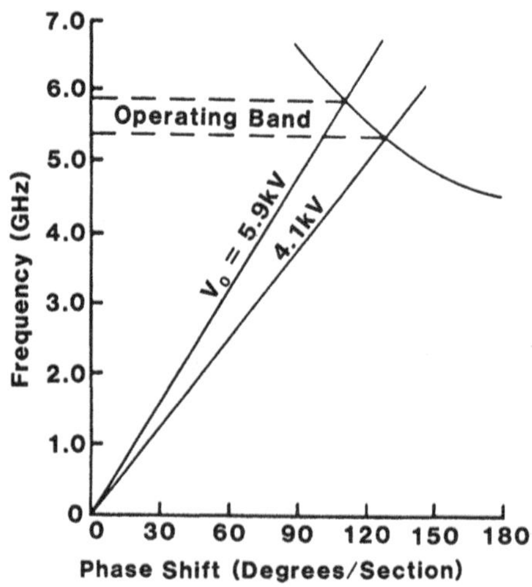

Figure 10.27 Dispersion Curve for a Backward-Wave Circuit

(From *Microwave Tube Manual* by Varian Associates, Air Force Publication Number T.O. 00-25-251, October 1979.)

It is clear from Fig. 10.27 that backward-wave amplifiers are synchronous for a given voltage at only one frequency. Thus, they are, in principle, voltage-tuned amplifiers. In practice, considerable bandwidth can be achieved depending on the circuit used and the design of the power supply.

10.6 HELIX TWT s

As was indicated at the beginning of this chapter, the TWT invented by Kompfner used a helix as a slow-wave circuit. During the formative years of TWT development, hundreds of other propagation structures were investigated, but the helix remained far superior to these in nearly every respect. The purpose of this section is to describe the characteristics of helix TWTs that have made them the mainstay of the TWT industry.

Many of the elements of the basic helix TWT were shown in Figs. 10.4, 10.6, 10.18, 10.19, and 10.20. For reference purposes in the following discussions, many of the geometrical characteristics of a helix that are of significance are given in Fig. 10.28. The dimensions shown are the radius of the helix, a, the pitch, p, and the diameter of the helix wire, d. The angle, Ψ, is the helix pitch angle. The path A—A will be referred to later in a discussion of the backward-wave oscillation (BWO) characteristics of a helix. Also, for reference purposes, the cross section of a periodic-permanent-magnet (PPM) focused helix TWT is shown in Figure 10.29. Various aspects of the helix, its support, and connections to it will be discussed in this section.

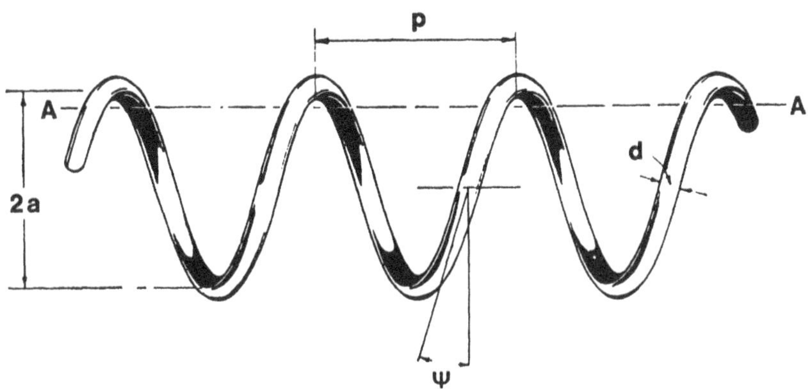

Figure 10.28 Illustration of Geometrical Characteristics of a Helix

(From *Power Travelling Wave Tubes* by J.F. Gittins, Published 1965 by American Elsevier, Inc.)

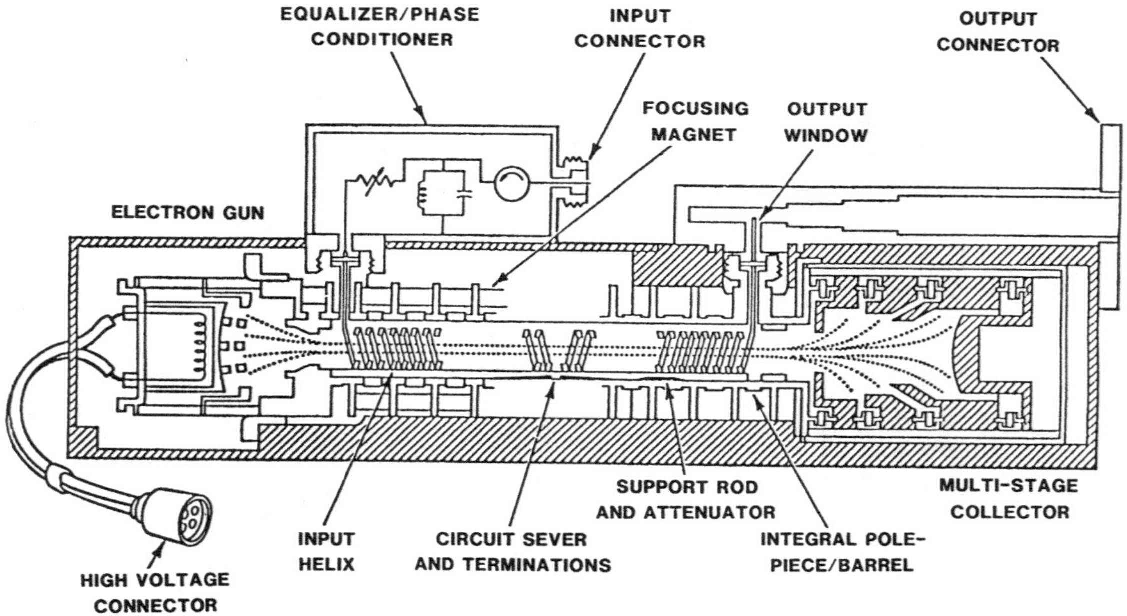

Figure 10.29 Cross Section of Helix TWT Developed for Electronic Countermeasures Purposes

(Courtesy of Varian Associates)

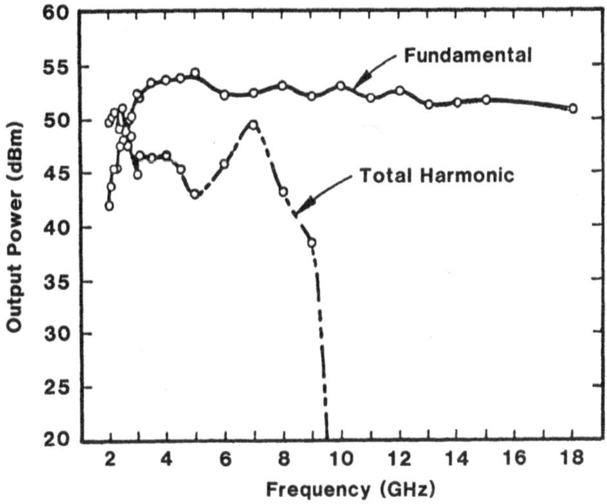

Figure 10.30 Saturated Output Power *versus* Frequency for the Varian VTU-6292A1 Helix TWT

(Courtesy of Varian Associates)

The extremely broad bandwidth characteristics of the helix is the single most important property of this slow-wave circuit. No circuit has ever rivaled the helix in bandwidth and most do not even come close. Helix TWTs can now be made with a bandwidth of over two octaves, as is shown in Fig. 10.30. Harmonic power indicated in Fig. 10.30 will be discussed later.

10.6.1 Helix Bandwidth

The broad bandwidth of the helix results from its low dispersion. As pointed out early in this chapter, the helical circuit can be thought of as being derived from a single-wire transmission line for which the dispersion is zero. When the transmission line is formed into a helix, the phase velocity becomes, approximately

$$v_p \approx c \sin \Psi$$

where c is the speed of light and Ψ is the helix pitch angle.

Unless corrective steps are taken, phase velocity varies somewhat with frequency for two reasons:

1. As frequency is reduced, the number of turns of the helix per wavelength is increased. As a result, the coupling of electric and magnetic fields between turns changes. For example, as shown in Fig. 10.31, when current flows in the same direction in adjacent turns, there is cancellation of the magnetic flux between turns. As a result, the flux generated per unit current (the inductance) decreases as the wavelength (and number of turns per wavelength) increases. Thus, as frequency decreases, inductance decreases and wave velocity increases. As the wave velocity around the helix increases, then, of course, the phase velocity increases. This is evident in Fig. 10.32, which is an ω-β diagram for a helix with a 10° pitch angle. The phase velocity at high frequencies is shown to be essentially constant, but increases at low frequencies. The effect of the dielectric material of the support rods for the helix is also evident in Fig. 10.32. Wave velocity is inversely proportional to the square root of dielectric constant. Thus, when dielectric material is added in the region surrounding the helix, the wave velocity decreases.
2. At high frequencies (and a relatively small number of turns per wavelength), the electric field is concentrated between turns of the helix. As the frequency is reduced and the number of turns per wavelength is increased, the electric field extends farther from the helix. Therefore, a metal shell surrounding the helix will have a greater effect on the electric field pattern at low frequencies than at high frequencies. As the shell is brought closer to the helix, the effect increases with the result that the wave velocity on the helix decreases. The decrease in velocity is greater at low frequencies than at high frequencies.

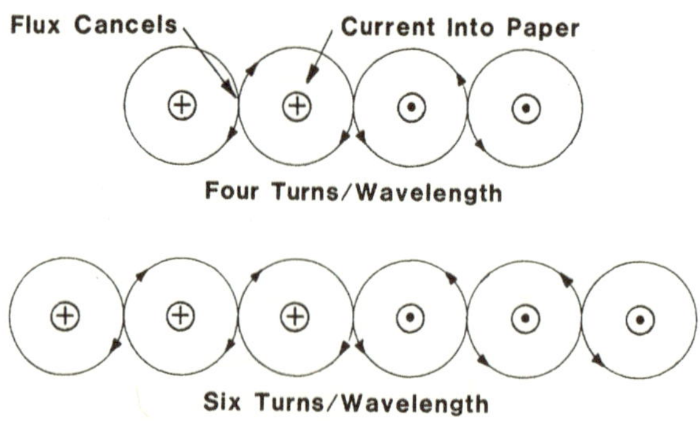

Figure 10.31 Illustration of Cancellation of Magnetic Flux Between Turns

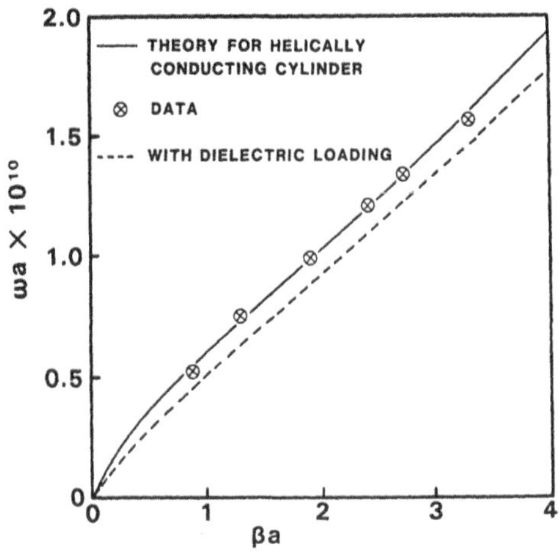

Figure 10.32 ω-β Diagram for a Wire Helix with a 10° Pitch Angle

(From *Power Travelling Wave Tubes* by J.F. Gittins, Published 1965 by American Elsevier, Inc.)

The increase in phase velocity resulting from the reduction in inductance at low frequencies can be partially offset by bringing the metal shell closer to the helix. Unfortunately, while the resulting decrease in dispersion is desirable, there is an accompanying decrease in circuit interaction impedance that is undesirable because gain and efficiency are decreased. The decrease in circuit impedance results from inductive coupling of the helix currents to the metal shell and the flow of current in the circumferential direction in the shell.

Ideally, if a metallic shell that conducted only in the axial direction could be placed around the helix, the effect of the shell on circuit impedance could be avoided because no circumferential currents could flow in the shell. Still, the shell would have a greater effect on the electric field at low frequencies than at high frequencies. As a result, control of the dispersion would be possible.

In practice, an axially conducting shell is approximated by one of the techniques, frequently referred to as anisotropic loading, illustrated in Fig. 10.33 [10.4]. One of the four configurations shown (labeled unloaded) is used for comparison purposes and has only the outer metal shell, which supports the helix through dielectric support rods. In the other three configurations, either solid, vane-shaped, or T-shaped metallic elements are used to provide the desired loading.

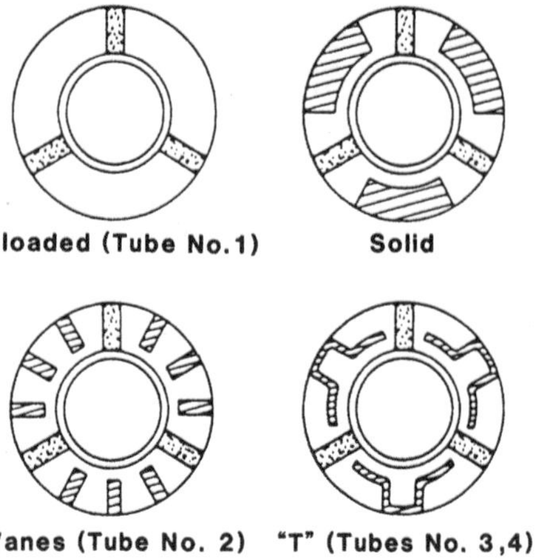

Figure 10.33 Techniques for Achieving Dispersion Control by Approximating an Axially Conducting Shell

(From J.L. Putz and M.J. Cascone, *Technical Digest,* 1979 IEDM.)

Experiments have been performed to provide a basis for the choice of the best loading scheme to use for controlling dispersion. Shown in Fig. 10.34 are phase velocity (in a sort of inverted ω-β diagram) and Pierce's velocity parameter as functions of frequency for three of the configurations in Fig. 10.33. Tests were performed with two T-shaped elements having different spacings from the helix. The results were obtained in tubes containing these configurations, which were nearly identical in all other respects.

With tube no.1, phase velocity increases and b decreases as frequency is reduced. In tube no.2, the vanes produced a virtually flat phase-velocity characteristic over nearly the entire frequency range. The T-shaped segments produced negative dispersion characteristics. That is, the phase velocity decreased at low frequencies.

The small-signal gain characteristics of the four tubes tested are shown in Fig. 10.35. The maximum gain for tube no.1 occurred in the 6 to 7 GHz range. This frequency corresponds to a wavelength such that there are about four helix turns per wavelength. This is in agreement with the prediction of the small-signal theory of helix TWTs. At higher frequencies, gain decreases because the field lines are concentrated closer to the helix and the component available for interaction with the beam decreases. At lower frequencies, gain decreases because the length of the helix, in terms of the number of wavelengths, decreases. Of the four tubes tested, tube no.4, which had a closer helix-to-loading-element spacing than tube no. 3, had the best small-signal bandwidth (in excess of two octaves).

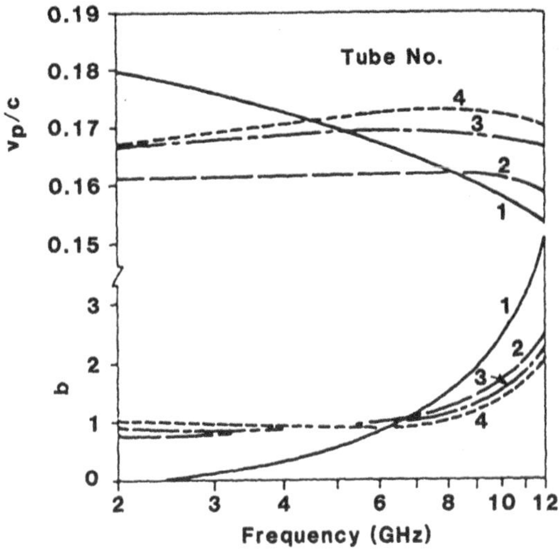

Figure 10.34 Normalized Phase Velocity and Pierce's Velocity Parameter as Functions of Frequency

(From J.L. Putz and M.J. Cascone, *Technical Digest*, 1979 IEDM.)

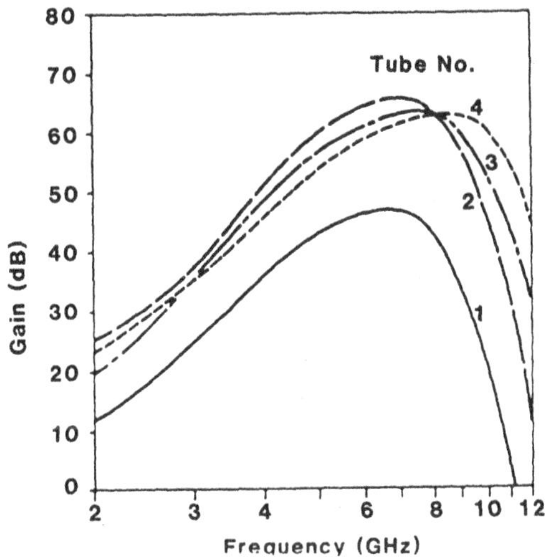

Figure 10.35 Small-Signal Gain for Various Loading Configurations

(From J.L. Putz and M.J. Cascone, *Technical Digest*, 1979 IEDM.)

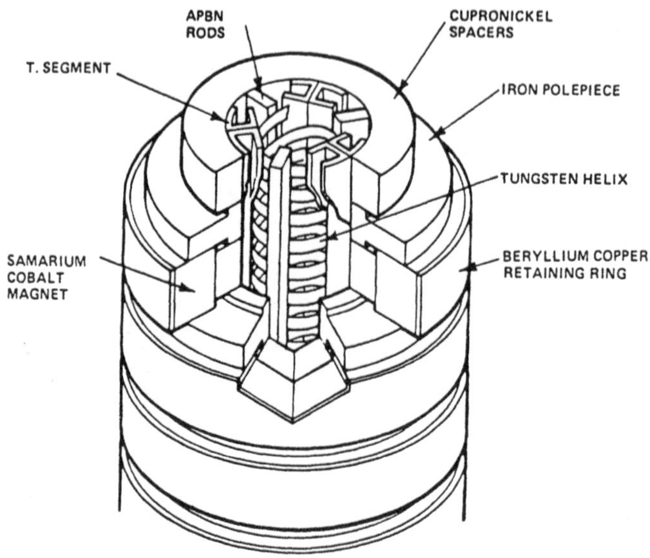

Figure 10.36 PPM Focused Helix TWT Containing T-Shaped Loading Elements.

(Courtesy of Varian Associates)

Shown in Fig. 10.36 is a cutaway view of a typical PPM-focused helix TWT containing T-shaped loading elements. In this example, the tungsten tape helix is supported by anisotropic pyrolytic boron nitride (APBN) rods. The reason for using APBN and the techniques for assembling the helix and supported-rod structure are discussed in the section on helix support techniques.

In Fig. 10.30, the harmonic power at low frequencies was shown to be very high. This results from the wide bandwidth of the tube and the fact that the velocity modulation and bunching processes are highly nonlinear. The extent of the nonlinearities was apparent in the electron density and velocity plots of Fig. 10.16. At the low end of the operating band of a broadband helix TWT, the second and even the third harmonic signals generated in the beam are within the bandwidth of the tube and are thus amplified. At the lower end of the band, harmonic power can rival fundamental power.

A technique known as harmonic injection is sometimes used to reduce harmonic output power in wideband helix tubes [10.5]. An indication of the effectiveness of this technique is illustrated in Fig. 10.37. When a second-harmonic signal of the proper phase and amplitude is injected, along with the fundamental signal, the harmonic output power drops and the fundamental power increases. This occurs when the injected second-harmonic signal is 180° out of phase with the second-harmonic signal generated by nonlinear processes inherent in the beam-circuit interaction process. With proper design of the input circuit, the cancellation process can be made reasonably noncritical.

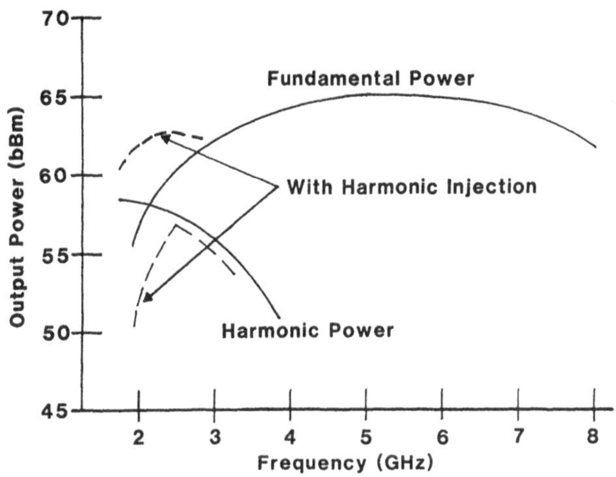

Figure 10.37 Affect of Harmonic Injection on Output Power for a Wideband Helix TWT

(From Hughes Aircraft Co., *TWT and TWTA Handbook.*)

10.6.2 Transitions

At the input to and output from a helix TWT, impedance transitions must be made to provide reasonable impedance matches. The impedance of the helix varies with frequency, as is shown in the example of Fig. 10.38 [10.6]. In this example, the helix structure was designed for operation from 4.8 to 9.6 GHz. At the center of the band, the impedance is seen to be about 50 ohms. At the low end of the band, the impedance increases to over 80 ohms, and at the high end it is about 35 ohms.

Ideally, a very broad-band impedance match to the helix could be made with the transition shown in Fig. 10.39. With this technique, a very gradual change in geometry is made so that the helix slowly changes into a coaxial line. As long as the overall transition is long compared with the wavelength of the signal being used, there will be an excellent impedance match.

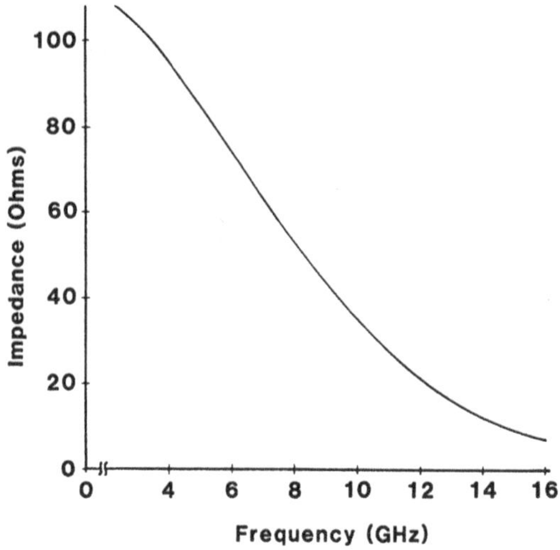

Figure 10.38 Typical Helix Impedance for an Octave-Bandwidth TWT

(From A.S. Gilmour *et al., IEEE Trans. Electron Devices,* Oct. 1979.)

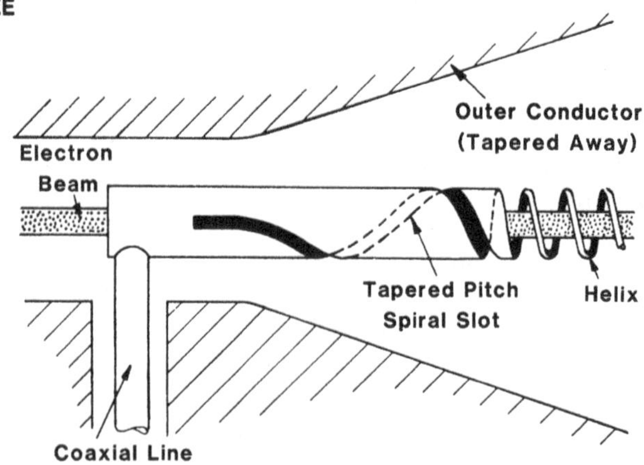

Figure 10.39 A Very Broad-Band Transition from a Helix to a Coaxial Line

(From *Power Travelling Wave Tubes* by J.F. Gittins, Published 1965 by American Elsevier, Inc.)

Unfortunately, transitions like that shown in Fig. 10.39 have the very serious disadvantage of being long. If used, they would increase the length and weight of a TWT. In practice, transitions are normally made after the signal is removed from the helix. For example, the center conductor of the coaxial line can be tapered as shown in Fig. 10.40

A problem with the configuration in Fig. 10.40 is that it requires a relatively long unsupported section of the center conductor of the coaxial line. Under severe shock and vibration conditions, the weld joint at the end of the helix may fail. Also, it should be noted that the last turn of the helix tends to be more poorly supported than other turns when the triangulation or pressure-insertion assembly techniques to be described in the following section are used. Thus, heat transfer from the last turn through the support rods may be relatively poor so that the last turn must be at least partially cooled by the center conductor of the coaxial line.

To help prevent failure of the weld and to provide adequate cooling, there should be a support for the center conductor located as close to the helix as possible. For example, in the output window of the TWT illustrated in Fig. 10.29, the insulating support for the center conductor is only a fraction of a wavelength from the connection to the helix. The material of this support is usually beryllium oxide, which is a very good thermal conductor. The beryllium oxide is brazed to the center conductor and to the outer conductor, and serves as part of the vacuum envelope. In the example shown in Fig. 10.29, a major portion of the impedance-matching transition is in the stepped taper of the waveguide located between the output window and the output connection.

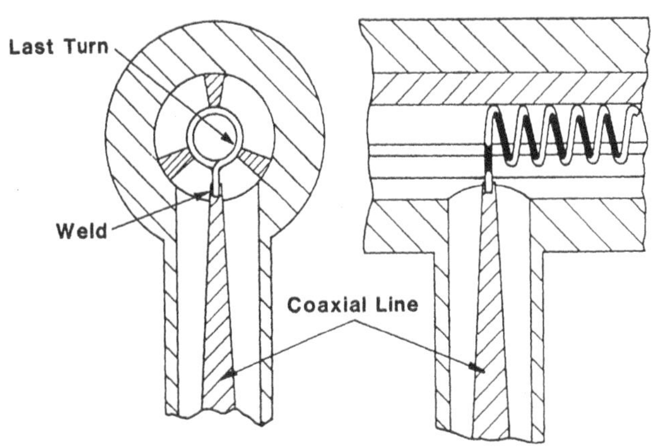

Figure 10.40 Transition Using Tapered Coaxial Line

10.6.3 Helix Support Techniques

The helix of a TWT is a relatively delicate structure. As an example, a cross-sectional view is shown in Fig. 10.41 for the helix of a 10 kW TWT designed to operate over the frequency range of 3.0 to 6.0 GHz. The helix is made of 0.010-in by 0.075-in tungsten tape. If the TWT is to be rugged and capable of operating at high average-power levels, this helix must be very well supported.

The use of metallic supports is out of the question because of their effect on the RF characteristics of the circuit. A dielectric cylinder completely surrounding the helix could conceivably provide excellent support and cooling. Unfortunately, the dielectric constants of suitable ceramic materials are high. As a result, their use in a cylinder would severely reduce the circuit interaction impedance of the helix. In fact, in order to minimize the loading effect of a ceramic support structure, it is necessary to minimize the amount of ceramic material used. As a result, thin, rectangular support rods like those shown in Figs. 10.36 and 10.41 are typically used. Even these load the helix to some extent, as is indicated by the example given in Fig. 10.42.

There are only two dielectric materials that are widely used for supporting the helices of TWTs. These are beryllium oxide (BeO) and APBN. These two materials have relatively high thermal conductivity (compared with other ceramics suitable for operation at high temperature in vacuum).

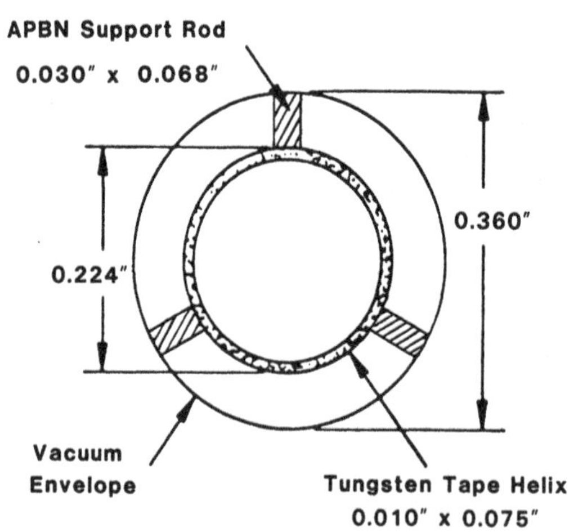

Figure 10.41 Cross Section of Helix of a 10 kW, 3.0 to 6.0 GHz TWT

(From Alan R. Jung, *Technical Digest*, 1978 IEDM.)

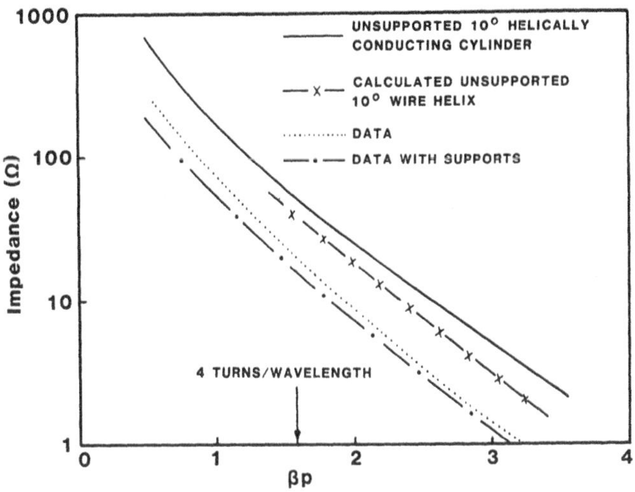

Figure 10.42 Coupled-Interaction Impedance of a Helix with and without Supports

(From *Power Travelling Wave Tubes* by J.F. Gittins, Published 1965 by American Elsevier, Inc.)

With the use of high thermal conductivity ceramics, the temperature drop across the support rods is not a severe problem in high-power helix TWTs. Instead, the major problems are the thermal resistances of the interfaces between the helix and the support rods and between the support rods and the shell (sometimes referred to as the barrel). In the example shown in Fig. 10.43, the thermal drops across these interfaces accounts for 70% of the total temperature difference between the hottest point on the helix and the barrel.

As a result of the large thermal drops at the support rod interfaces, the helix and support rod structures in high-power helix TWTs must be assembled using techniques that minimize the thermal interface resistances. Basically, there are three techniques that are used for assembling helix and support-rod structures. These are

- Triangulation
- Pressure insertion
- Brazing

With the triangulation technique, the metal barrel that is to surround the helix is slightly distorted into a triangular shape. The helix with three equally spaced support rods is then inserted into the barrel and the distorting forces on the barrel are removed. The higher the resulting forces are at the interfaces

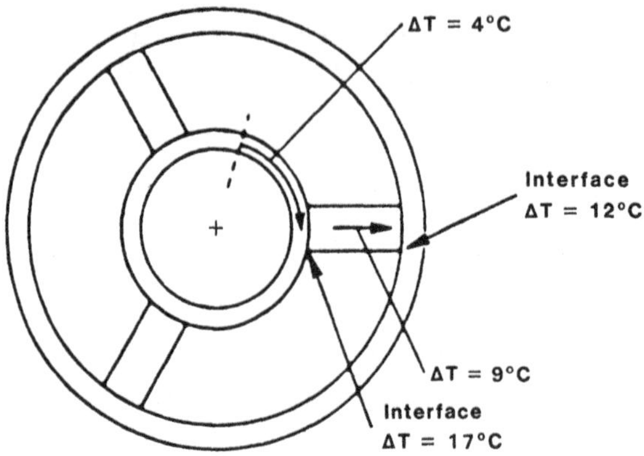

Figure 10.43 Temperature Drops Calculated for a Tungsten Helix and APBN Rods Dissipating 1 Watt

(From A. Scott and M.J. Cascone, *Technical Digest,* 1978 IEDM.)

between the support rods and the helix or barrel, the lower are the thermal resistances. A limit is reached, however, when the forces are high enough to distort the helix excessively. For helix strength, the helix material normally used is tungsten or molybdenum.

It should be pointed out that with the triangulation technique the barrel is not at all like that shown in Fig. 10.36. Instead, it is a relatively thin metallic tube over which the magnetic pole pieces are slid after the helix has been inserted. With this technique, there is a relatively poor thermal interface between the barrel and the pole pieces.

With the insertion technique, the helix and rod assembly is inserted into a barrel structure like that shown in Fig. 10.36. In some cases, the barrel is slightly expanded by heating before the helix and rod assembly is inserted. This is normally referred to as the hot-insertion technique. In other cases, the helix and rod assembly are machined with tolerances for an interference fit and then inserted under high pressure into the cold barrel. The temperatures given in Fig. 10.43 are for a pressure-inserted structure. As with the triangulation technique, the helix materials used are tungsten or molybdenum.

The most effective way of reducing the thermal interface resistances is to braze the interfaces. While the brazing technique sounds attractive, it is difficult to implement in practice. If rods which have been metallized on the surfaces to be brazed are used, there are difficulties with thermal expansion problems. It should be kept in mind that the support rods must be brazed to every turn of the helix. There may be several tens of contact points of one support rod with a helix and several support rods (usually three). Every contact point of every rod must be brazed for proper protection of the helix. After brazing, the metallization of the support rods must be removed in the regions between turns of the helix.

With the brazing technique, a wider choice of helix materials is possible than with the triangulation or pressure-insertion techniques. In addition to tungsten or molybdenum, copper can be used. While tungsten and molybdenum can be copper plated to reduce RF losses, only the use of pure copper throughout the helix leads to the maximum possible transfer of heat and the ultimate loss reduction.

Finally, it should be pointed out that in brazed structures the helix is mechanically bonded to the dielectric support material. There is reduced probability, therefore, that helix failure will result from overheating of the helix and the resulting possible loss of thermal contact between the helix and the dielectric.

An alternative to the use of brazed rods is the block-supported structure illustrated in Fig. 10.44 [10.7]. Individual ceramic blocks are brazed to the helix and, usually, to the barrel. The use of ceramic blocks eliminates the thermal expansion problems associated with rods.

Additional advantages of the block-supported structure are reduced dielectric loading (and higher interaction impedance) and a reduced effect of the dielectric material on phase velocity. As an example of the capability made possible by the block-supporting technique, Thomson-CSF of France produces a 1.5-kW average power, 6-GHz TWT for communications service.

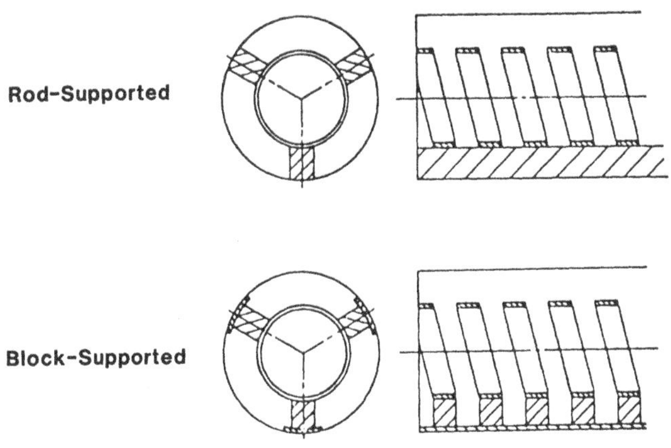

Figure 10.44 Comparison of Rod-Supported and Block-Supported Helix Structures

(From Yukio Hiramatsu, *Technical Digest*, 1979 IEDM.)

Comparisions of the advantages of various techniques are often misleading because test conditions differ. Still, because of the continuing emphasis on higher and higher powers at higher and higher frequencies, it is important to attempt to compare the triangulation, pressure-insertion, and brazing techniques. This has been done and the results are summarized in Fig. 10.45 [10.8]. The base-plate temperature for the test fixture in these tests was 100° C. Thus, the temperature rise of the brazed helix (which was copper) was about two-thirds of that for the insertion (shrunk) technique and about one-half of that for the triangulation technique.

It is important to note that the results in Fig. 10.45 were obtained by passing a dc current through the helices. RF losses and at least a small amount of additional heating are therefore, not included. The additional heating may be larger in the metallized, brazed helix than in the other two.

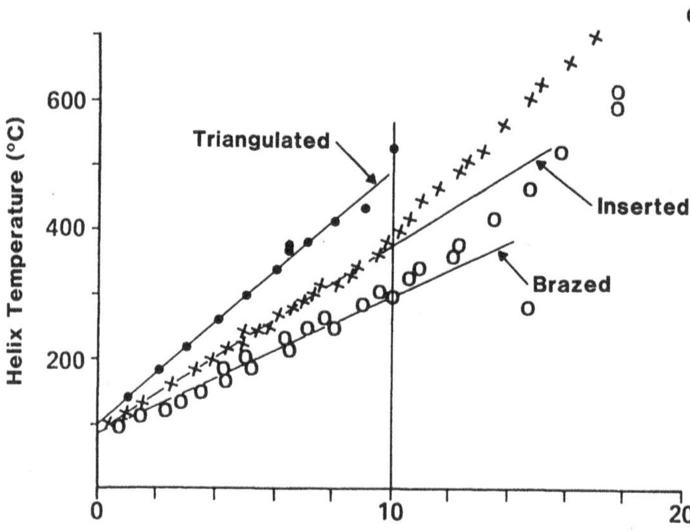

Figure 10.45 Comparison of Thermal Characteristics of Triangulated, Pressure-Inserted (Shrunk), and Brazed Helices

(From T. Mosser, W. Pinger, D. Zavadil, 1980 Microwave Power Tube Conference.)

10.6.4 MM-Wave Helix TWTs

In general, TWT dimensions are scaled with wavelength. At frequencies in the mm-wave range, dimensions become extremely small and miniaturized helices and support rods are used. For example, rectangular support rods having cross-sectional dimensions of .010-in by .020-in, or smaller, are common. As helix assemblies become smaller, heat dissipation problems become more severe, and so the thermal management techniques described in the previous sections become even more important than they are at lower frequencies.

The brazing technique, which offers excellent thermal capabilities, is very difficult to implement in the mm-wave range. As a result, at present, the pressure insertion technique is widely used and average power levels are limited to a few tens of watts at best.

Two new techniques are being investigated that may make possible significant improvements in average power capability. One of these is the use of support rods fabricated from type II-A diamond, which has a thermal conductivity several times that of copper. The other is the use of a plasma-spraying technique for coating the entire length of the outer surface of a helix with BeO.

Diamond and plasma-sprayed BeO have been investigated for use in pressure-fitted assemblies. It is predicted that in either case average power levels as high as 200 W may be achievable at frequencies as high as 50 GHz [10.9].

10.6.5 High Power Operation

If corrective actions are not taken, a major problem that limits the power capability of helix TWTs is backward-wave oscillation (BWO). The circumstances leading to BWO were discussed in the section on dispersion. In particular, in connection with Fig. 10.26, the phase conditions leading to backward-wave interaction were discussed.

At a frequency such that there are two turns per wavelength or 180° per turn (the pi-point), a backward-wave interaction occurs, which is essentially identical to that illustrated by Fig. 10.26. Consider, for example, a portion of the electron beam traveling close to the helix (such as along path A-A in Fig. 10.28). In regions between wires, electrons interact with the electric field. Adjacent to the wires, the electrons are shielded from the field. In the forward direction, when the transit angle of the electrons from turn to turn is 180°, the electrons maintain synchronism with the forward traveling wave. For a wave traveling in the reverse direction at a frequency such that there are two turns per wavelength, electrons with a transit angle from turn to turn of 180° will also remain in phase with the field. Thus, a backward-wave interaction can take place.

If the backward-wave interaction is sufficiently intense, oscillations occur. Whether or not a tube will oscillate in the BWO mode is dependent on a number of factors including beam-current level, beam focusing (and, in particular, how close the beam is to the helix), and reflection coefficient at the collector end of the tube.

Because the frequency of backward-wave oscillations is such that the beam has a 180° transit angle from turn to turn, the BWO frequency is tunable with beam voltage. In fact, tubes designed for operation in the BWO mode can be voltage tuned over a reasonably broad range of frequencies. Helix BWOs are discussed in Section 10.6.7.

High power TWTs are especially prone to BWO, primarily because of the high beam currents. When these oscillations occur, there is a reduction in power at the desired frequency of operation and so the oscillations are said to produce in-band power holes. In addition, the BWO signal may induce oscillations or may intermodulate with the main signal to produce spurious signals inside the operating band of the tube.

One method of suppressing backward-wave oscillations is the introduction of frequency-sensitive attenuation into the helix circuit. A technique developed by Varian Associates for producing frequency-sensitive loss is illustrated in Fig. 10.46 [10.10]. Meander-line circuits are placed on the support rods in the regions between turns of the helix. The circuits are fabricated using standard microwave integrated circuit (MIC) technology. The total length of each line is equal to half the wavelength at the BWO frequency so that the lowest resonant frequency is the BWO frequency.

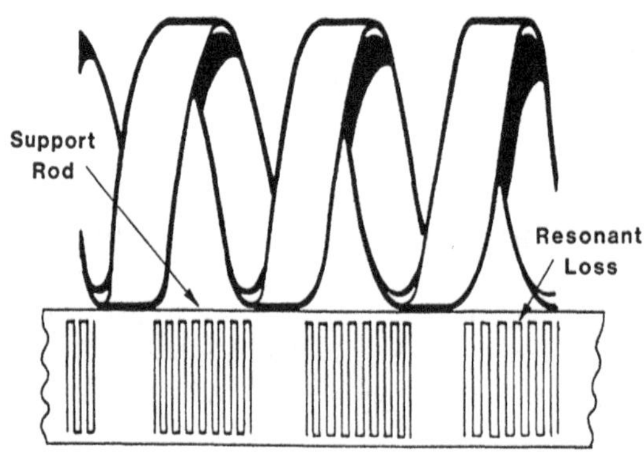

Figure 10.46 Method of Producing Frequency-Sensitive Attenuation through the Use of Resonant Loss

(From A. Scott and M.J. Cascone, *Technical Digest*, 1978 IEDM.)

The amount of loss that can be introduced with the resonant loss technique is very large, as shown by the example given in Fig. 10.47. The loss pattern shown was used in a 10 kW tube designed for operation in the 3.0 to 6.0 GHz range. The pi-mode (BWO) frequency was about 8.0 GHz. Power output as a function of frequency for this 10 kW tube is shown in Fig. 10.48. Notice the very sharp drop in power at the high end of the band where resonant loss becomes effective.

In another technique for suppressing BWO, developed by Raytheon, the output helix is constructed with a step change in the pitch and, thus, in the phase velocity as is shown in Fig. 10.49. The change in pitch is on the order of 20%, which effectively separates the helix into two sections. The slow section is designed to produce the desired gain for the growing wave. Backward-wave oscillations induced in this section propagate through the step and onto the fast section of the helix. Here, the phase velocity is sufficiently out of synchronism with the beam velocity so that energy from the backward wave is actually coupled back into the beam, thus attenuating the backward wave.

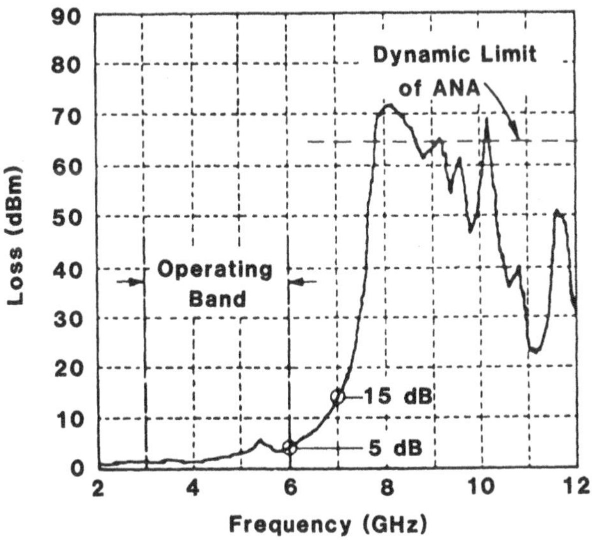

Figure 10.47 Example of Insertion Loss for Resonant Loss on Three Support Rods

(From Alan R. Jung, *Technical Digest,* 1978 IEDM.)

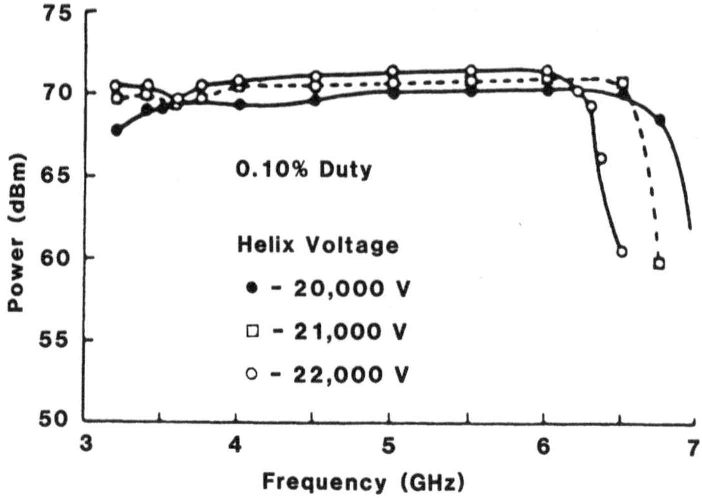

Figure 10.48 Saturated Power Output of 10 kW Helix TWT with Resonant Loss

(From Alan R. Jung, *Technical Digest,* 1978 IEDM.)

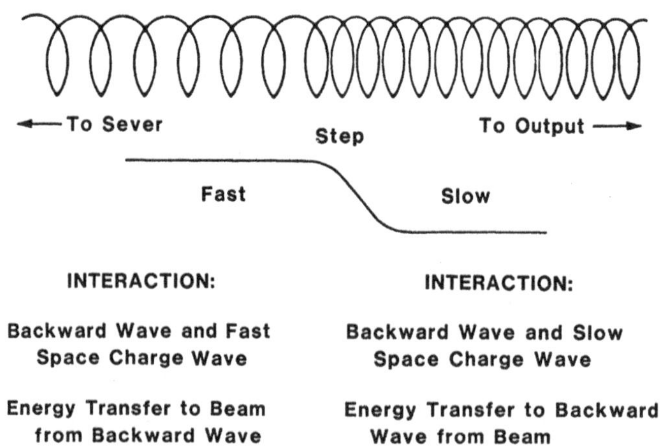

Figure 10.49 BWO Suppression Using Step Change in Phase Velocity

(From A. Scott and M.J. Cascone, *Technical Digest,* 1978 IEDM.)

In the theoretical analysis of the velocity step technique, it is the space-charge wave and its fast and slow components that provide the mechanism for energy coupling to and from the backward wave. The velocities of the space-charge waves were indicated in Fig. 10.14.

In Fig. 10.50, estimates are shown of the peak output power that may be attainable using BWO suppression techniques. Notice that peak power levels in excess of 100 kW appear to be possible in PPM-focused tubes using resonant loss.

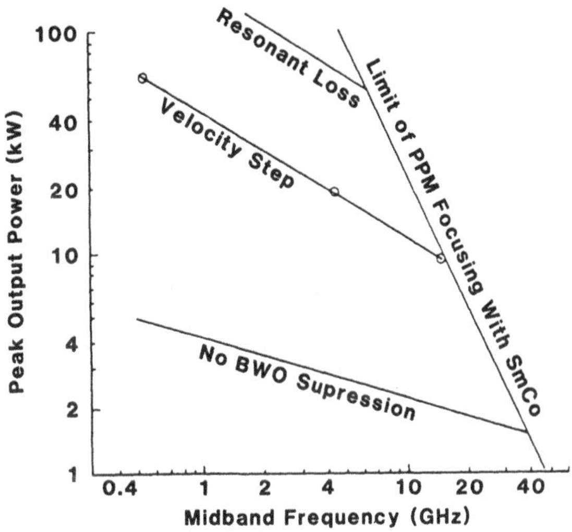

Figure 10.50 Obtainable Peak Power as a Function of Midband Frequency for Helix TWTs with Various Oscillation-Suppression Techniques

(From Alan R. Jung, *Technical Digest,* 1978 IEDM.)

10.6.6 Ring-Bar TWTs

The development of the resonant loss and velocity step techniques described in the previous section, for suppression of backward-wave oscillations has occurred during the past decade. Before these techniques were available, the ring-bar circuit illustrated in Fig. 10.51 was developed to make possible high-power operation of a helix-type circuit. Basically, the ring-bar circuit is derived from the contra-wound helix circuit also shown in Fig. 10.51 and is far easier to fabricate.

To understand the operation of these circuits and how they suppress backward-wave interactions, it is helpful to examine the bifilar helix shown in Fig. 10.52 and to compare qualitatively its characteristics with those of a single helix. The pitch of the helices making up the bifilar helix is the same as that of the single helix. For the moment, interactions between the two helices of the bifilar and single helices are assumed to be the same. As a result, for a given frequency of operation, the voltage distributions indicated in Fig. 10.52 for both the single helix and the bifilar helix are the same. For the BWO discussion, these have been chosen so as to be those associated with two turns per wavelength.

For electrons near the helix moving to the right (designated by the lower case *e*) and the backward wave moving to the left, the backward-wave interaction is as described in the previous section. That is, electrons that are initially accelerated (top left, Fig. 10.52) experience essentially no forces 180° later (middle left) and then are accelerated again 360° later (bottom left). Similarly, and not shown in Fig. 10.52, electrons that are initially decelerated experience no forces 180° later and then are decelerated 360° later. As a result, there is bunching and gain, even though the wave on the helix moves in the opposite direction from the electrons.

For the bifilar helix shown in Fig. 10.52, it is assumed that the electrons and waves have the same velocities as for the single helix. In this case,

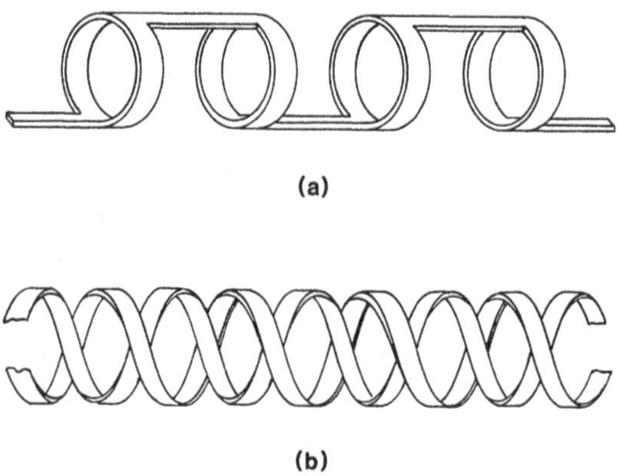

Figure 10.51 Ring-Bar (a) and Two-Tape Contra-Wound Helix (b) Circuits

(From Hughes Aircraft Co., *TWT and TWTA Handbook.*)

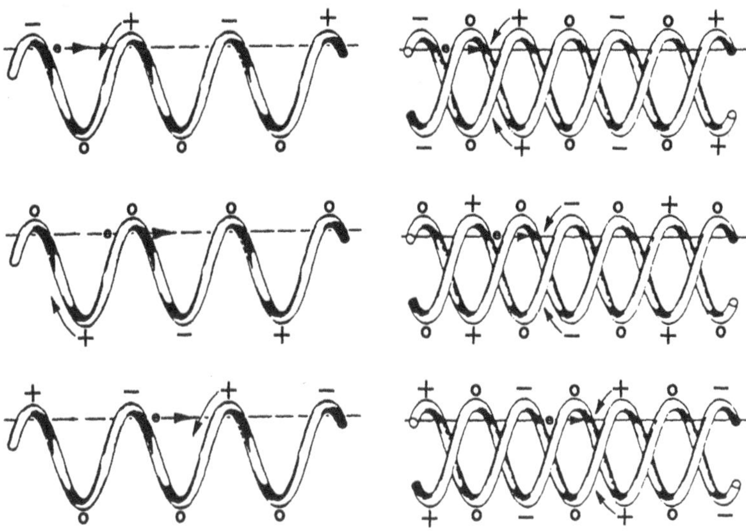

Figure 10.52 Backward-Wave Interaction on a Single and a Bifilar Helix for Two Turns per Wavelength.

however, electrons that are initially accelerated (top right) are decelerated 180° later (center right) and then accelerated again 360° later (bottom right). Because, on each cycle, the accelerating and decelerating forces tend to cancel, there is no net gain that can lead to backward-wave oscillations. As a result, higher operating voltages, currents, and RF power levels are possible with the bifilar helix than with the single helix (with no suppression techniques being used).

The same arguments apply to the contra-wound helix and to the ring-bar circuit, so these circuits can operate without BWO at higher power levels than can the single helix with no BWO suppression. The ring-bar structure has been analyzed by Birdsall and Everhart [10.11]. Some of the parameters used in their analysis are shown in Fig. 10.53. The interactions between the equivalent right-hand-wound and left-hand-wound helices result in considerable dispersion, as is indicated by the ω-β diagram for a ring-bar circuit shown in Fig. 10.54. Thus, while operating power level is high for a tube using a ring-bar circuit, the bandwidth is generally limited to about 10% to 20%. Figure 10.55 contains a typical plot indicating the power and bandwidth capabilities of the ring-bar circuit.

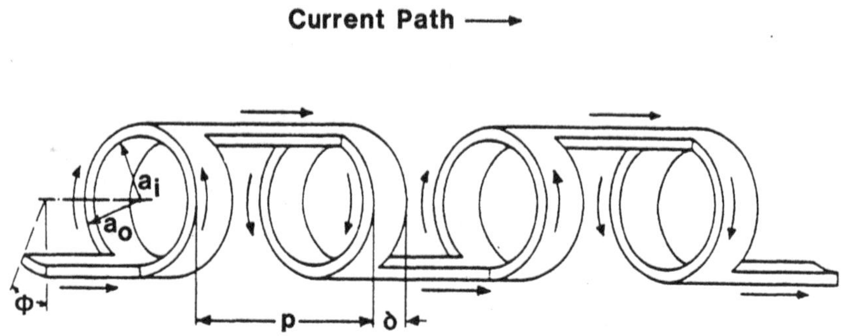

Figure 10.53 Quantities Used in Analysis of Ring-Bar Structures

(From *Power Travelling Wave Tubes* by J.F. Gittins, Published 1965 by American Elsevier, Inc.)

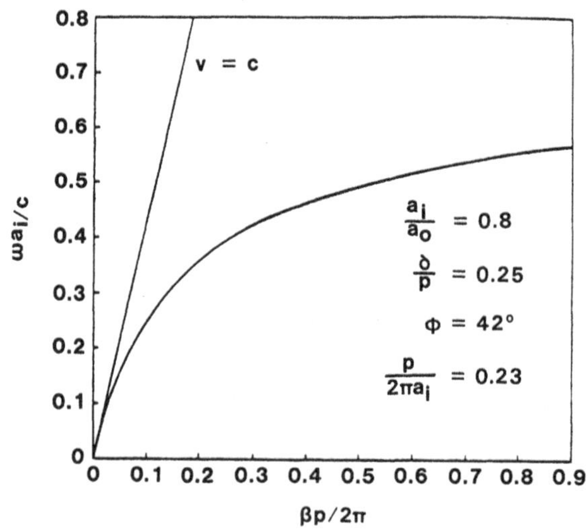

Figure 10.54 ω-β Diagram for Ring-Bar Structure

(From *Power Travelling Wave Tubes* by J.F. Gittins, Published 1965 by American Elsevier, Inc.)

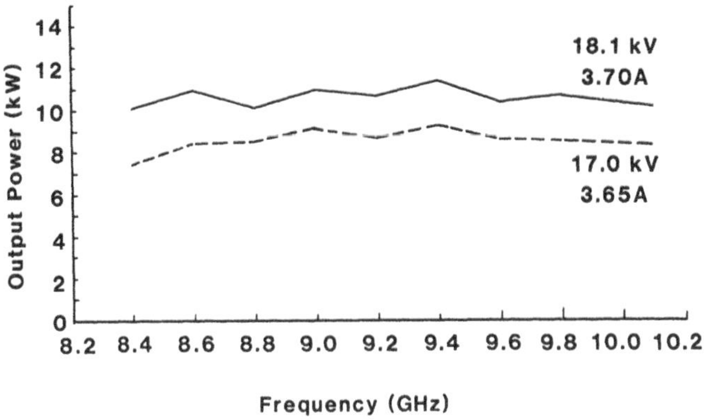

Figure 10.55 Output-Power Characteristic for a Ring-Bar TWT

(From Hughes Aircraft Co., *TWT and TWTA Handbook.*)

10.6.7 Helix BWOs

As was indicated in Section 10.6.5, backward-wave oscillations serve a useful purpose in voltage-tuned signal generators. Helix-type backward-wave oscillators (BWOs) are rapidly tunable (in less than 1 μs) over a bandwidth up to an octave. Devices operating at frequencies up to 50 GHz are available.

Shown in Fig. 10.56 are the essential elements of a BWO. Most significant, perhaps, is the electron-gun assembly with a ring-shaped cathode for generating a hollow electron beam. The use of a hollow electron beam traveling close to the helix structure enhances the backward-wave interaction described in Section 10.6.5.

Power outputs are typically tens of milliwatts as indicated by the power curve for a Varian 33-to-50 GHz BWO shown in Fig. 10.57. The corresponding voltage tuning characteristic and helix modulation coefficient are given in Fig. 10.58.

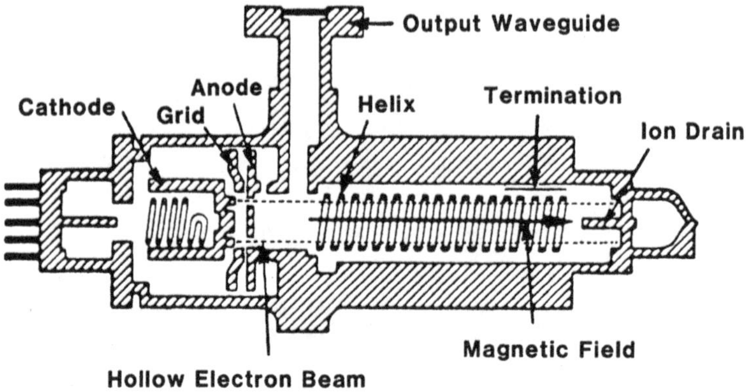

Figure 10.56 Helix-Type Backward Wave Oscillator

(Courtesy of Varian Associates)

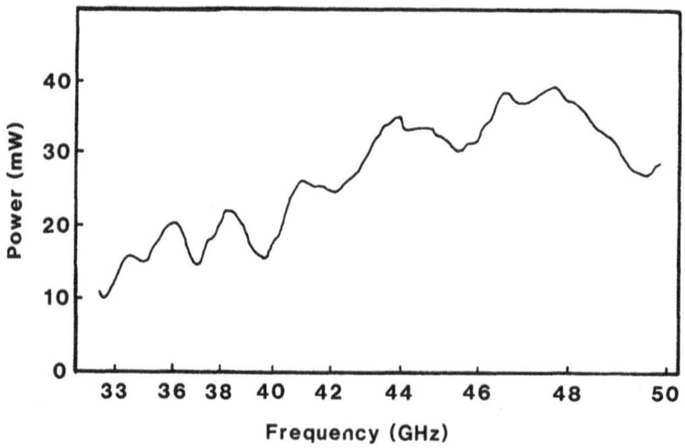

Figure 10.57 Power Output for Varian Helix BWO

(Courtesy of Varian Associates)

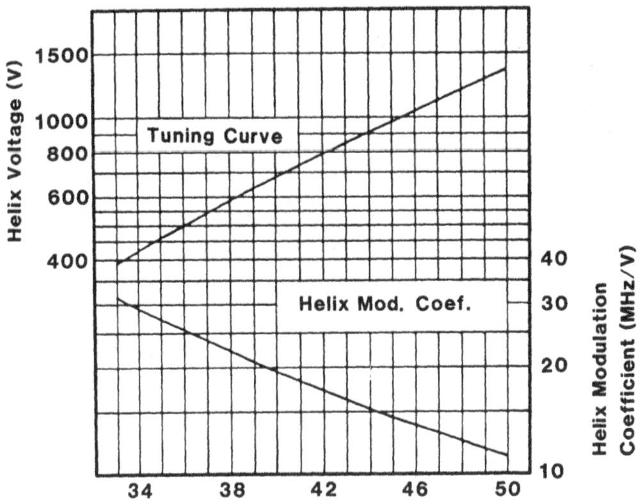

Figure 10.58 BWO Voltage Tuning Characteristic and Helix Modulation Coefficient

(Courtesy of Varian Associates)

10.6.8 Dual-Mode Operation

The purpose of a dual-mode TWT is to provide two different RF power levels (separated by as much as 10 dB) efficiently in a single device. This is highly desirable in airborne electronic countermeasures systems, for example, where volume and weight are of extreme importance. In these systems, the use of two tubes, each of which is designed for operation at one of the power levels, often results in excessive system weight and volume.

There are basically three ways in which a single tube can be efficiently operated at two power levels:

1. Beam current and voltage are held constant and RF drive level is changed. Efficiency is obtained by using a multistage depressed collector to recover beam power. This technique has only recently been demonstrated and is discussed further in Chapter 12.
2. A complex tube, basically consisting of two tubes in one envelope, is used. A drawing of a device of this type is shown in Fig. 10.59. The tube consists of a conventional TWT in series with a very short, low gain (10 dB), nearly lossless device. In the low-power mode, the beam

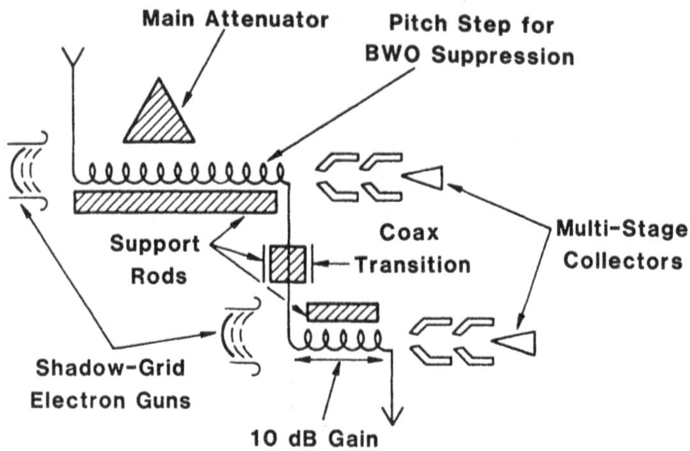

Figure 10.59 Dual-Beam, Dual-Mode TWT

(Courtesy of Varian Associates)

in the 10 dB device (sometimes referred to as a transparent tube) is turned off and the signal from the conventional TWT passes through it with no amplification and little loss. In the high-power mode, the transparent tube is turned on to provide an additional 10 dB of gain. This device has the advantage of providing a full 10 dB of pulse-up capability (from the CW to the pulse mode) and the disadvantage of being more complex and heavier than the true single-tube device.

3. Beam current is varied with a gridded electron gun. (Beam voltage is not changed because of the weight and complexity of a modulator and because the beam velocity must remain near synchronism with the circuit phase velocity).

The third technique (changing beam current) for achieving dual-mode operation is very difficult to implement when the power separation between modes must be more than a few dB. One difficultly encountered is that of focusing both the low-power and the high-power beams with the same magnet structure (usually PPM because of the requirement for light weight).

Assuming constant efficiency, beam power and beam current must vary in direct proportion to RF power. If beam diameter is held constant, then beam-charge density varies directly with beam current. Finally, the radial electric field in the beam is proportional to charge density and, therefore, to power level. If the PPM structure is designed for the low-power mode, the magnetic field will be inadequate to offset the electric field in the high-power mode. Of course, the opposite situation exists if the PPM structure is designed for the high-power mode. This problem may be alleviated by changing the beam diameter when the beam current is changed. An electron gun for changing beam diameter with beam current is shown in Fig. 6.27.

The other difficultly encountered when beam current is changed is that beam impedance changes and this changes Pierce's gain parameter, C. As a result, the RF characteristics of the tube change with beam current, making it extremely difficult to obtain the desired performance at two power levels that differ by a large amount.

10.7 COUPLED CAVITY TWTs

In coupled cavity TWTs [10.12], several tens of klystron-like cavities are used for the slow-wave structure. Except for sever regions, these cavities are electromagnetically coupled together so that an RF signal can propagate along the structure. Coupling techniques take on widely differing forms and are a primary subject for discussion in this chapter. If an electron beam is passed through the structure and if it has a velocity near the phase velocity, then amplification can occur.

With appropriate choices of cavity dimensions and cavity coupling techniques, electron-beam velocity can be high, and so operation at high peak-power levels is possible. In addition, coupled-cavity structures are mostly metallic (copper and iron) and have geometries such that thermal resistances are very low. As a result, high average powers are possible.

Many different coupling techniques are available for transmitting a signal from one cavity to the next. These lead to one or the other of the two phase-velocity characteristics illustrated in Fig. 10.60. In Fig. 10.60(a), the field in cavity no. 2 is the same as that in cavity no. 1, except for a phase delay of roughly 90°. Thus, an electron beam leaving cavity no. 1 at ωt must arrive at cavity no. 2 about 90° later at ωt_2 to remain in phase with the signal. The ω-β diagram for this type of coupling is also shown in Fig. 10.60(a). Circuits with this type of coupling are referred to as fundamental forward-wave circuits.

The other phase-velocity characteristic frequently encountered in coupled-cavity circuits is shown in Fig. 10.60(b). The field in cavity no. 2 is 180° out of phase with that in cavity no. 1 plus an additional delay of approximately 90°. An electron beam leaving cavity no. 1 at ωt_1 must arrive at cavity no. 2 about 270° later at ωt_2 to remain in phase with the signal. Circuits of this type are fundamental backward-wave circuits. The reason for this is readily apparent from Fig. 10.60(b). The ω-β diagram shows that the group velocity is negative when the electron transit angle is 90°.

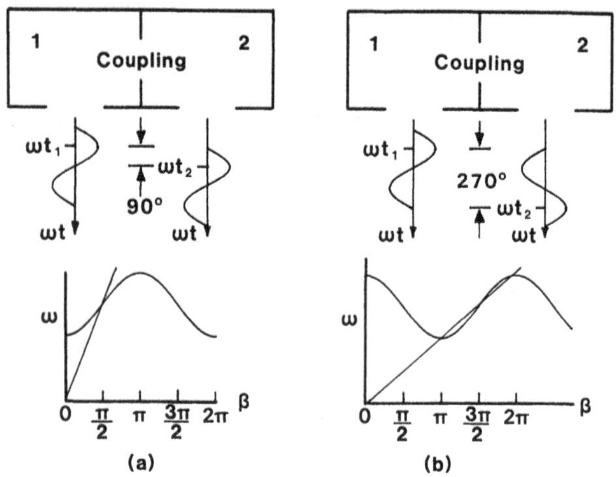

Figure 10.60 Fundamental Forward-Wave (a) and Fundamental Backward-Wave (b) Coupling

10.7.1 Fundamental Forward-Wave Operation

The operation of coupled-cavity TWTs with fundamental forward-wave coupling may be described with the aid of Fig. 10.61. In this example, the signal is coupled from cavity to cavity with inductive coupling loops. The magnetic field in one cavity induces a current in the coupling loop. The current flows through the aperture between cavities and induces a magnetic field in the adjacent cavity. Except for a phase delay of roughly 90° associated with the flow of current from one cavity to the next, the magnetic fields in adjacent cavities are in phase.

The operation of the first cavity is identical to that described at the beginning of the klystron chapter. The voltage across the first gap modulates the electron-beam velocity so that an ac component of current develops as the beam moves toward the second cavity. The beam induces a signal into the second cavity and the voltage across the second gap, V_2, results from the combination of the induced signal with the signal fed into the second cavity from the first cavity by the coupling loops.

The voltage, V_2, enhances the velocity modulation of the beam, thereby increasing the ac current in the beam. This process is repeated at subsequent gaps with each gap voltage being the combination of the signal induced by the beam and the sum of all the components fed forward from previous gaps.

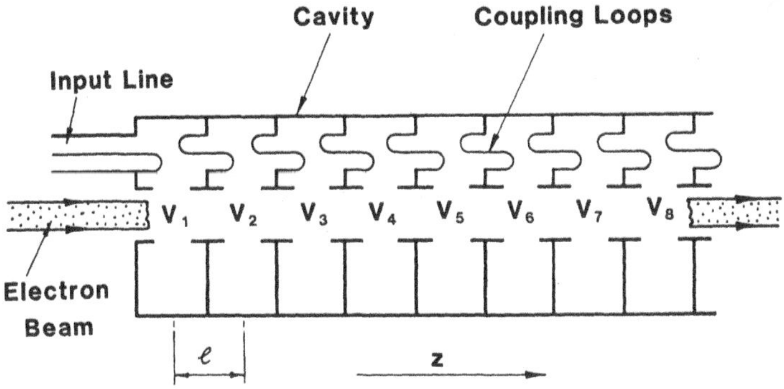

Figure 10.61 Circuit for Illustrating Fundamental Forward-Wave Operation

(From *Power Travelling Wave Tubes* by J.F. Gittins, Published 1965 by American Elsevier, Inc.)

The ac beam current at any point is the sum of all the currents produced by previous gaps.

The vector-diagram approach used by Gittins is helpful in illustrating how circuit voltages combine with beam-induced voltages to produce signal growth. For example, Fig. 10.62 shows the buildup of circuit voltage for the synchronous case with the phase shift from gap to gap being 90°. The quantity g contains Pierce's gain parameter and is the ratio of the magnitude of the voltage induced in a gap to that fed forward along the circuit from the previous gap.

The relative phases of the voltages shown in Fig. 10.62 are as they would appear to an observer traveling at the same velocity as the beam. Waves traveling at the beam velocity are shown by stationary vectors. If a wave is traveling faster than the beam, the vector rotates in a counterclockwise direction, and if a wave is traveling slower than the beam the vector rotation is clockwise. Thus, Fig. 10.62 shows the circuit wave traveling more slowly than the beam, as is predicted by Pierce's small-signal theory.

It will be noted in Fig. 10.62 that the voltage resulting from the induced beam current is initially 90° out of phase with the cavity voltage. The reason for this is apparent from the Applegate diagram in Chapter 9, which showed the formation of electron bunches at alternate zeros of the modulating voltage. Thus, the ac current in the beam lags the modulating voltage by 90° (plus the electron transit angle).

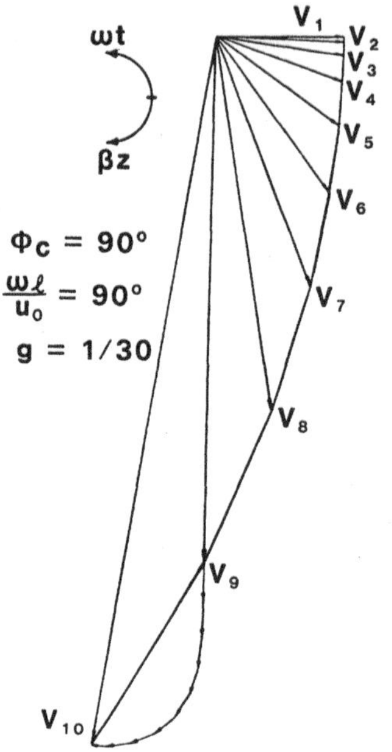

Figure 10.62 Vectors Showing Buildup of Circuit Voltage when Beam and Natural Circuit Wave Velocities are Equal (Synchronous Case)

(From *Power Travelling Wave Tubes* by J.F. Gittins, Published 1965 by American Elsevier, Inc.)

10.7.2 Fundamental Forward-Wave Circuits

The centipede structure shown in Fig. 10.63 is an implementation of the coupling-loop scheme that was illustrated in Fig. 10.61 for achieving fundamental forward-wave operation. As many coupling loops as possible are used to maximize coupling and, therefore, operating bandwidth. As shown in Fig. 10.63 (b), the height of each cavity must be adequate to accommodate the large coupling loops required to maximize coupling. Since the electron-beam transit angle from cavity to cavity is only about 90°, the electron beam velocity must be high, so the centipede structure is suitable for high-power operation.

The clover-leaf structure shown in Fig. 10.64 is another commonly used fundamental forward forward-wave circuit. The four-lobed cavity distorts the magnetic field lines into a clover-leaf pattern. Radial components of the field are coupled from cavity to cavity by radial slots. By rotating each cavity

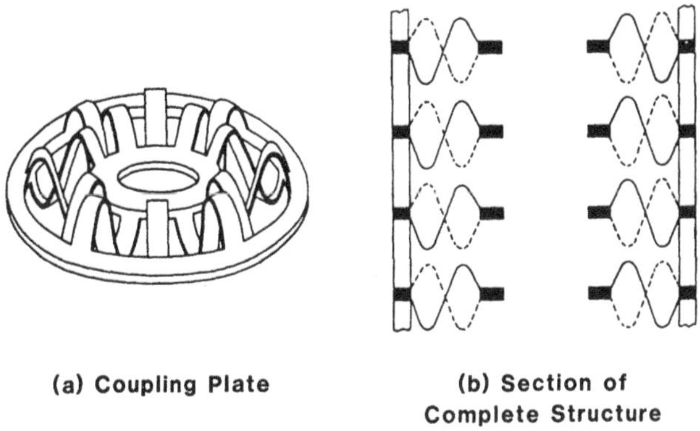

Figure 10.63 Centipede Structure

(From *Power Travelling Wave Tubes* by J.F. Gittins, Published 1965 by American Elsevier, Inc.)

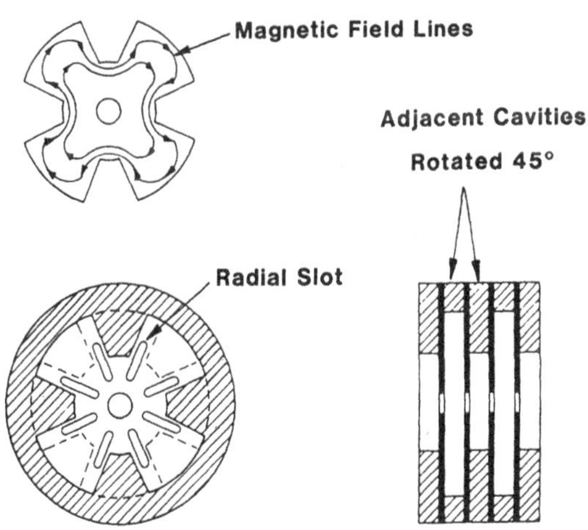

Figure 10.64 Clover-Leaf Structure

(From *Power Travelling Wave Tubes* by J.F. Gittins, Published 1965 by American Elsevier, Inc.)

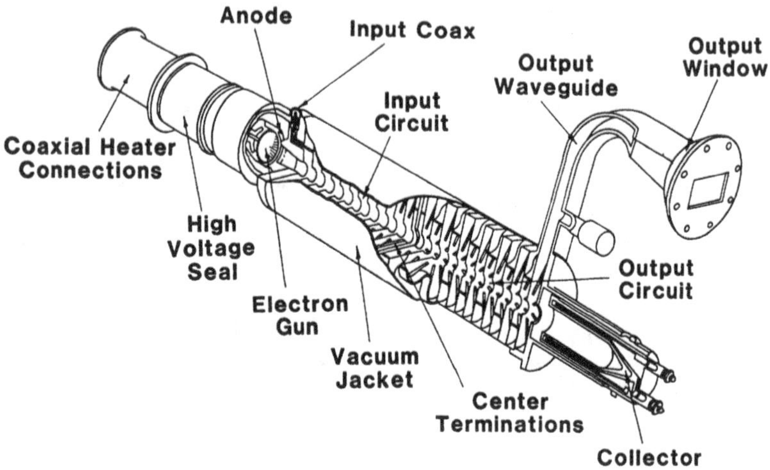

Figure 10.65 Example of Use of Clover-Leaf Structure in Varian VA-125 High-Power TWT

(From *Power Travelling Wave Tubes* by J.F. Gittins, Published 1965 by American Elsevier, Inc.)

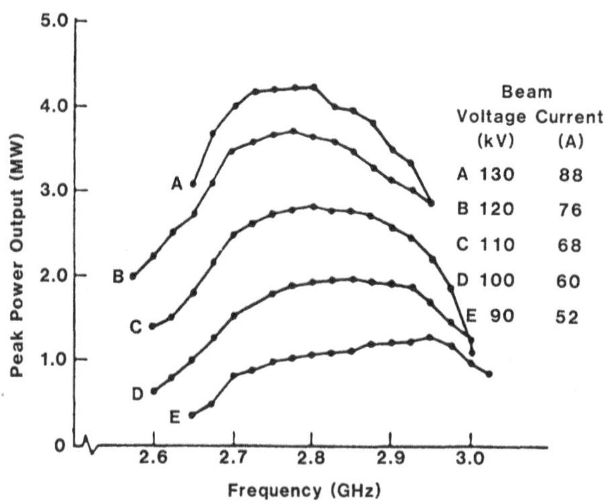

Figure 10.66 Power Output for VA-125A TWT

(From Armand Straprans, *et al., Proc. IEEE*, March 1973.)

45° with respect to adjacent cavities, the fields in adjacent cavities are in phase except for a delay of approximately 90° associated with coupling the signal from cavity to cavity.

The clover-leaf structure is widely used in high-power TWTs. An example, shown in Fig. 10.65, is the Varian VA-125 for which the power output is given in Fig. 10.66. Note that output powers in excess of 4 MW are obtained.

The clover-leaf structure is also used for the output circuit in twystrons (described in the following chapter) at power levels as high as 10 MW. Bandwidth is limited to about 10% because of limitations imposed by the coupling between cavities.

10.7.3 Fundamental Backward-Wave Operation

If the coupling loops of Fig. 10.61 are reversed as shown in Fig. 10.67, then fundamental backward-wave coupling occurs. The magnetic field in a cavity induces a current in the coupling loops to the cavity on the right. That current then induces a field in the adjacent cavity, which is 180° out of phase with the cavity on the left, plus an additional phase shift of roughly 90° associated with the flow of current from cavity to cavity. The total phase from cavity to cavity is, therefore, about 270°.

If the beam velocity is high enough so that the electron transit angle is about 90°, then backward-wave amplification is possible. For forward-wave amplification, the beam velocity must be slow enough so that the electron transit angle from cavity to cavity is about 270°.

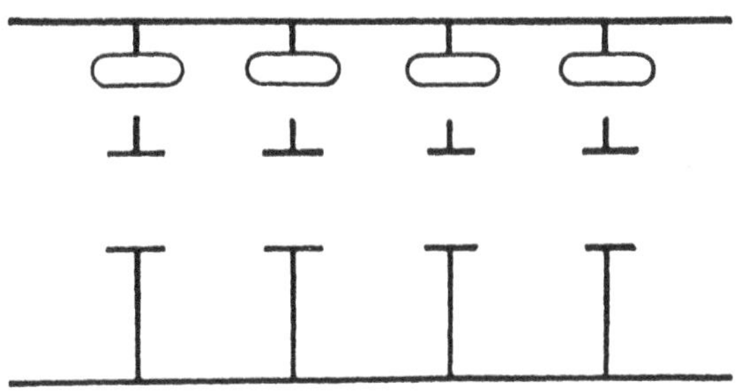

Figure 10.67 Configuration of Coupling Loops for Fundamental Backward-Wave Operation

(From *Power Travelling Wave Tubes* by J.F. Gittins, Published 1965 by American Elsevier, Inc.)

10.7.4 Fundamental Backward-Wave Circuits

Two commonly used fundamental backward-wave circuits are illustrated in Fig. 10.68. In each of these, apertures are used to couple the fields from one cavity to the next. The staggered slot structure in Fig. 10.68 (a) can be regarded as a folded waveguide. From this point of view, it is clear that as a wave propagates upward in one cavity, through the slot and downward in the adjacent cavity, the direction of the electric field reverses. Thus, there is basically a 180° phase shift from cavity to cavity plus an additional phase delay associated with the time that it takes the wave to propagate from one gap to the next.

Alternatively, in a manner similar to that described in connection with Fig. 10.67, the magnetic field in any of the cavities in Fig. 10.68 induces current flow in the cavity walls. At the edges of the coupling slots, these currents induce magnetic fields in the adjacent cavities. Because the field establishing the currents is 180° out of phase with the field established by the currents, there is basically a 180° phase shift from cavity to cavity. In addition, there is a phase shift associated with the inductive delay in coupling the signal from cavity to cavity.

In general, because the electron transit angle from cavity to cavity is on the order of 270° (three times that for fundamental forward-wave circuits), the beam velocity and power for fundamental backward-wave circuits are lower than those for fundamental forward-wave circuits. Output powers from one to several hundred kilowatts are typical.

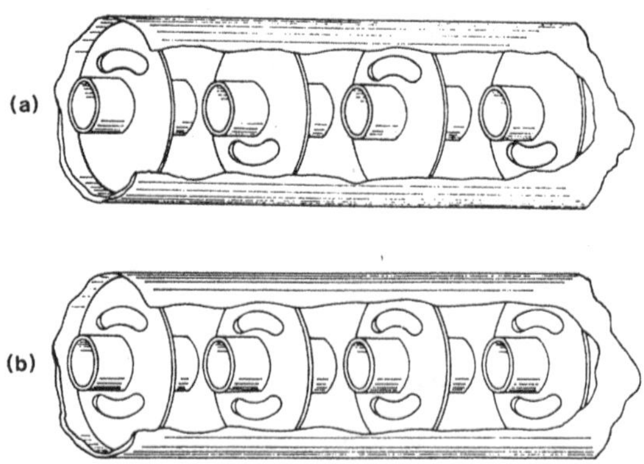

Figure 10.68 Staggered Slot (a) and Aligned Slot (b) Fundamental Backward-Wave Circuits

(From *Power Travelling Wave Tubes* by J.F. Gittins, Published 1965 by American Elsevier, Inc.)

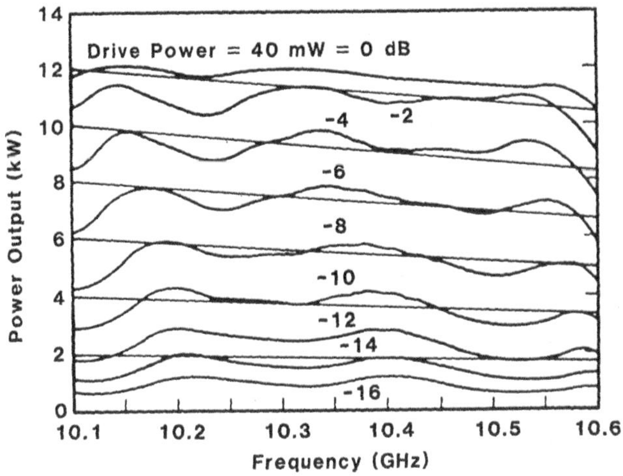

Figure 10.69 Example of Power Capability of CW Coupled-Cavity TWT

(From Armand Staprans, *et al., Proc. IEEE,* March 1973.)

A specific example of the capability of a device with a fundamental backward-wave circuit is shown in Fig. 10.69. The tube utilizes confined-flow solenoid focusing and a depressed collector. The overall efficiency for this tube is in the 20 to 30% range.

Through the use of velocity resynchronization techniques (to maintain synchronism at the output end of the tube), overall efficiency is maximized. This is usually accomplished by reducing the cavity spacing near the output end of the tube. As an alternative, the beam velocity can be increased as has been demonstrated with the technique illustrated in Fig. 10.70. The resulting interaction efficiency was over 50% as is indicated by Fig. 10.71. The combination of this voltage-jump technique with modern multistage depressed collector technology should produce extremely high overall efficiencies. The obvious difficulty in implementation is the insulation necessary between sections of the body of the tube and between the tube and the solenoid and RF circuits.

Coupled-cavity structures are ideally suited for use in TWTs that are focused with periodic permanent magnets. As shown in Fig. 10.72, the iron pole pieces can be made integral with the cavity structure. The pole pieces are copper plated, and copper spacers are used to form cavities of the required height. Liquid cooling channels may also be included in the coupled-cavity structure. In some cases, the coolant is channeled around the drift tubes between cavities to carry away the heat resulting from interception of the electron beam.

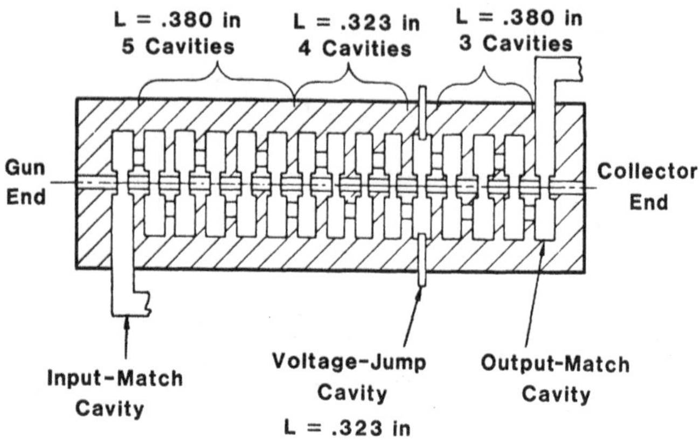

Figure 10.70 High-Efficiency Voltage-Jump TWT

(From John T. Mendel, *Proc. IEEE,* March 1973.)

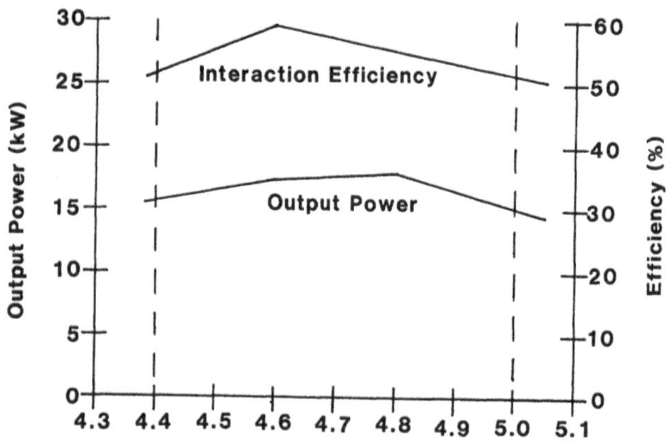

Figure 10.71 Efficiency and Output Power for Voltage-Jump TWT

(From John T. Mendel, *Proc. IEEE,* March 1973.)

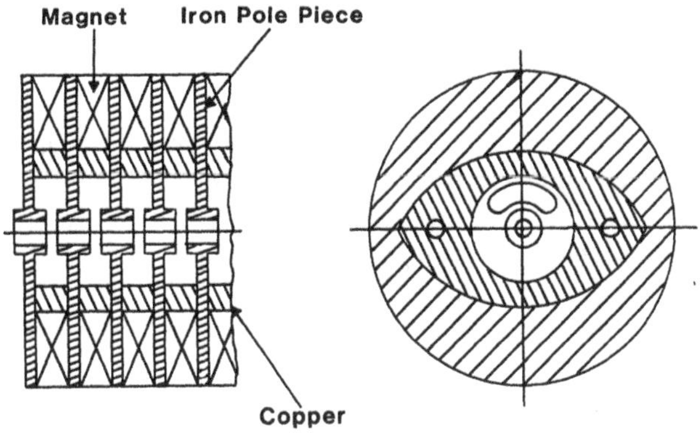

Figure 10.72 Coupled-Cavity Circuit with Integral Periodic-Permanent-Magnet Focusing

(From Hughes Aircraft Co., *TWT and TWTA Handbook.*)

When periodic-permanent magnets (PPM) are used, the electron beam is not as well focused as it is when a solenoid is used. As a result, the percentage of the beam current intercepted by the circuit is relatively high, and so the average power capability is relatively low. The reduction in average power capability of PPM-focused tubes compared with solenoid-focused tubes is typically a factor of four or more.

10.7.5 Analytical Considerations

In Pierce's small-signal theory, a continuous interaction of the beam with the circuit is assumed. In coupled-cavity structures, the interaction is periodic rather than continuous. In order to describe adequately the various effects, such as backward-wave interactions and band-edge oscillations, it is necessary to develop accurate equivalent circuits for the coupled-cavity structures [10.13].

As an example, the development of the equivalent circuit for the staggered slot circuit is illustrated in Fig. 10.73. In the first step of the circuit development, the cavity and coupling slots (a) are represented by inductances and capacitances (b). Losses can be included through the use of resistors, but they are not considered here. Then, the circuits for individual cavities are coupled together (c) and converted into a relatively simple transmission line (d), (e), and (f). Resonant circuits associated with the cavity and with the coupling slots are carefully preserved in the development of the equivalent circuits, and so ω-β and impedance characteristics in both the cavity and the slot can be derived.

The ω-β diagram in Fig. 10.74 and the impedance characteristic in Fig. 10.75 were derived from the circuit model in Fig. 10.73. In these figures, ω_c is the resonant frequency of the cavity and ω_s is the resonant frequency of the coupling slots. In determining the sizes of the inductances, the circulating current in each cavity is divided into three parts. Two of these link the cavity to preceding and following cavities and the third represents current which does not couple. The fraction of the current taking part in the coupling is k. The value of k_c is related to the ratio of the inductance of the coupling slot to the inductance of the cavity.

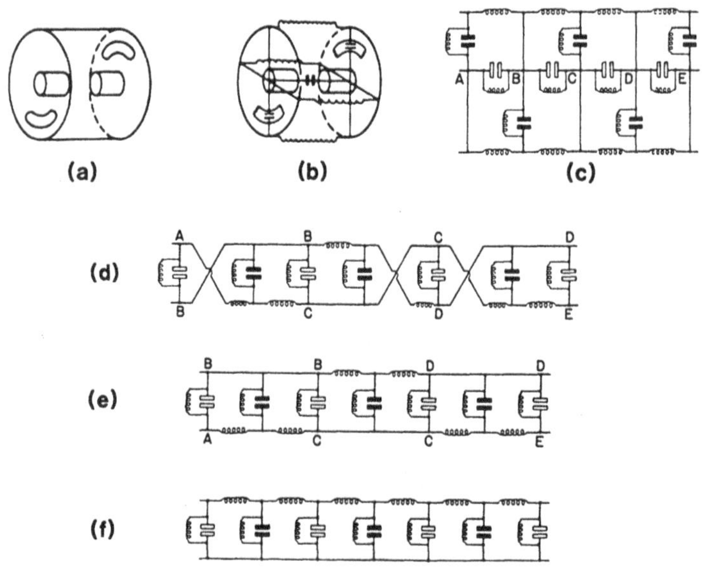

Figure 10.73 Development of Equivalent Circuit Model for the Staggered-Slot Coupled-Cavity Circuit

(From *Power Travelling Wave Tubes* by J.F. Gittins, Published 1965 by American Elsevier, Inc.)

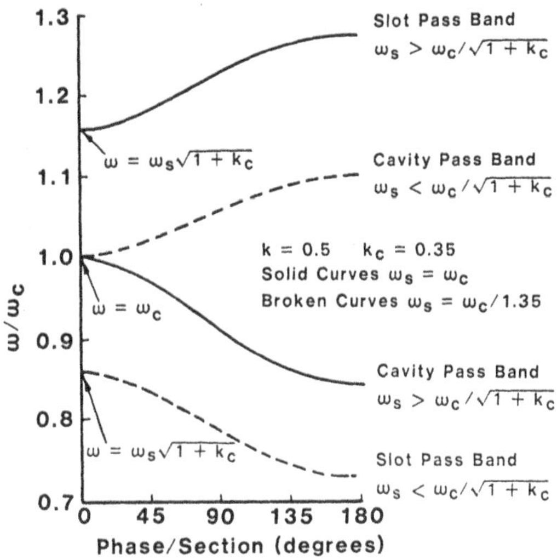

Figure 10.74 ω-β Diagram for Circuit Model of Staggered-Slot Coupled-Cavity Structure

(From *Power Travelling Wave Tubes* by J.F. Gittins, Published 1965 by American Elsevier, Inc.)

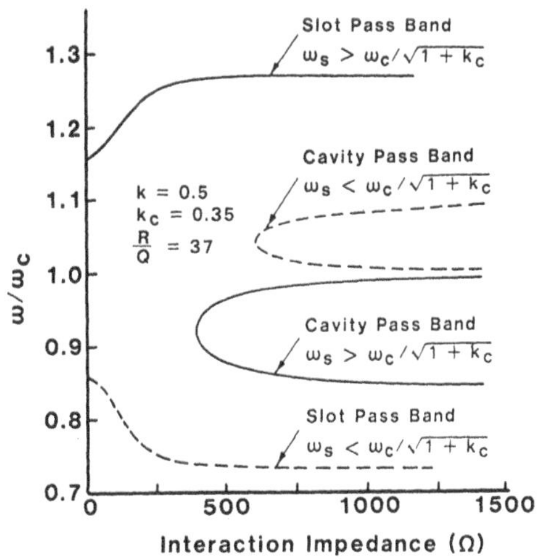

Figure 10.75 Interaction Impedance Derived from Circuit Model for Staggered-Slot, Coupled-Cavity Structure

(From *Power Travelling Wave Tubes* by J.F. Gittins, Published 1965 by American Elsevier, Inc.)

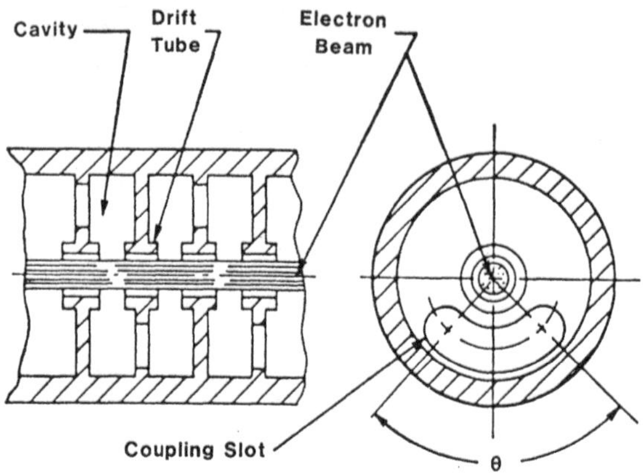

Figure 10.76 Coupling Slot Between Cavities

(From Hughes Aircraft Co., *TWT and TWTA Handbook.*)

The resonant frequency of the coupling slot, ω_s, is dependent on the size of the coupling aperture between cavities. Referring to Fig. 10.76, when the kidney angle θ is small, the slot resonant frequency is high, and the slot passband is well above the cavity passband as indicated in Fig. 10.74. As the angle θ is increased and the slot resonant frequency decreased, it is possible for the cavity passband to be higher in frequency than the slot passband.

Normally, the cavity passband is used for interaction with the electron beam, and such was the case for the results given in the previous section. It is possible for the slot passband to be used, and tubes designed to operate in this mode have existed for many years [10.14]. An important property of the slot passband is that the interaction impedance at the upper cut-off frequency is zero (see Fig. 10.75). This results from the vanishing axial electric field component on the axis as is indicated in Fig. 10.77. As a result, no special oscillation-suppression techniques are required at this frequency.

Attempts have been made [10.15] to coalesce the passband and slot band in an effort to achieve octave-bandwidth operation. While the ω-β diagram indicates near synchronism of the circuit wave and the beam over the desired bandwidth, two difficulties are in-band backward-wave instability and a large gain variation with frequency.

Once the circuit model of Fig. 10.73 with appropriate component values has been established for a coupled-cavity structure, the analysis of the interaction with the electron beam can be carried out. The steps of the

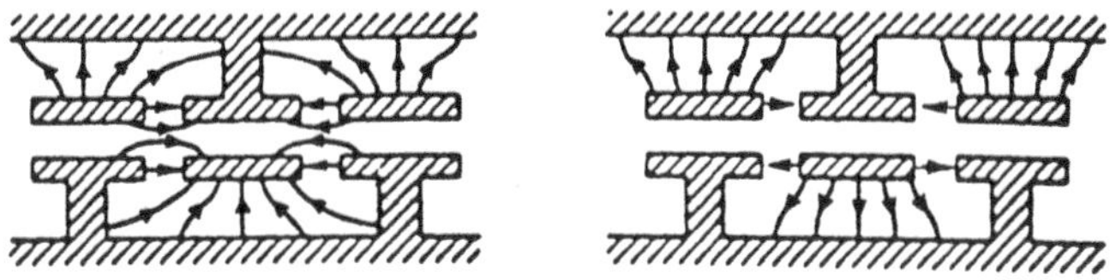

π Phase Shift   0 (2π) Phase Shift

Figure 10.77 Electric Field Patterns for the Slot-Band Cut-Off Frequencies

(From J.R. Frey and I. Tammaru, *Technical Digest,* 1981 IEDM.)

analysis are conceptually the same as those used by Gittins in the vector-diagram description of coupled-cavity TWT operation (see Fig. 10.62, for example). That is, the electron beam induces current into the appropriate elements (cavity gaps) of the equivalent circuit, and the electric field of the circuit excites current and velocity modulation of the beam. The shapes of the electric fields of the circuit and the space-charge fields in the beam must be accurately determined so that the analysis can properly predict tube performance.

10.7.6 Terminations and Transitions

The need for severs in TWTs is described in Section 10.4. Terminations are required at the severs to minimize reflections. The RF power that a termination is required to dissipate can be appreciable in coupled-cavity TWTs. As a result, the termination is normally a lossy ceramic material which is capable of high-temperature operation and is appropriately bonded to the surrounding metallic structures to achieve the required heat dissipation capability. Examples of terminations for a clover-leaf structure and for a staggered-slot structure are shown in Figs. 10.78 and 10.79, respectively.

Two types of transitions are shown for the high-power coupled-cavity TWT in Fig. 10.65. At the input where the power level is low, a coaxial line is used to couple to a cavity and at the output, a transition from the cavity to a waveguide is used. The use of a coaxial line and a coupling loop is common at low power levels and has been described in the Chapter 9. A coaxial line and coupling loop for a coupled-cavity structure are illustrated in Fig. 10.80. Also shown is the side view of a waveguide transition. In coupling a cavity to a waveguide, it is often necessary to use inductive impedance-matching elements as shown in Figs. 10.81 and 10.82. These elements were described in Chapter 2 and serve to offset the capacitive reactance of the cavity gap.

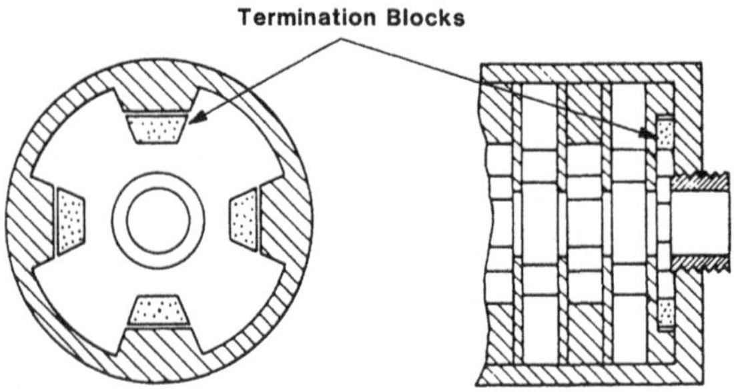

Figure 10.78 Termination for a Clover-Leaf Structure

(From *Power Travelling Wave Tubes* by J.F. Gittins, Published 1965 by American Elsevier, Inc.)

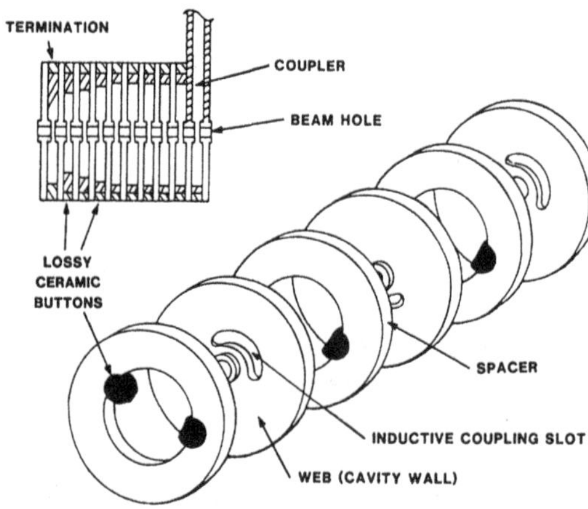

Figure 10.79 Termination and Lossy Ceramic Buttons (for Producing Tapered Loss Pattern) in a Staggered-Slot Structure.

(From Hughes Aircraft Co., *TWT and TWTA Handbook.*)

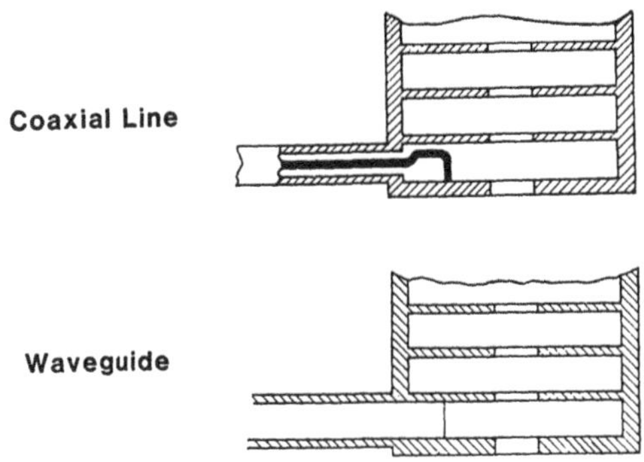

Figure 10.80 Coaxial-Line and Waveguide Transition to Coupled-Cavity Structures

(From *Power Travelling Wave Tubes* by J.F. Gittins, Published 1965 by American Elsevier, Inc.)

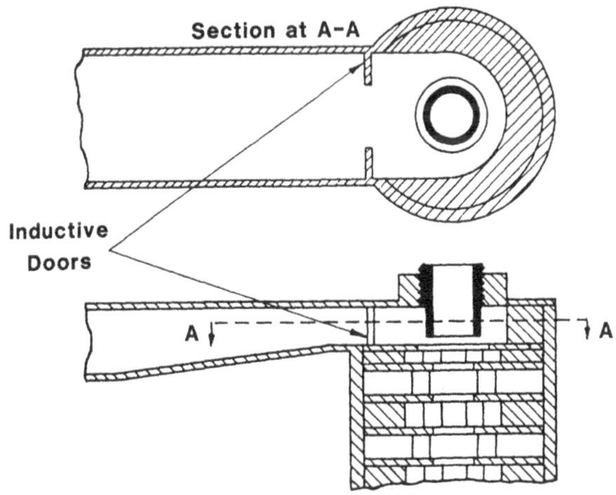

Figure 10.81 Transition from Waveguide to a Coupled-Cavity Structure

(From *Power Travelling Wave Tubes* by J.F. Gittins, Published 1965 by American Elsevier, Inc.)

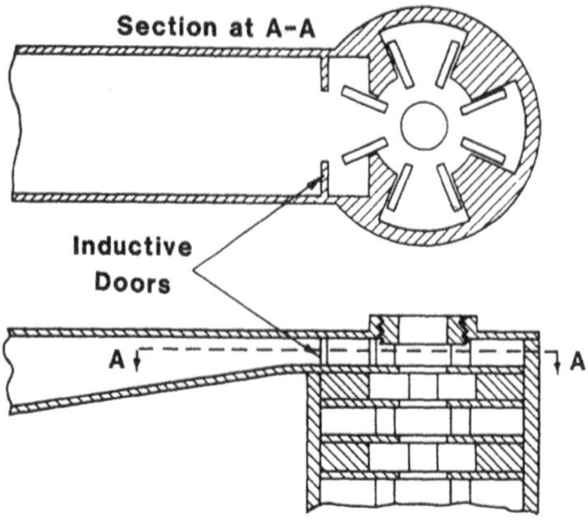

Figure 10.82 Transition from Waveguide to Clover-Leaf Structure

(From *Power Travelling Wave Tubes* by J.F. Gittins, Published 1965 by American Elsevier, Inc.)

REFERENCES

[10.1] R. Kompfner, "The invention of the traveling-wave tube," from lecture series at the University of California at Berkely, published by San Francisco Press, 1963.

[10.2] Recommended for further reading:

J.R. Pierce, *Traveling Wave Tubes.* Princeton, N.J.: Van Nostrand, 1950.

M. Chodorow and C. Susskind, *Fundamentals of Microwave Electronics.* New York: McGraw-Hill, 1964.

J.F. Gittins, *Power Traveling-Wave Tubes.* New York: American Elsevier, 1965.

Reich, Ordung, Krauss, and Skolnik, *Microwave Theory and Techniques.* Princeton, NJ: Van Nostrand, 1953.

A.S. Gilmour, Jr., "Basic principles and some aspects of the electromagnetic compatibility of microwave tubes," *Inter-System Electromagnetic Compatibility*, New York: Holt, Reinhart and Winston, 1972.

[10.3] P. Hess, "Traveling-wave tube large signal theory with application to amplifiers having dc voltage tapered with distance," Ph.D. dissertation, University of California at Berkeley, 1960.

[10.4] J.L. Putz and M.J. Cascone, "Effective use of dispersion shaping in broadband helix TWT Circuits," *Technical Digest,* 1978 International Electron Devices Meeting, pp. 422-424.

[10.5] J.T. Mendel, "Helix and coupled-cavity traveling-wave tubes," *Proc. IEEE,* vol. 61, March 1973, pp. 280-298.

[10.6] A.S. Gilmour, Jr., M.R. Gillette and J.T. Chen, "Theoretical and experimental TWT helix loss determination," *IEEE Trans. Electron Devices,* vol. ED-26, Oct. 1979, pp. 1581-1588.

[10.7] Y. Hiramatsu, "Study of block-supported helix circuits for high-power TWTs," *Technical Digest,* 1979 International Electron Devices Meeting, pp. 416-417.

[10.8] T. Mosser, W. Pinger, and D. Zavadil, "High-power brazed helix fabrication," 1980 Microwave Power Tube Conference.

[10.9] O. Sauseng, A.E. Manoly, and A. Hall, "Thermal properties and power capability of helix structures for millimeter waves," *Technical Digest,* 1978 International Electron Devices Meeting, pp. 534-537.

[10.10] Alan R. Jung, "10 kW and up, from a helix TWT?" *Technical Digest 1978 International Electron Devices Meeting*, pp. 530-553.

[10.11] Birdsall and Everhart, *IRE Trans. on Electron Devices,* vol. ED-3, 1956, p. 190.

[10.12] Recommended for further reading:

John T. Mendel, "Helix and Coupled-Cavity Traveling-Wave Tubes," *Proc. IEEE,* vol. 61, no. 3, March 1973, pp. 280-298.

Armand Straprans, Earl William McCune, and Jack A. Ruetz, "High-power linear-beam tubes," *Proc. IEEE,* vol. 61, no. 3, March 1973, pp. 299-327.

[10.13] H.J. Curnow, "A general equivalent circuit for coupled-cavity slow-wave structures," *IEEE Trans. Microwave Theory Tech.*, vol. MTT-13, Sept. 1965, pp. 671-675.

[10.14] J.R. Frey and I. Tammaru, "A coupled-cavity TWT operating in the inverted slot mode," *Technical Digest,* 1981 International Electron Devices Meeting, pp. 504-506.

[10.15] Arthur Karp and W. Revis Ayers, "Design concepts for an octave-bandwidth coupled-cavity TWT," *Technical Digest,* 1978 International Electron Devices Meeting, pp. 546-549.

Chapter 11
Hybrid Tubes

In hybrid tubes, klystron and coupled-cavity TWT technologies are combined to achieve enhanced bandwidth, power capability, or operation at high frequencies [11.1].

Linear-beam amplifier tubes can be considered to consist of a driver section and an output section. The output section must have a bandwidth at least as wide as that of the driver section. In both the extended interaction klystron (EIK) and the twystron, the driver section consists of cavities that are stagger-tuned in a manner somewhat like that in an intermediate-frequency (IF) amplifier. The output section is a coupled-cavity structure with an appreciable bandwidth.

Conceptually, the primary difference between the EIK and the twystron is as was indicated in Figs. 9.4 and 9.5. That is, in the EIK, each element of the driver section may consist of multiple cavities, whereas in the twystron more conventional klystron cavities are used.

11.1 EXTENDED INTERACTION KLYSTRON (EIK)

An extended interaction structure consists of heavily coupled cavities as shown in Fig. 11.1. In one mode of operation, the fields in all cavities are in phase and the electron transit angle between gaps is 2π radians. The other common mode of operation is similar to the fundamental backward-wave mode of coupled-cavity TWT operation. That is, the fields in adjacent cavities are approximately 180° out of phase.

In comparing an extended interaction resonator with a single-gap output cavity, the bandwidth is, of course, increased. Perhaps more important is the fact that the power-handling capability is increased. This is because the total output voltage, which must be applied across a single gap in a conventional klystron, is divided among the output cavities of the EIK. The high-power capability of the EIK was perhaps best illustrated by a 1 MW average-power X-band device that used a five-cavity resonator like that shown in Fig. 11.1. [11.2]

An example of the bandwidth that can be achieved in an EIK is in the VA 812C, the frequency response for which is shown in Fig. 11.2. The bandwidth for this tube is over 10%. The efficiency of nearly 40% is not as high as can be achieved in a narrowband tube because of the high perveance beam used (to improve beam-loading characteristics and bandwidth) and because of compromises made in output-section impedance and tuning of cavities in the driver section. Table 11.1 lists the operating characteristics for the VA-812C.

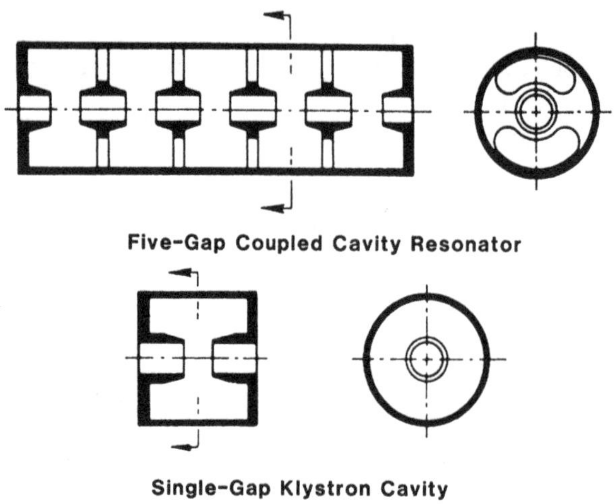

Figure 11.1 Comparison of a Five-Gap Extended Interaction Structure to a Single-Gap Klystron Cavity

(From Armand Staprans, *Proc. IEEE*, March 1973.)

TABLE 11.1
VA-812C Operating Characteristics
(From Armand Staprans, *Proc. IEEE*, March 1973)

Frequency	400 – 450 MHz
Peak Beam Voltage	145 kV
Perveance	2.5×10^{-6} A/V$^{3/2}$
Peak Power Output	8 MW
Average Power Output	30 kW
Pulse Length	6 μs
Efficiency	40%
Electronic Bandwidth	12%
Gain	30 dB

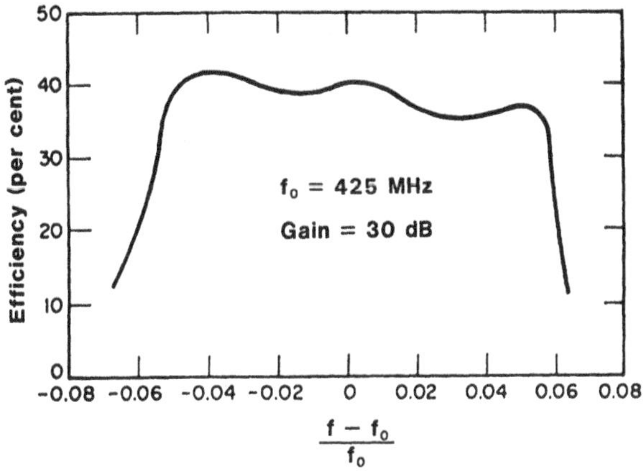

Figure 11.2 Frequency Response of Broadband Klystron Amplifier, VA-812C

(From Armand Staprans, *Proc. IEEE*, March 1973.)

11.2 EXTENDED INTERACTION OSCILLATOR

The extended interaction oscillator (EIO) is a single-cavity device as shown in Fig. 11.3. Within the cavity, there is a segmented drift tube. The cavity and drift tube form a resonator that functions very much like a TWT coupled-cavity structure in which the cavity-to-cavity coupling is extremely strong. Thus, an electron beam entering the resonator is velocity modulated and, if the resulting bunches pass the gaps in the drift tube in synchronism with the RF fields, then the signal level in the cavity is reinforced. At sufficiently high beam currents, oscillations are sustained and significant RF power levels can be generated.

As with most klystrons and TWTs, EIOs are designed for minimum interception of the electron beam on the RF structure. As a result, at high frequencies, these devices are capable of operation at higher power levels than reflex klystrons or magnetrons in which the beam current is collected on the RF structure.

As shown in Fig. 11.4, the output power of the EIO is a function of beam velocity and, therefore, beam voltage. In addition, electronic tuning also occurs when beam voltage is varied because of the change in the rate at which electron bunches pass the gaps in the segmented drift tube.

As indicated in Fig. 11.3, the anode of the electron gun in the EIO is electrically isolated from the body of the tube. As a result, beam current can be controlled independently from beam voltage. Thus, as indicated in

Fig. 11.5, amplitude modulation with constant frequency is possible if the beam voltage and anode voltage are properly controlled. Similarly, as shown in Fig. 11.6, frequency may be varied while power level is held constant by appropriate control of voltages.

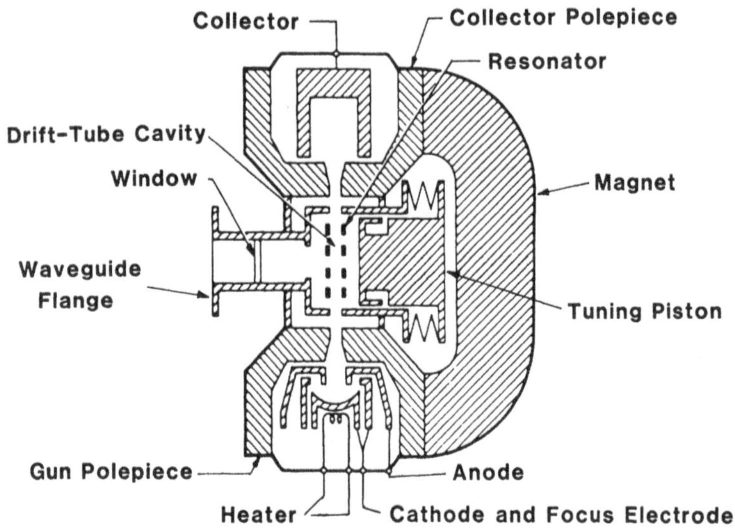

Figure 11.3 Tunable Extended Interaction Oscillator

(From *Microwave Tube Manual* by Varian Associates, Air Force Publication Number T.O. 00-25-251, October 1979.)

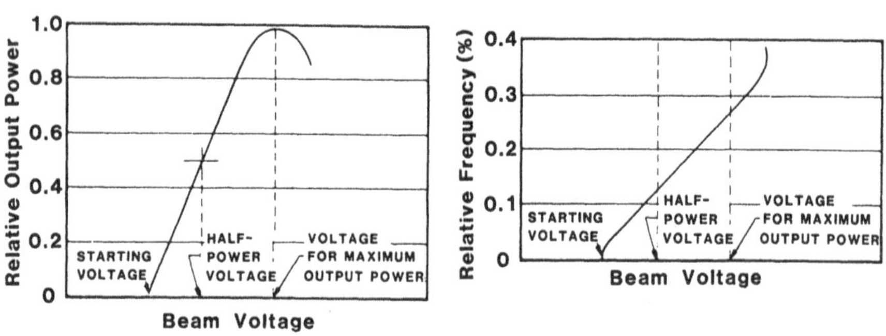

Figure 11.4 Output Power and Frequency Variation with Beam Power

(From *Microwave Tube Manual* by Varian Associates, Air Force Publication Number T.O. 00-25-251, October 1979.)

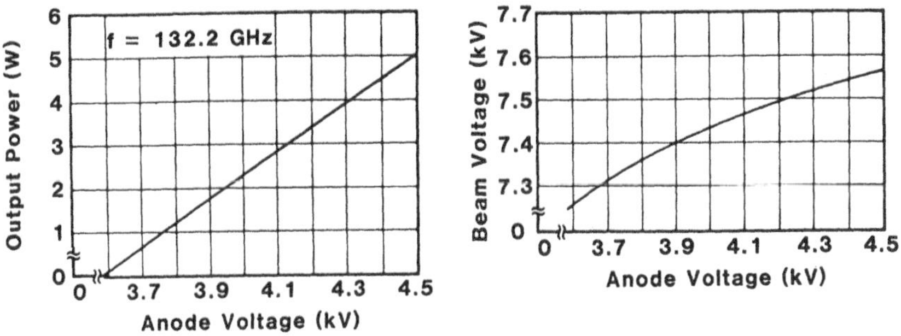

Figure 11.5 Amplitude Modulation with FM Suppressed

(From *Microwave Tube Manual* by Varian Associates, Air Force Publication Number T.O. 00-25-251, October 1979.)

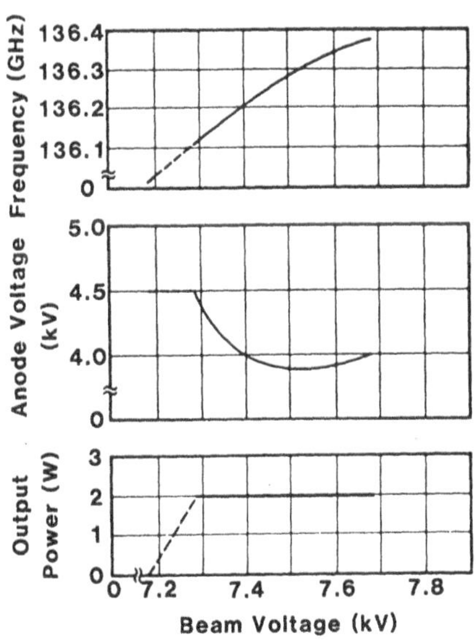

Figure 11.6 Frequency Modulation with AM Suppressed

(From *Microwave Tube Manual* by Varian Associates, Air Force Publication Number T.O. 00-25-251, October 1979.)

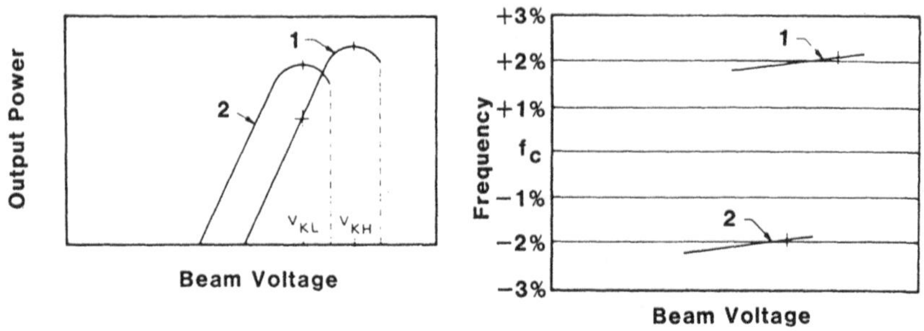

Figure 11.7 Output Power and Frequency *versus* Beam Voltage for Various Mechanical Tuner Settings

(From *Microwave Tube Manual* by Varian Associates, Air Force Publication Number T.O. 00-25-251, October 1979.)

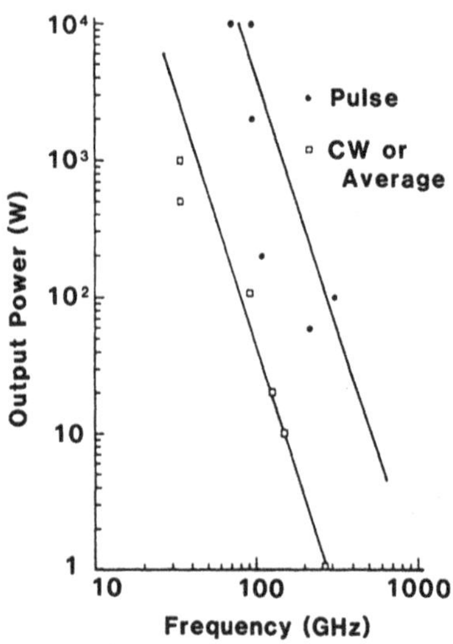

Figure 11.8 Achieved Power Outputs of CW and Pulsed Extended Interaction Oscillators

(From Varian of Canada Data Sheet Entitled "Extended Interaction Oscillator Selection Guide.")

In Fig. 11.3, a tuning piston is shown in the resonator structure. As indicated in Fig. 11.7, frequency may be changed several percent by mechanically positioning the tuner.

The range of power capabilities for extended interaction oscillators is shown in Fig. 11.8. EIOs are available for operation at frequencies up to 260 GHz. Pulsed EIOs are capable of power outputs of several kilowatts at 40 GHz varying down to 100 W at over 200 GHz.

11.3 TWYSTRONS

In a twystron, a klystron-like driver section is used in conjunction with (usually) a clover-leaf output section. With this combination, the low gain of the TWT section at band edges (for an example of typical effect on power output, see Fig. 10.66) is offset by the gain response of the stagger-tuned klystron driver section. The overall result, as Fig. 11.9 indicates, can be a very flat gain characteristic over a bandwidth of up to 15%. The resulting power output as a function of frequency is also very flat as shown by the example in Fig. 11.10.

Table 11.2 contains a listing of the characteristics of several twystrons. Peak powers up to 7 MW are listed and higher powered devices have been built. Average powers up to about 30 kW are listed and can be increased about an order of magnitude before circuit losses and interception present fundamental limitations.

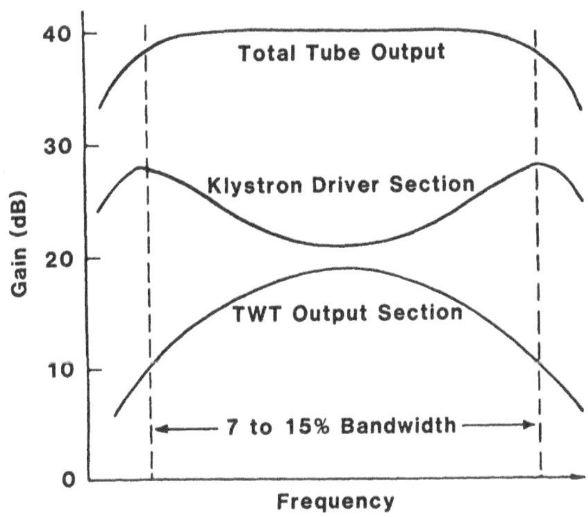

Figure 11.9 Gain of a Twystron Amplifier

(From *Microwave Tube Manual* by Varian Associates, Air Force Publication Number T.O. 00-25-251, October 1979.)

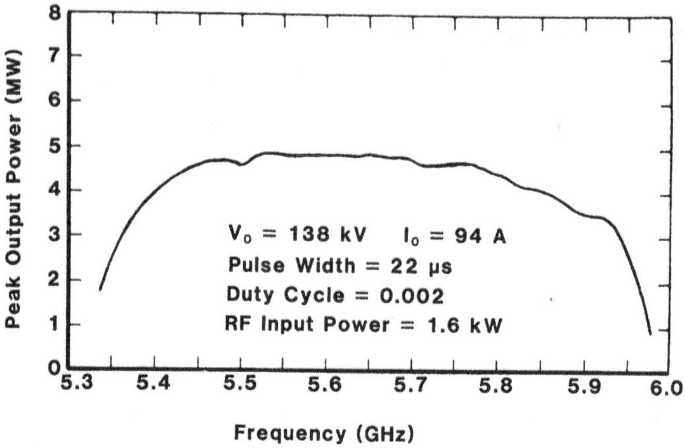

Figure 11.10 Power Output of Twystron Amplifier VA-146

(From *Microwave Tube Manual* by Varian Associates, Air Force Publication Number T.O. 00-25-251, October 1979.)

Efficiencies listed in Table 11.2 are about 30%. At midband, the efficiency is closer to 40% and recent results with multistage depressed collectors show that efficiencies near 50% or higher should be possible.

Finally, the various elements of a typical twystron are illustrated in Fig. 11.11.

TABLE 11.2
Representative Twystron Characteristics
(From *Microwave Tube Manual* by Varian Associates, Air Force Publication Number T.O. 00-25-251, October 1979.)

Tube Type	*VA-145*	*VA-915*	*VA-146*	*VA-145LV*
Frequency Bandwidth (GHz)	2.7–2.9 2.9–3.1 3.0–3.2	3.1–3.6	5.4–5.9	3.1–3.5
Peak Output Power (MW)	3.5	7.0	4.0	1.0
Average Output Power (kW)	7.0	28.0	10.0	1.0
Pulsewidth (μs)	10	40	20	50
Efficiency (%)	35	30	30	30
Beam Voltage (kV)	117	180	140	80
Beam Current (A)	80	150	95	45
Drive Power (kW)	0.3	3.0	2.0	1.0

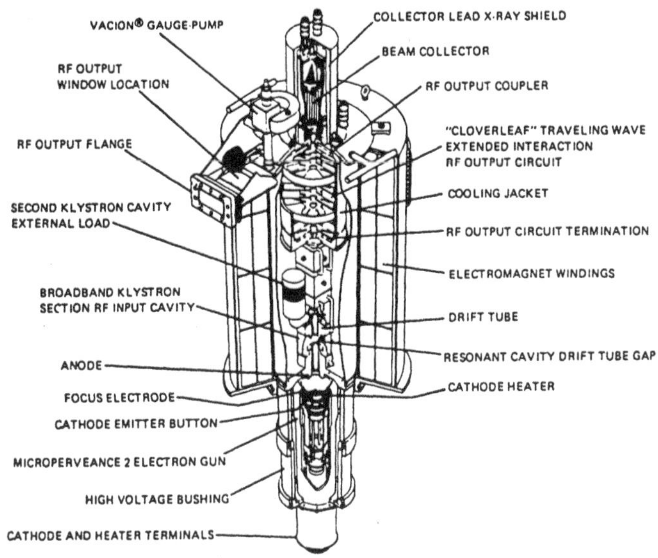

Figure 11.11 Typical Multimegawatt Twystron

(From *Microwave Tube Manual* by Varian Associates, Air Force Publication Number T.O. 00-25-251, October 1979.)

REFERENCES

[11.1] Recommended for Further Reading:

Armand Staprans, Earl William McCune, and Jack A. Ruetz, "High-power linear-beam tubes," *Proc. IEEE*, vol. 61, no. 3, March 1973, pp. 299–327.

Technical Manual, Installation, Operation, Maintenance, Care and Handling, General, Microwave Tubes, Magnetron, Tubes, Electron Tubes, by Varian Associates. Air Force Publication T.O. 00-25-251, October 1979.

[11.2] W. Luebke and G. Caryotakis, "Development of a one megawatt CW Klystron," *Microwave Journal*, vol. 9, No. 8, Aug. 1966.

Chapter 12

Efficiency Enhancement in Linear Beam Tubes

A primary consideration in selecting microwave tubes for many applications is the overall efficiency in converting input power to output power at the desired frequency.

One very important part of overall efficiency is electronic efficiency which is the efficiency of conversion of power in the electron beam to RF power. Electronic efficiency has received considerable attention since the invention of microwave tubes. In this chapter, it will be assumed that the electronic efficiency has been established in the design of the RF characteristics of a tube. Then, with electronic efficiency fixed, techniques will be investigated for enhancing overall efficiency.

12.1 POWER FLOW

The various elements of power flow in a linear beam tube are illustrated in Fig. 12.1. Power is supplied for the heater, the electron beam, and the solenoid. The RF input power is normally at least 30 dB below the RF output power and is assumed to be negligible.

Power is lost when electrons are intercepted by the circuit and by RF losses within the tube. In addition, some of the RF power that is generated is lost because it is not at the desired frequency (harmonics, intermodulation products, *et cetera*).

Finally, after the RF output is removed from the tube, the beam leaving the RF circuit (the spent beam) contains power that is delivered to the collector. Some of this power is converted to heat by electrons striking the collector. If the collector is designed properly, the electrons can be slowed down before they strike the collector surfaces, and the collector can thereby recover power from the spent beam [12.1].

Now, using the various elements of power identified in Fig. 12.1, it is straightforward to derive power flow and efficiency relationships. The overall tube efficiency, η_{ov}, is simply the ratio of the RF output power to the total input power, P_{in}. That is,

$$\eta_{ov} = \frac{P_{out}}{P_{in}} \tag{12.1}$$

where

$$P_{in} = P_o + P_h + P_{sol} - P_{rec} \tag{12.2}$$

or

$$P_{in} = P_{ot} - P_{rec} \tag{12.3}$$

where

$$P_{ot} = P_o + P_h + P_{sol} \tag{12.4}$$

is the total power associated with generating and focusing the electron beam. As a result,

$$\eta_{ov} = \frac{P_{out}}{P_{ot} - P_{rec}} \tag{12.5}$$

The power recovered by the collector may be written as

$$P_{rec} = \eta_{coll} P_{sp} \tag{12.6}$$

where η_{coll} is the efficiency of the collector. The spent beam power is beam power from the gun less the power removed from the beam as it passes through the tube. That is,

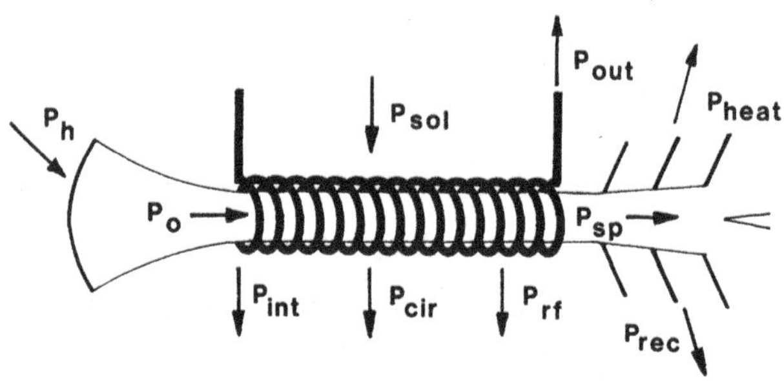

P_h	= heater power	P_{sol}	= solenoid power
P_o	= beam power from gun	P_{out}	= rf output
P_{int}	= interception loss	P_{sp}	= power in spent beam
P_{cir}	= circuit losses	P_{heat}	= heat in coll.
P_{rf}	= harm's, intermod, etc.	P_{rec}	= power rec. by coll.

Figure 12.1 Power Flow in a Linear Beam Tube

$$P_{sp} = P_o - P_{int} - P_{cir} - P_{RF} - P_{out} \tag{12.7}$$

so

$$P_{rec} = \eta_{coll}\,(P_o - P_{int} - P_{cir} - P_{RF} - P_{out}) \tag{12.8}$$

Two additional efficiency terms are useful in completing the analysis. One is the electronic efficiency, η_e, which is the efficiency of conversion of beam power to RF power at the desired frequency. That is,

$$P_{out} + P_{cir} = \eta_e P_o \tag{12.9}$$

Values for η_e range from a few percent to well over 50%.

The other efficiency is the circuit efficiency, η_{cir}, which is the efficiency of the RF circuit in delivering the generated RF power at the desired frequency to the output connector of the tube. Thus,

$$P_{out} = \eta_{cir}(P_{out} + P_{cir}) = \eta_{cir}\eta_e P_o \tag{12.10}$$

Values for η_{cir} are usually in the 75 to 90% range.

Now, by combining (12.10) and (12.8) with (12.5), the overall efficiency may be written as

$$\eta_{ov} = \frac{\eta_{cir}\eta_e}{\dfrac{P_{ot}}{P_o} - \eta_{coll}\left(1 - \eta_e - \dfrac{P_{int} + P_{RF}}{P_o}\right)} \tag{12.11}$$

The plot of this equation in Fig. 12.2 is useful in illustrating the relative significances of the various efficiencies and powers.

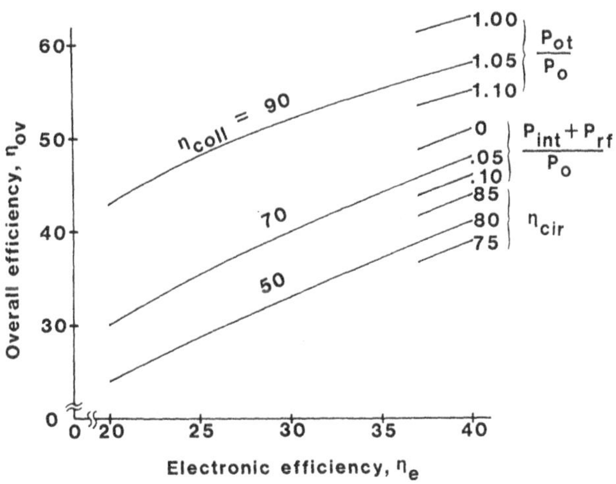

Figure 12.2 Variations of Overall Efficiency with the Various Elements of Efficiency and Power

In Fig. 12.2, the effect of P_{ot} is shown on the 90% collector efficiency plot. This same range of variations would apply to other η_{coll} plots, but is only shown for the 90% plot. Similarly, the effects of variations in $P_{int} + P_{RF}$ are shown on the 70% η_{coll} plot, but would apply to other collector efficiency plots. The effects of variations in circuit efficiency are shown on the 50% η_{coll} plot but would apply to other η_{coll} plots.

Perhaps the most striking aspect of Fig. 12.2 is the strong dependence of overall efficiency on collector efficiency, especially for high values of collector efficiency. This leads directly to the next section of this chapter, which deals with the design of efficient collectors.

The importance of circuit efficiency is also evident in Fig. 12.2. Notice that a 10% increase in η_{cir} increases η_{ov} about five percentage points. This leads to considerations of circuit efficiency, which are discussed in the section following collector efficiency.

Finally, in considering Eq. (12.11) and Fig. 12.2, it should be noted that heater power and RF losses (harmonics and intermodulation products) are normally small. Thus, P_{ot} contains primarily the power penalty resulting from the use of a solenoid for focusing. Increases in η_{ov} that result from decreases in solenoid power (and P_{ot}) can readily be traded off against the decrease in η_{ov} that results from an increase in P_{int} (interception power), which, in turn, normally results from a decrease in solenoid power.

12.2 DEPRESSED COLLECTORS

A collector for a linear beam tube is shown in Fig. 12.3. As the beam passes through the pole piece, the magnetic focusing field is removed, and so space charge forces cause the beam to expand as it enters the collector.

With the collector at the same potential as the body of the tube, electrons strike the collector at relatively high velocity. Electron energy is converted to heat at the collector surface. With a collector like that shown in Fig. 12.3, a relatively large percentage (normally 50% or more depending on electronic efficiency) of the power input to the tube is dissipated in heat in the collector.

By reducing (depressing) the voltage on the collector below the body potential, the velocities of the electrons striking the collector and the heat generated in the collector are reduced. As a result, the collector recovers some of the power in the spent electron beam.

A typical circuit for operating the collector at a depressed potential is shown in Fig. 12.4. The cathode-to-anode voltage is V_0 and the cathode-to-collector voltage is V_c, where V_c is less than V_0. Thus, the collector is depressed with respect to the anode.

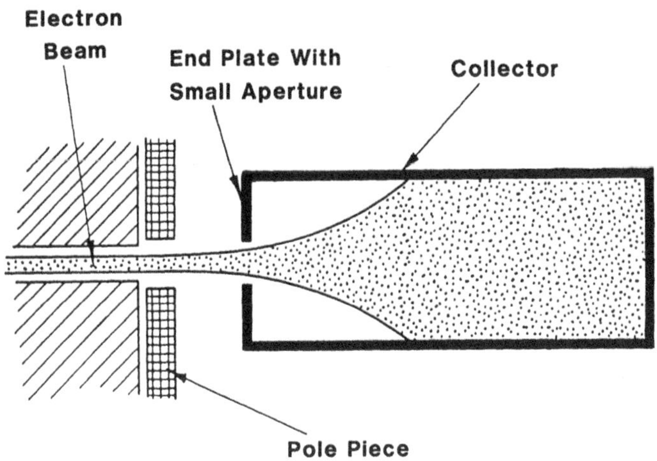

Figure 12.3 Collector for Linear Beam Tube

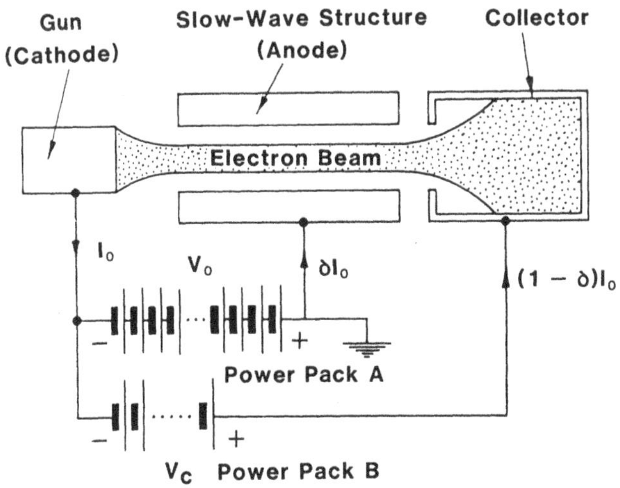

Figure 12.4 Circuit for Depressed Collector Operation

(From *Power Travelling Wave Tubes* by J.F. Gittins, Published 1965 by American Elsevier, Inc.)

It is important to understand that even though the anode supply (power pack A) supplies the beam voltage, the only current that it must supply is the interception current, which is a small fraction, δ, of the total beam current, I_0. The voltage supplied by the collector supply (power pack B) is somewhat less than V_0, but the current $(1-\delta)\ I_0$, is much larger than that from the anode supply. As a result, the power from the collector supply is much larger than that from the anode supply. Within limits, the power supplied by the collector supply is reduced as the collector potential is depressed. Thus, while the tube engineer thinks of collector depression as a technique for recovering power, the tube user looks at it as a technique for reducing power input, rather than for recovering power because the power was not supplied to the tube in the first place.

It should be noted that the use of a depressed collector significantly reduces the power regulation requirement for a tube. This is because only the anode voltage (and power) must be tightly regulated in order to regulate beam power and phase. Within limits (to be discussed in the following paragraphs), the collector voltage and, with it, the major portion of the power to the tube can be unregulated.

The operating potential of a depressed collector power supply like that shown in Fig. 12.4 is highly dependent on the distribution of electron velocities in the electron beam. The slower electrons will be reflected by the collector if the potential is depressed too far. With increasing collector voltage, as shown in Fig. 12.5, collector current will decrease and body current will increase. Excessive body current will result in damage to the RF structure. Single-element collectors like those illustrated in Figs. 12.3 and 12.4 are capable of recovering roughly 30 to 40% of the power in a spent beam (highly dependent on the tube type).

To recover a large fraction (over 80%) of the power in a spent beam, and, at the same time, prevent electrons from being reflected, a collector must contain electric fields shaped so that all electrons are nearly stopped, regardless of velocity, just as they strike the collector electrodes. In addition, the fields must be shaped to prevent secondary electrons from being accelerated (thus wasting energy) or leaving the collector.

To better understand the approach to the solutions of these problems, it is helpful to examine the electron energy spread in spent beams. Shown in Fig. 12.6 are energy distribution curves for the electron beams in a high efficiency klystron and in a helix TWT. The horizontal axis is the spent beam energy normalized with respect to the initial beam energy (before modulation). The vertical axis is the fraction of the beam current having more than a particular value of energy. For example, in the UHF klystron, which is a highly efficient tube with an intense interaction between the beam and the circuit, only about 25% of the beam current enters the collector with an

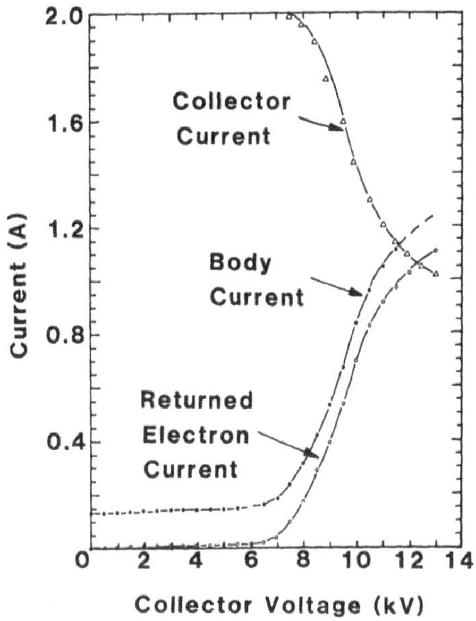

Figure 12.5 Affect of Collector Voltage of Body Current

(From *Microwave Devices and Circuits* by Samuel Y. Liao, Copyright 1985 by Prentice Hall, Inc., Reproduced with permission.)

energy greater than 40% of the initial beam energy. Thus, a collector electrode depressed 40% (of the cathode-to-anode voltage) below the anode voltage could collect no more than 25% of the beam current. The remaining 75% of the beam current would be reflected.

For the helix TWT, in which the beam circuit interaction is weak, a collector depressed 40% could collect the entire beam. The fraction of the original beam power (from the gun) recovered by the collector would be $I(V)/I_0 \times 0.4V/V_0$, or 0.4. The power remaining in the beam to be dissipated in heat in the collector would be the shaded area under the curve (depending on perveance) above a normalized energy of $0.4V/V_0$.

To recover a large fraction of the power represented by the shaded area in Fig. 12.6, several additional electrodes would be required. Also, the electric field distribution would have to be such that the electrons were collected by the correct electrodes.

For example, an electrode depressed to 0.6 V_0 could collect nearly all of the electrons in the helix TWT. A small percentage (less than 5%) would be repelled, but could be collected by an electrode at 0.4 V_0. An electrode depressed to 0.8 V_0 could collect about 85% of the current for the 1.0 microperveance case. The remaining 15% would be reflected to electrodes at potentials closer to the anode potential.

12.3 MULTISTAGE DEPRESSED COLLECTORS

The preceding discussion provides the basis for the design of efficient collectors. Modern, highly efficient collectors use several electrodes at different potentials to collect selectively electrons at low energy levels. These collectors are called multistage depressed collectors (MDCs).

The research and development for the collectors to be described has been the work of Dr. H.G. Kosmahl and his associates at the NASA Lewis Research Center in Cleveland, Ohio. The original impetus for the NASA work was the extremely high premium on efficiency resulting from the high cost of power generation and heat reflection on satellites using microwave tubes. Now, the NASA collector technology is being put to use in a broad range of applications where high microwave power is required and prime power is severely limited. An example is airborne electronic countermeasures systems [12.2].

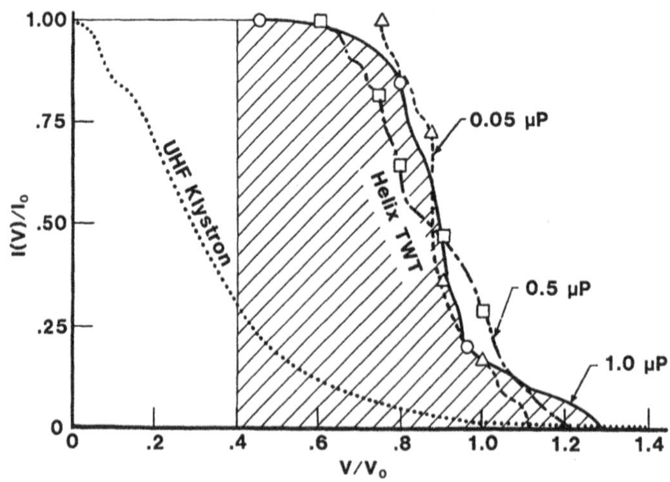

Figure 12.6 Beam Energy Spread for a High Efficiency UHF Klystron and for a Helix TWT (at Various Perveances and an Electronic Efficiency of about 11%)

(Adapted from H.G. Kosmahl, *Proc. IEEE*, November 1982.)

The collector-electrode configurations for the NASA MDCs are illustrated in Figs. 12.7 and 12.8. The first NASA collectors were derived from mathematical solutions to the problem of both slowing down and dispersing the electron beam entering the collector. As shown in Fig. 12.7, the equipotential surfaces were designed to be concave toward the aperture where the beam entered the collector. This forced electrons to move radially outward.

Collector electrodes were designed to coincide with the equipotential surfaces that were derived mathematically. The aperture sizes in the electrodes where chosen so that electrons would land on electrodes at the appropriate voltage. Since most beam electrons landed on the back sides of the electrodes, secondary electrons were emitted into a negative potential region and, therefore, were immediately returned to the electrode surfaces.

In all microwave tubes, some electrons are out of phase with the RF voltage and are accelerated, rather than decelerated, by the interaction process. These electrons leave the interaction region with energies greater than the original beam energy. This was shown in Fig. 12.6 by the curves at $V/V_0 > 1.0$. In high-power klystrons, this situation is particularly severe and some electrons are accelerated to as much as twice the original beam energy.

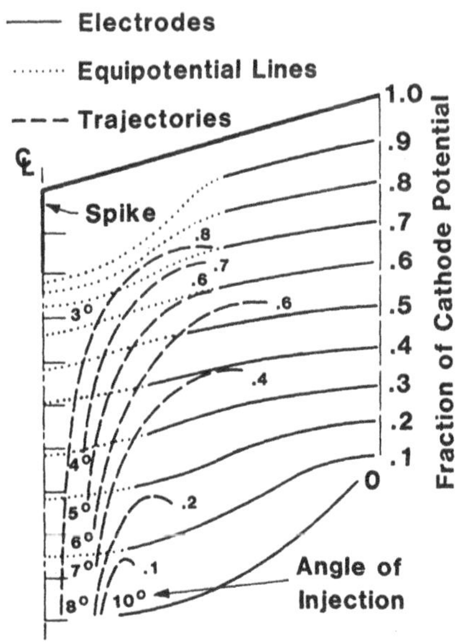

Figure 12.7 Axisymmetric MDC with Equipotential Surfaces Determined Mathematically

(From H.G. Kosmahl, *Proc. IEEE,* November 1982.)

In the NASA collector in Fig. 12.7, a sharp spike is shown at the back of the collector. This spike helps shape the fields at the back of the collector to disperse high-energy electrons. In addition, the surface area presented to high-energy electrons entering the collector on axis is minimized. Thus, the generation of high-energy secondary electrons is minimized.

The efficiencies of collectors patterned after the design in Fig. 12.7 is very high (> 85%) and the current reflected from the collector is virtually zero. Complexity and size are disadvantages.

The collector shown in Fig. 12.8 was developed during a joint NASA/USAF program. Collector size was reduced significantly with a small compromise in efficiency so that MDCs could be used in airborne applications where size and weight are important considerations. In these collectors, called individual lens collectors, each electrode and its associated fields are optimized to produce maximum collector efficiency while minimizing the current resulting from reflected electrons.

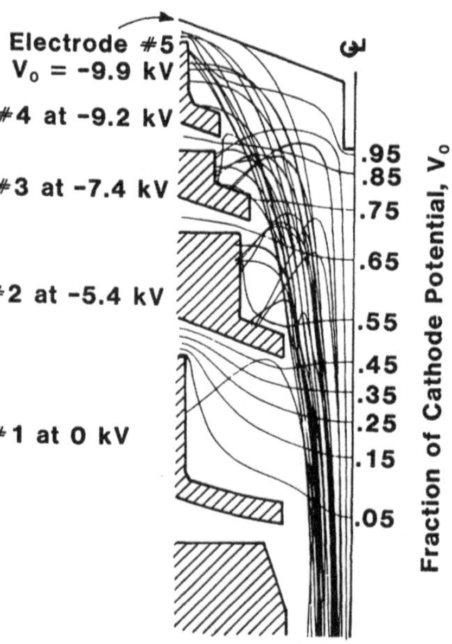

Figure 12.8 Equipotential Surfaces and Electron Trajectories in "Individual Lens" Collector

(From H.G. Kosmahl, *Proc. IEEE,* November 1982.)

In order for an MDC (either the mathematically designed version or the individual lens version) to operate efficiently, the electron beam entering it must not diverge rapidly. Rapid divergence leads to large radial components of electron velocities and to energy components that cannot be recovered by the collector.

To reduce the divergence of a beam it is necessary to expand its diameter so that space-charge forces are reduced. A magnetic refocusing system, one version of which is shown in Fig. 12.9, is used for this purpose. An additional benefit of the refocusing system is an improvement in the laminarity of the beam entering the collector.

Improvements in overall efficiency that have been demonstrated through the use of a MDC on a dual-mode helix TWT for electronic countermeasures applications are illustrated in Fig. 12.10. An overall efficiency improvement by a factor of three of more was demonstrated across a bandwidth of an octave. The collector efficiency was nearly constant at about 85% as shown in Fig. 12.11.

A very interesting application for MDCs is in tubes where efficient operation is necessary in the linear (small-signal) region. As shown in Fig. 12.12, the efficiency of the MDC increases significantly as the input signal level is reduced. For the example in Fig. 12.12, the MDC efficiency exceeds 97% when the drive signal level (and basic efficiency) are reduced to zero.

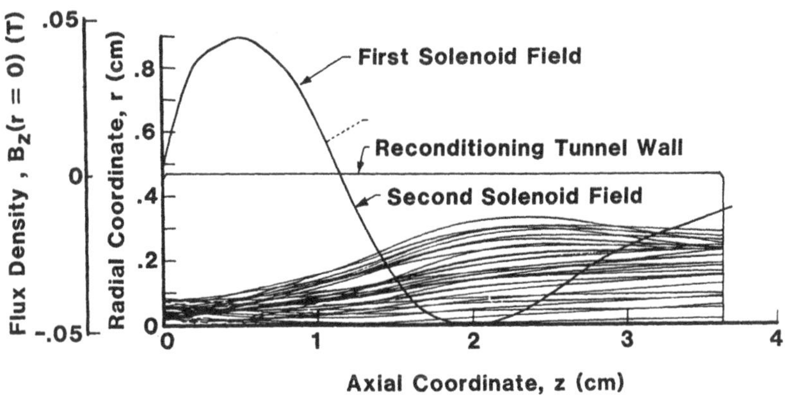

Figure 12.9 Magnetic Refocusing System for Reducing Beam Divergence and Improving Laminarity

(From H.G. Kosmahl, *Proc. IEEE,* November 1982.)

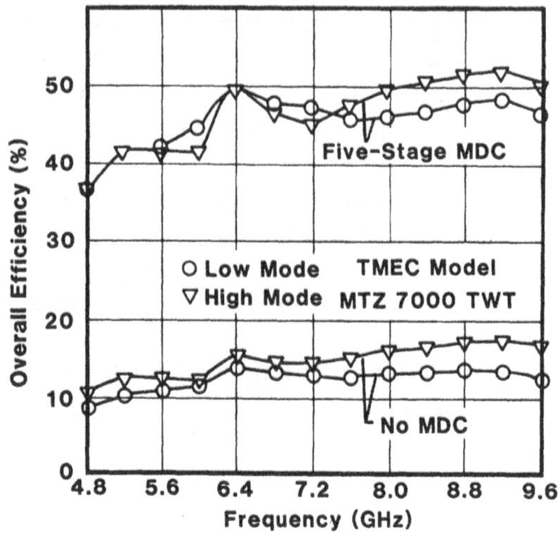

Figure 12.10 Overall Efficiency of a Dual-Mode ECM Helix TWT with and without a MDC

(Adapted from H.G. Kosmahl, *Proc. IEEE*, November 1982.)

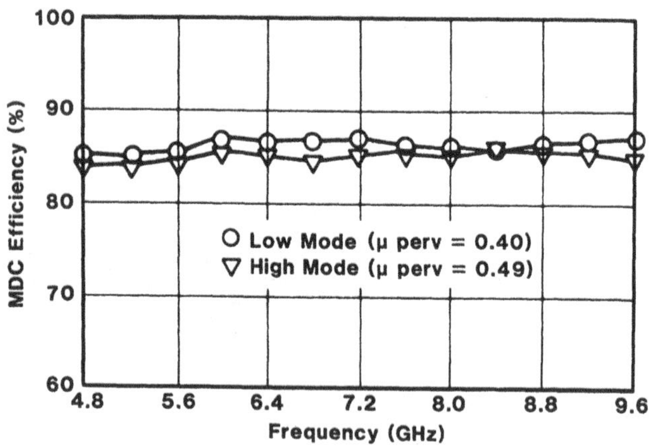

Figure 12.11 Collector Efficiency at Saturation

(Adapted from H.G. Kosmahl, *Proc. IEEE*, November 1982.)

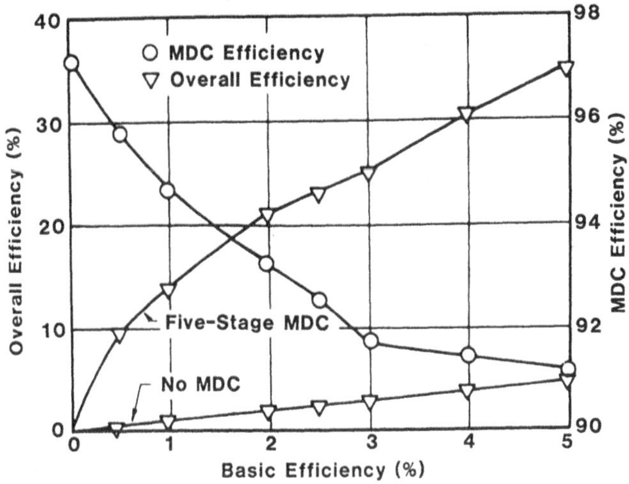

Figure 12.12 Overall and MDC Efficiency *versus* Basic (Electronic) Efficiency in the Small-Signal Region

(Adapted from H.G. Kosmahl, *Proc. IEEE*, November 1982.)

At a drive signal level low enough so that the basic (electronic) efficiency and the overall efficiency (without MDC) is 1%, the addition of the MDC increases the overall efficiency to 14%. At a basic efficiency of 4%, the MDC increases the efficiency to 30%. Thus, using a MDC, efficient operation over a wide range of drive levels is possible. Table 12.1 lists computed and measured efficiencies for a wide range of TWT operating conditions. Of particular interest is the high degree of agreement between the computed and measured collector efficiencies.

A typical power supply configuration for a MDC is shown in Fig. 12.13. As was the case for the single element depressed collector, the cathode-to-anode (and helix) supply must be regulated, but it must supply only the power resulting from interception current. The tapped power supply for the collector is essentially unregulated. This is because variations in voltages to collector elements may cause some electrons to move from one collector element to another, but the overall effect on efficiency is negligible.

TABLE 12.1
Analytical *versus* Experimental MDC/TWT Performance
(Adapted from H.G. Kosmahl *Proc. IEEE*, November 1982.)

Model	*Microper veance*	*Electronic efficiency % (experi-mental)*	*Operation Mode*	*MDC Stages*	*Collector Efficiency:*		*Overall Efficiency %*
					Computed	*Measured*	*Measured*
TMEC 5897C	0.47	17	Saturation	5	84.9	84.2	41.6
TMEC 5897C	0.47	17	Saturation	3	81.6	81.9	39.0
TMEC 5897C	0.47	3.7	Linear	5	89.7	87.4	18.6
TMEC 5897C	0.47	7.7	Linear	5	86.7	86.4	30.1
TMEC MTZ 7000	0.49	19.5	Saturation, high mode	5	81.4	84.3	48.0
	0.40	16	Saturation, low mode		84.4	85.5	47.6

TABLE 12.1 (cont'd)

Model	*Microper veance*	*Electronic efficiency % (experi-mental)*	*Operation Mode*	*MDC Stages*	*Collector Efficiency: Computed*	*Collector Efficiency: Measured*	*Overall Efficiency % Measured*
TMEC MTZ 7000	0.49	19.5	Saturation, high mode	5	70.4	77.6	42.0
	0.40	2.4	Saturation, low mode		91.8	91.9	18.5
TMEC MTZ 7000	0.40	16	Saturation, low mode	4	83.8	81.7	42.7
TMEC MTZ 7000	0.40	3.6	Linear	5	88.7	90.4	25.5
Varian VTG 6336 A1	1.23	24.4	Saturation	4	78.4	76.0	45.6

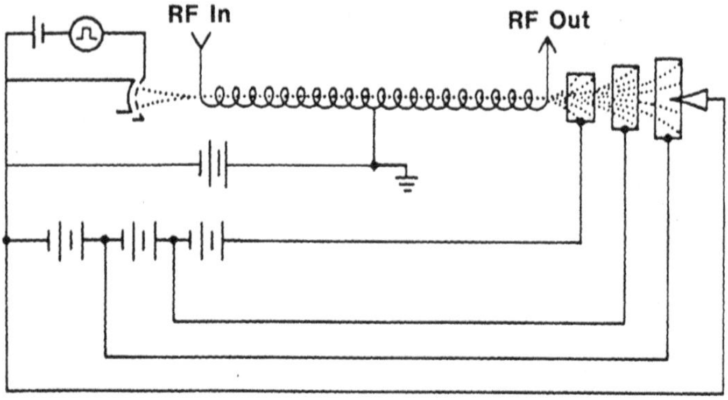

Figure 12.13 TWT Power Supply Operation with a Multistage Depressed Collector

12.4 CIRCUIT EFFICIENCY [12.3]

As was illustrated in Fig. 12.2, circuit efficiency has a significant effect on overall efficiency. This is shown more clearly in Fig. 12.14. As an example of the importance of circuit efficiency, assume a collector efficiency of 85%. Assume further that a lossy helix such as rough tungsten (circuit efficiency ~ 70%) is replaced by a properly finished copper-plated helix (circuit efficiency ~ 90%). The increase in overall efficiency is 10%.

Circuit efficiency is primarily dependent on the effective resistance of the circuit material. At high frequencies, dielectric losses are also important. To illustrate the effects of circuit resistivity, calculations were performed of losses in the helix structure for a TWT. The model for the analysis is shown in Fig. 12.15. A helically conducting sheath helix was assumed. Allowance was made for dielectric support rods, and the resistivity of the barrel as well as that of the helix were taken into consideration.

Nominal values for the various helix parameters were as follows:

helix pitch	= 0.141 (18 T.P.I.)
helix radius	= 0.12 cm (0.047 in)
barrel-to-helix diameter ratio	= 2.0
helix rsistivity	= 2.5×10^{-6} Ω cm
barrel resistivity	= 2.5×10^{-6} Ω cm
effective relative dielectric constant of support rods (ϵ_2)	= 1.5

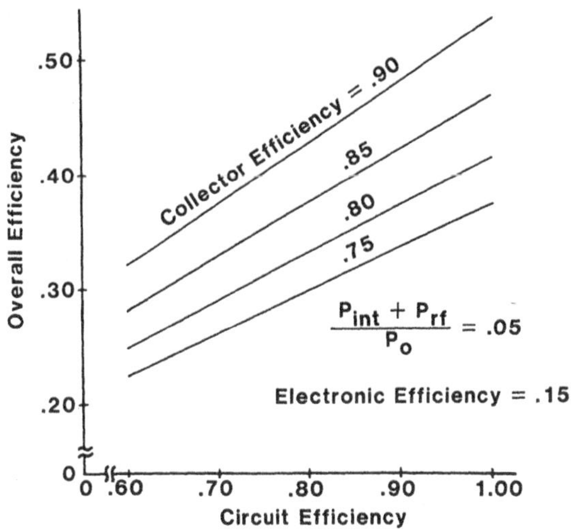

Figure 12.14 Effect of Circuit Efficiency on Overall Efficiency

(From A.S. Gilmour, Jr., *et al., IEEE Trans. Electron Devices*, October 1979.)

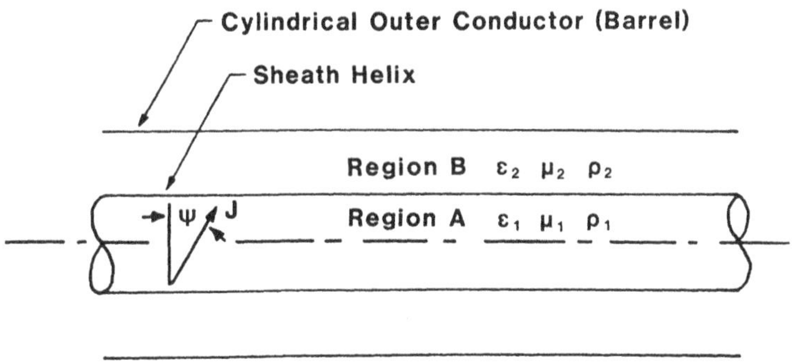

Figure 12.15 Helically Conducting Helix Model Used in Analysis

(From A.S. Gilmour, Jr., *et al., IEEE Trans. Electron Devices*, October 1979.)

The effect of variations in helix resistivity with the other parameters is shown in Fig. 12.16. Notice that loss increases roughly as the square-root of helix resistivity, as is expected for skin-effect resistance (see Ch. 2). Comparisons of measured and theoretical loss for a copper-plated helix are given in Fig. 12.17.

The dashed curves of Fig. 12.17 are calculated loss curves based on the bulk resistivity of copper (1.724×10^{-6} Ω-cm) and based on the best fit to experimental data at the frequencies where reflections in the measurement circuitry were small and a resistivity of 4×10^{-6} Ω-cm was assumed. The circled data points are measured losses and the solid curve is the calculated loss for a resistivity of 4×10^{-6} Ω-cm, taking into consideration reflections in the measurement circuitry.

The excellent fit of the theoretical curves to data when a resistivity of 4×10^{-6} Ω-cm is assumed leads to the conclusion that 4×10^{-6} Ω-cm was the effective value of the resistivity of the copper surface of the helix. This value is more than twice the bulk resistivity of copper.

The reason for the high effective resistivity is indicated in Fig. 12.18, which contains photomicrographs of the surface of the copper-plated helix [12.4].

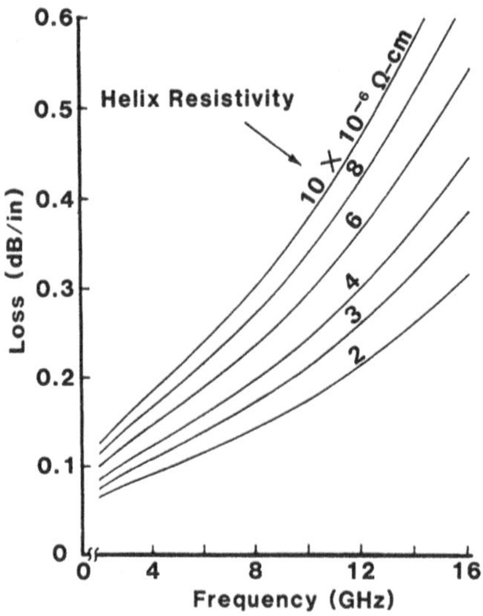

Figure 12.16 Loss of Functions of Frequency and Helix Resistivity

(From A.S. Gilmour, Jr., *et al.*, *IEEE Trans. Electron Devices*, October 1979.)

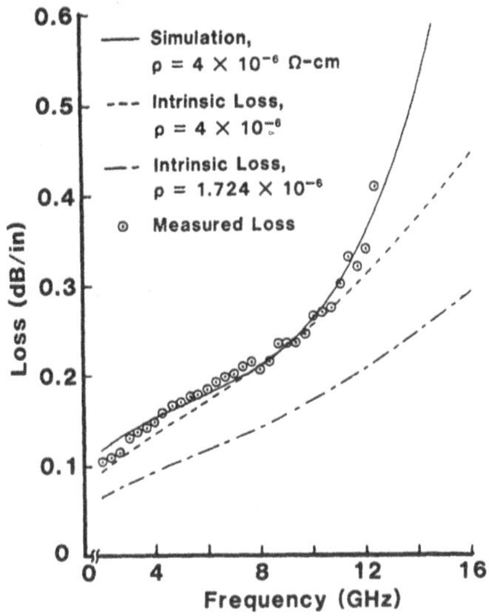

Figure 12.17 Measured and Theoretical Loss as a Function of Frequency for a Copper-Plated Helix

(From A.S. Gilmour, Jr., *et al., IEEE Trans. Electron Devices*, October 1979.)

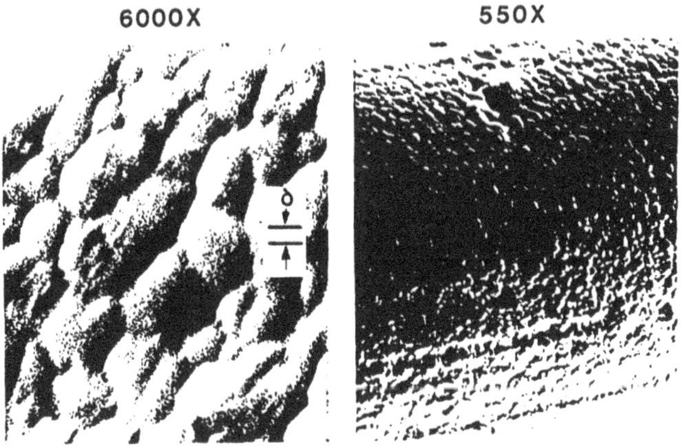

δ ≈ Skin Depth at 10 GHz

Figure 12.18 Photomicrographs of Copper-Plated Helix

(From A.S. Gilmour, Jr., *et al., IEEE Trans. Electron Devices*, October 1979.)

The skin depth at 10 GHz was small compared with the sizes of the irregularities on the orange-peel textured copper surface. As a result, the path length for the RF current and the effective resistance were much larger than they would have been for a smooth surface. To minimize resistance and loss, it is essential to properly polish the surface of the helix.

The analysis showed that loss increased as helix pitch and radius increased. Furthermore, the amount of the increase scaled with the increase in current path length as expected. A similar effect caused by the relative effective dielectric constant of the helix support rods is illustrated in Fig. 12.19. As the dielectric constant is increased, signal velocity decreases and loss increases (by the square root of dielectric constant).

Finally, the analysis showed that the resistivity of the metallic barrel surrounding the helix had essentially no effect on loss.

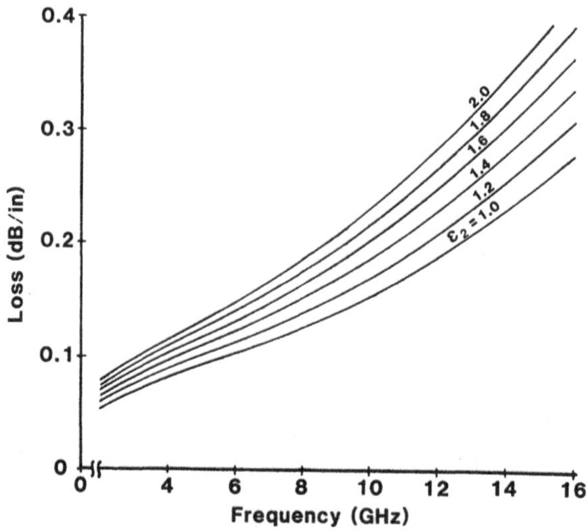

Figure 12.19 Loss as Functions of Frequency and Effective Relative Dielectric Constant of Support Rods

(From A.S. Gilmour, Jr., *et al.*, *IEEE Trans. Electron Devices*, October 1979.)

REFERENCES

[12.1] H.G. Kosmahl, "Modern Multistage Depressed Collectors — A Review," *Proc. IEEE,* Vol. 70, Nov. 1982, pp. 1325–1334.

[12.2] H.G. Kosmahl and P. Ramins, "Small-Size 81 to 83.5 Percent Efficient 2- and 4-Stage Depressed Collectors for Octave-Bandwidth High Performance TWT's," *IEEE Trans. Electron Devices*, Vol. ED-24, Jan. 1977, pp. 36–44.

[12.3] A.S. Gilmour, Jr., M.R. Gillette, and J.T. Chen, "Theoretical and Experimental TWT Helix Loss Determination," *IEEE Trans. Electron Devices*, Vol. ED-26, Oct. 1979, pp. 1581–1588.

[12.4] A.S. Gilmour, Jr., "SEM Analysis of TWT Helix Samples," *IEEE Trans. Electron Devices*, Vol. ED-24, June 1977, pp. 774–775.

Chapter 13

Crossed-Field Tubes

In crossed-field tubes, the dc electric field is perpendicular to the magnetic field. The general motion of the electron beam is perpendicular to both fields that is

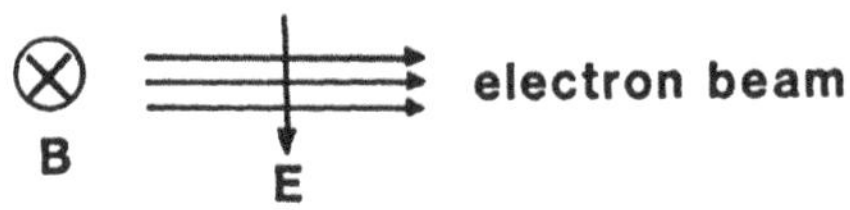

This is in contrast to linear beam tubes where the general direction of beam motion is aligned with the fields:

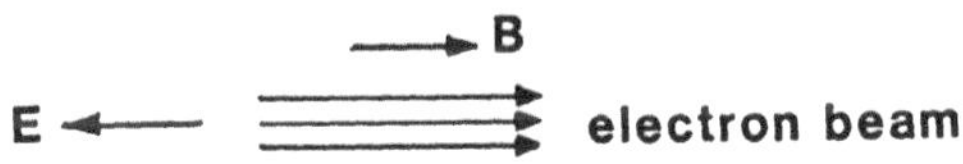

Crossed-field tubes are referred to as "M-type" tubes after the French TPOM, which in English means *tubes for propagation of waves in a magnetic field.*

Many types of crossed-field tubes have been developed and can be categorized as indicated in Fig. 13.1. With the exception of injected-beam crossed-field amplifiers, these devices are in use in deployed systems. Injected-beam crossed-field amplifiers, while highly developed have not found application and will not be discussed.

13.1 HISTORY OF THE MAGNETRON [13.1, 13.2]

The earliest reports of magnetron development date back to 1921. These early magnetrons used anodes that were cylindrical in shape. Efficiencies and power capabilities were very low. In 1935, Posthumous deduced theoretically that in magnetrons with multisegment anodes and large magnetic fields, RF power should be generated with high efficiency. He then demonstrated an efficiency of about 50% in a magnetron operating at a wavelength of 50 cm (600 MHz).

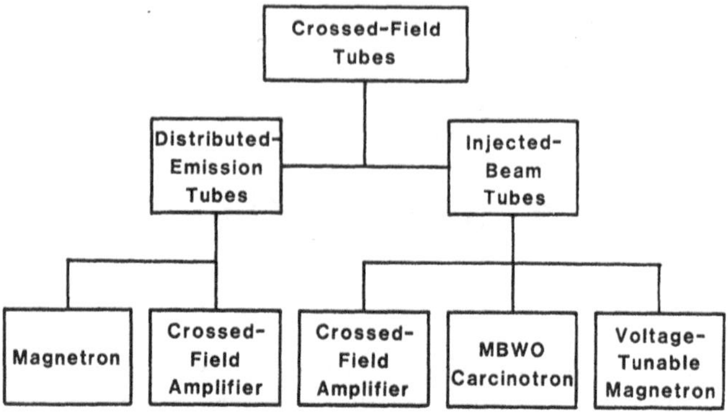

Figure 13.1 Family Tree of Crossed Field Tubes

(From *Microwave Tube Manual* by Varian Associates, Air Force Publication Number T.O. 00-25-251, October 1979.)

In 1936, Samuel proposed the hole-and-slot resonator arrangement for use in a magnetron anode at high frequencies, but apparently did not reduce its use to practice.

By about 1938, a magnetron with a copper anode and hole-and slot resonators had been built by Russian scientists Alekseroff and Malearoff. A tungsten helix cathode was used and the power generated was about 100 W CW at a wavelength of 10 cm.

In late 1939, Boot and Randall at the University of Birmingham (UK), who were apparently unaware of the work of Posthumous, Samuel, and the Russians, set about to combine the cylindrical-anode-magnetron work with the klystron's advantage of using resonators for confining the RF fields.

By early 1940, a magnetron with an anode containing six resonant cavities had been built and operated. By May 1940, an experimental radar containing a pulsed 10-cm cavity magnetron was in operation and by September 1940 a submarine periscope could be detected at a range of seven miles.

At some point during the early development of the magnetron, an oxide-coated cathode was supplied by Warnecke and Ponte of CSF (France) to replace the carburized thoriated-tungsten cathode being used at the time. The peak power capability increased by about an order of magnitude to 100 kW.

In an article [13.2] by J.J Coupling (a pseudonym of J.R. Pierce) concerning the magnetron, the following opening statement appeared:

> The two most important weapons of the recent war were the atomic bomb, and radar . . . and it was Maggie, the magnetron, that made Allied radar incomparably superior to the Nazi's boat.

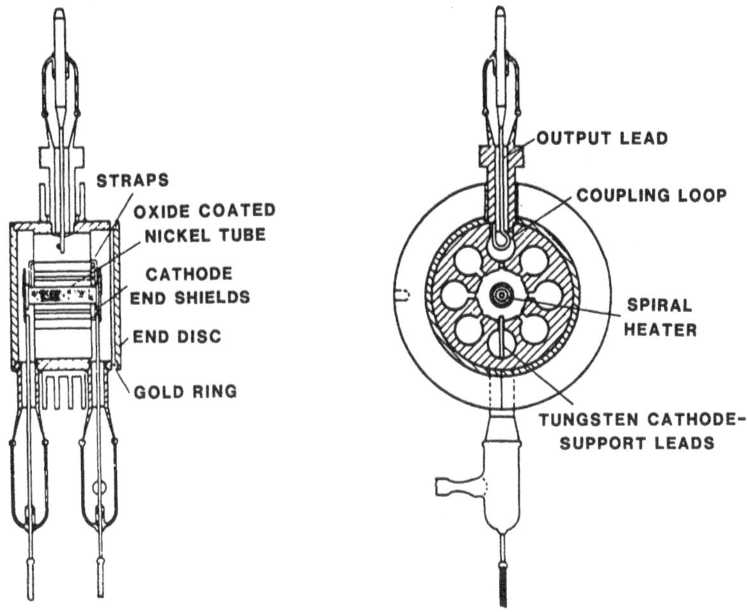

Figure 13.2 Early Sealed-Off Pulsed Cavity Magnetron

(From H.A.H. Boot and J. T. Randall, *IEEE Trans. Electron Devices*, July 1976.)

13.2 MAGNETRONS [13.3]

An early pulsed cavity magnetron is shown in Fig. 13.2. The anode contains eight resonators and is concentric with an oxide-coated cathode. A permanent magnet (not shown) was used for applying a magnetic field parallel to the cathode surface. Details of the electron motion about the cathode, the resonator operation and the interactions of the electrons with the RF field, are given in the following sections.

13.2.1 Electron Motion [13.4]

Before attempting to describe the motions of electrons in a magnetron in the presence of RF fields, it is necessary to examine their trajectories is crossed dc fields.

First of all, in the absence of an electric field, an electron with initial velocity u_0 will follow a circular trajectory of radius u_0/ω_c and will rotate at the cyclotron frequency ω_c as shown in Fig. 13.3. The value of ω_c is ηB, where η is the electron charge to mass ratio and B is the magnetic flux density. When an electric field is applied, then, of course, the electron velocity is changed, but the frequency of rotation remains fixed at ω_c. For an electric field in the minus-y direction, the electron velocity steadily decreases as the electron moves in the minus-y direction, and so the radius of curvature of the electron trajectory decreases. When the y-directed motion of the electron reverses, its velocity increases, and so the radius of curvature increases. The resulting electron trajectory is the cycloid shown in Fig. 13.4. As the electric field (or the initial velocity of the electron) is varied, the various trajectories shown in Fig. 13.5 are followed.

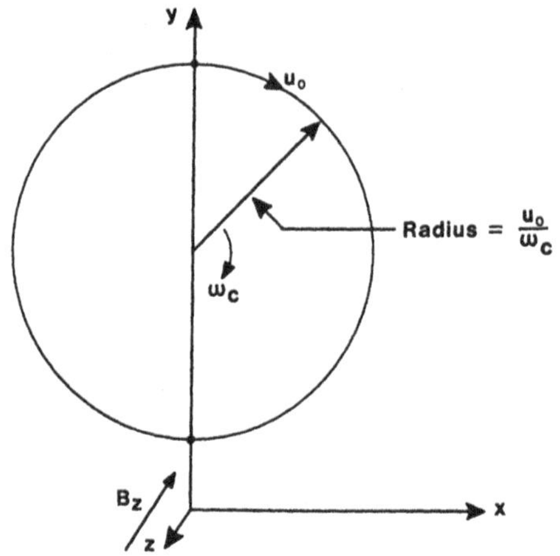

Figure 13.3 Electron Orbit in a Magnetic Field in the Absence of an Electric Field

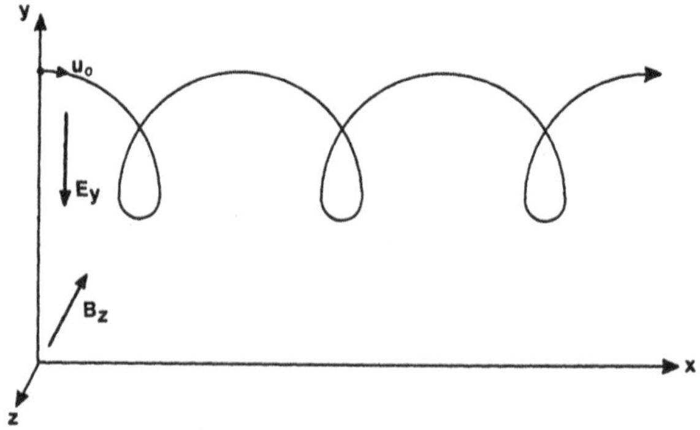

Figure 13.4 Cycloidal Electron Trajectory in the Presence of a y-Directed Electric Field

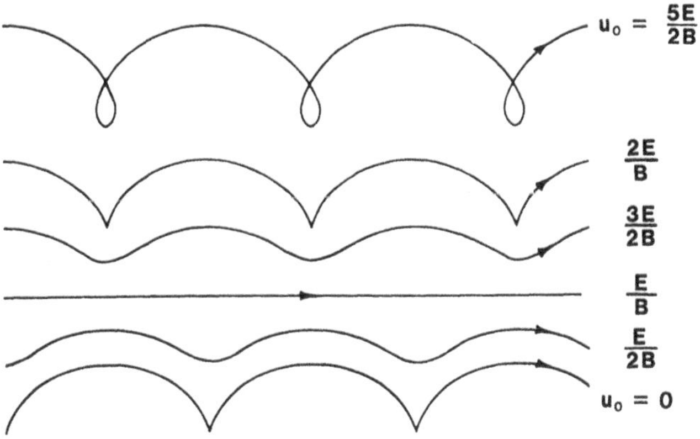

Figure 13.5 Electron Trajectories for Various Initial Velocities or Electric Field Levels

The most significant aspect of Fig. 13.5 is that the electron acquires an average velocity component perpendicular to both the electric and magnetic fields. In a circular configuration, like a magnetron, electrons follow trajectories similar to those shown in Fig. 13.5 around the center electrode.

In particular, an electron leaving the center electrode at essentially zero velocity will follow a trajectory like that shown in Fig. 13.6. If the magnetic field is reduced or the electric field is increased, the trajectory will move closer to the outer positive electrode.

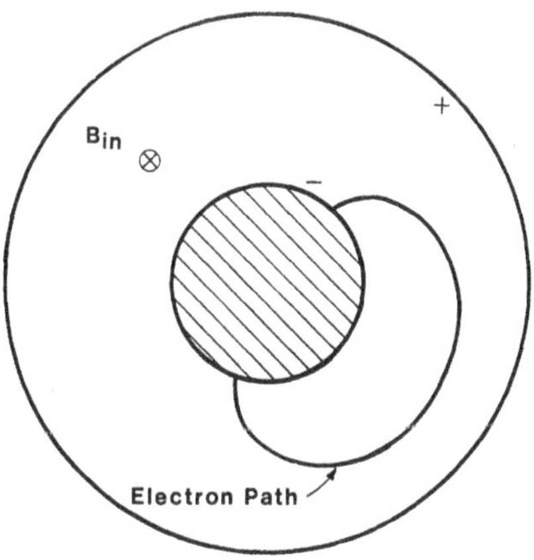

Figure 13.6 Electron Trajectory in Circular Electrode Configuration

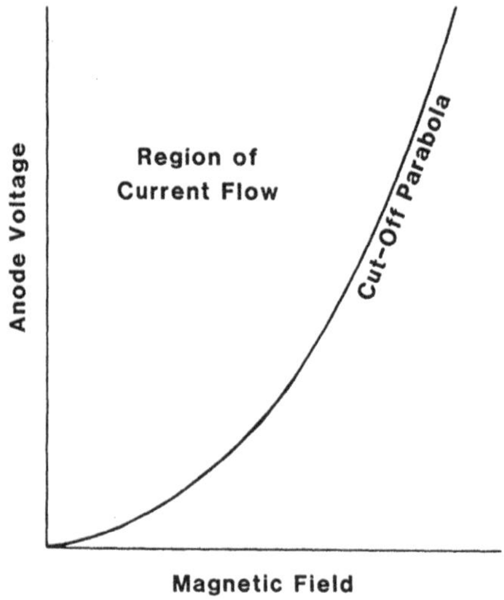

Figure 13.7 Hull Cut-Off Condition

There is a critical combination of magnetic and electric fields known as the Hull cut-off condition, shown in Fig. 13.7, at which electrons just touch the outer electrode. In a cylindrical diode, that combination of fields would result in the flow of electron current from the cathode to the anode.

In addition to the dc field effects, RF electric fields and space-charge forces combine to alter the trajectories indicated in Figs. 13.4, 13.5, and 13.6. The RF circuit and field patterns for a cavity magnetron are described in the following section.

13.2.2 Cavity Structure

Each cavity of the hole-and-slot resonator structure in a magnetron of the type shown in Fig. 13.2 can be visualized as a parallel LC circuit, as indicated in Fig. 13.8. The equivalent circuit for an eight-cavity resonator is eight LC circuits in parallel (Fig. 13.9).

The magnetic and electric field patterns for each cavity are as indicated in Fig. 13.10. In the desired mode of operation, some of the magnetic flux lines of each cavity link the adjacent cavities and the fields in adjacent cavities are 180° out of phase. This is referred to as the π-mode of operation. The resulting field configurations in an eight-cavity structure are shown in Fig. 13.11

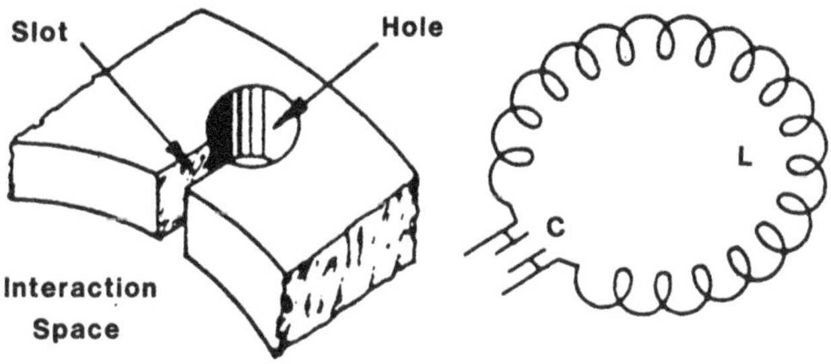

Figure 13.8 Equivalent Circuit of Hole-and-Slot Cavity

(From *Microwave Devices* by James T. Coleman, Copyright 1982 by Reston Publishing Company, Inc., Reproduced with permission.)

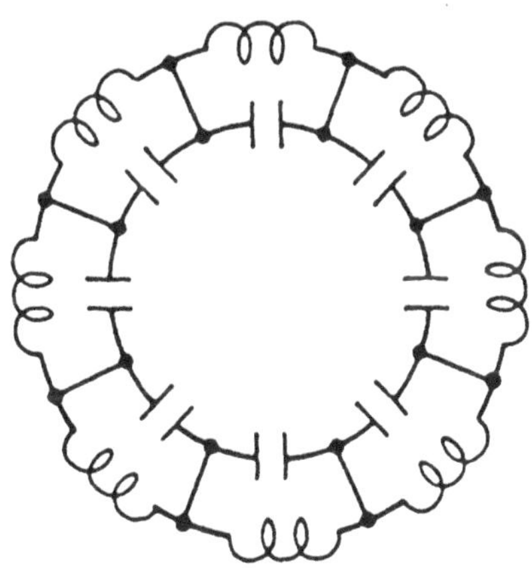

Figure 13.9 Equivalent Circuit for Eight-Cavity Resonator

Modes and field patterns other than those associated with the π-mode can exist and result in magnetron operation at frequencies other than the desired frequency. These other modes and the precautions taken in attempts to prevent their occurrence are discussed in Section 13.2.4.

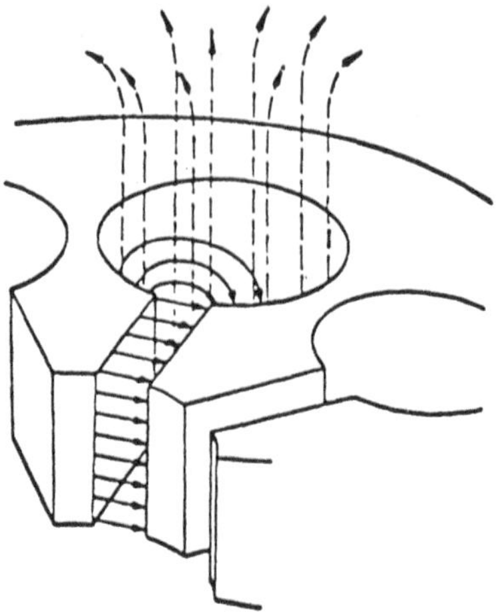

Figure 13.10 Magnetic (Dashed Lines) and Electric (Solid Lines) Field Patterns in Hole-and-Slot Cavity

(From *Microwave Magnetrons* by George B. Collins, Copyright 1948 by McGraw-Hill, Inc., Reproduced with permission.)

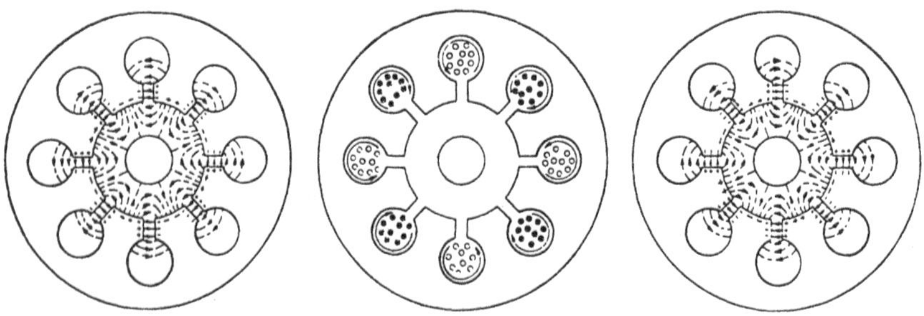

(a) (b) (c)

Figure 13.11 Charges, RF Fields, and Currents for π-Mode in an Eight-Cavity Resonator (The Phase of (a), (b), and (c) Differs Progressively by $\pi/2$)

(From *Microwave Magnetrons* by George B. Collins, Copyright 1948 by McGraw-Hill, Inc., Reproduced with permission.)

13.2.3 Electron Interaction

In considering the interactions of electrons with combined dc and RF electric fields, it is initially helpful to review one aspect of the electron trajectories shown in Fig. 13.5. That is, as is shown again in Fig. 13.12, electrons with velocity E/B move in a straight line. Those electrons with $u_0 > E/B$ are deflected downward in the direction of the electric field toward the negative electrode. Those electrons with $u_0 < E/B$ are deflected upward in the direction opposite to the electric field toward the positive electrode.

The deflections indicated in Fig. 13.12 correspond to those occurring in a magnetron and depend on the position of the electrons relative to the RF field pattern. As an example, consider Fig. 13.13. An electron leaving the cathode at point A enters an accelerating field and is driven back into the cathode with appreciable energy. This electron extracts energy from the RF field and the only useful purpose that it serves is to contribute to the emission process by generating secondary electrons.

An electron leaving the cathode at B enters a decelerating field and moves toward the anode. With the correct frequency of operation, this electron will remain in synchronism with the decelerating field as it passes successive anode segments and will eventually strike the anode. In moving from the cathode to the anode, this electron acquires dc energy, which it gives up to the RF field. The paths of individual electrons may be traced in a relatively straightforward manner, as indicated in Fig. 13.13.

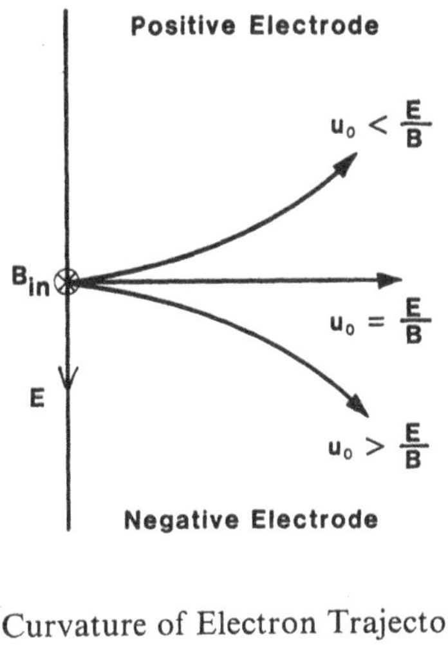

Figure 13.12 Initial Curvature of Electron Trajectories

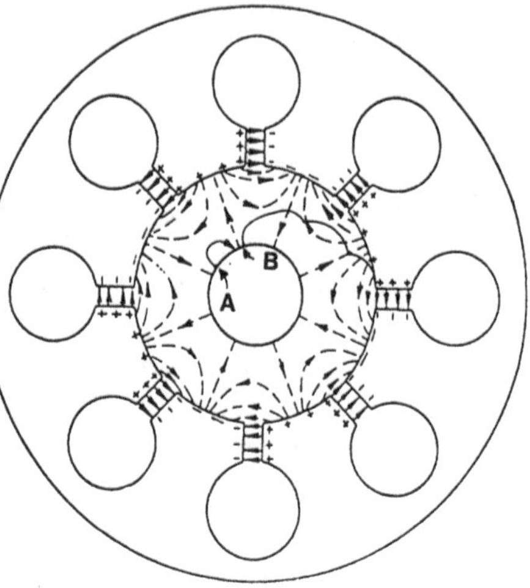

Figure 13.13 Paths Followed by Electrons in a Magnetron

(From *Microwave Magnetrons* by George B. Collins, Copyright 1948 by McGraw-Hill, Inc., Reproduced with permission.)

In an operating magnetron, a space-charge cloud of electrons surrounds the cathode, thus complicating the analysis of the trajectories. In the absence of RF fields, the electrons form a rotating sheath or hub around the cathode. When RF fields are present, those electrons in accelerating regions are forced back toward the cathode, thereby producing a depression in the space-charge hub. In the decelerating regions, the electrons move toward the anode and produce projections on the space charge hub.

As a result, the rotating space-charge cloud is distored into the spoke-like pattern shown in Fig. 13.14. Electron current flows through the spokes from the cathode to the anode. the spoke-like cloud rotates with an angular velocity such that it remains in synchronism with the alternating RF voltage on the anode segments. As the spokes pass the slots between anode segments, the RF fields provide a focusing action that tends to further bunch the electrons. As is indicated in Fig. 13.15, an electron at position a on the leading edge of a spoke is in a region where the radial component of the RF field is in opposition to the dc field. As a result, from $u = E/B$ its angular velocity will be reduced so that it will fall back into the spoke. An electron at point b will experience an increased field and will be accelerated into the spoke. This same focusing action helps maintain synchronism between the rotating spokes and the alternating field so that the operating conditions of the magnetron (magnetic field, voltage, et cetera) are not as critical as they would appear to be at first.

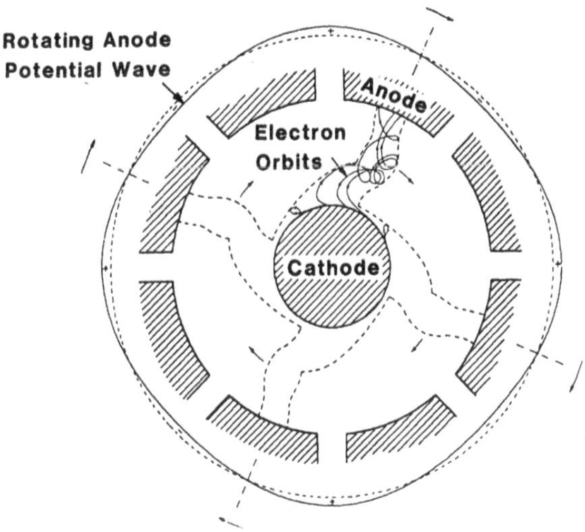

Figure 13.14 Spoke-like Electron Cloud in Eight-Cavity Magnetron

(From H.A.H. Boot and J. T. Randall, *IEEE Trans. Electron Devices*, July 1976.)

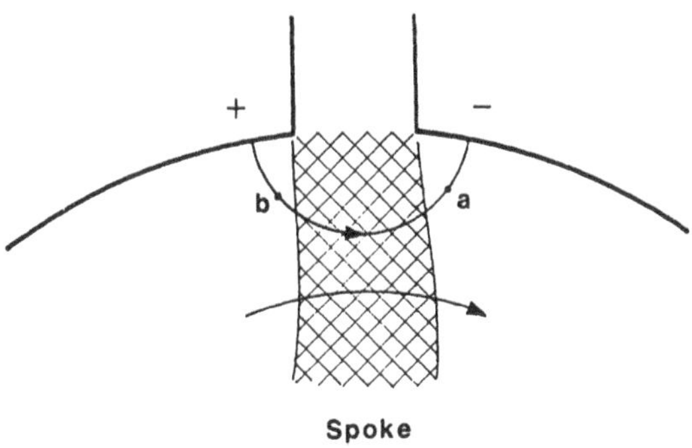

Figure 13.15 Focusing Action of RF Fields on Space Charge

13.2.4 Moding

In his science-fiction article on magnetrons, Pierce leads into a discussion of moding by commenting on the various problems faced by early magnetron engineers and the humorous names that they applied to the peculiarities of performance encountered.

> The Civil War Effect, or North-South Effect had to do with differences in performance when the direction of the magnetic field was reversed. The differences proved to be due to asymmetries in the tubes, and disappeared when the tubes were made more acurately. The Blowing Out of Brains Effect had simply to do with the melting of a copper end cap which covered the end of the anode block of a small magnetron to form part of the vacuum envelope. The Gnawing Away of the Anode Effect was similar. The Gardarens Swine Effect has to do with the particularly precipitous descent of certain contours on a perfomance chart, and the Valley of the Shadow was a region of low efficiency lying close to a high-efficiency region known as the Heights of Abraham.
>
> Undoubtedly the worst effect was one perhaps too serious for a fancy name. It was called, simply, moding. An engineer in using an early magnetron would be adjusting the load, that is, the impedance into which the magnetron fed power, in order to get more power. Suddenly, the power, which had been rising, would drop, and he would find the tube to be oscillating feebly at some new and unwanted frequency a little removed from the old. Or, he would be gradually raising the voltage to get more power, and the same thing would happen. The tube was said to be operating in another mode of oscillation.

Moding occurs because the RF circuit of a magnetron consists of many resonators coupled together. An analogous situation exists when several identical pendulums are suspended from a common support which is not rigidly held in place. For example, in Fig. 13.16, three pendulums are shown along with the three forms of motion (modes of oscillation) that they can have. The frequencies at which the pendulums swing back and forth are different for each mode.

Similarly, a magnetron may have several modes of oscillation which depend upon the coupling between cavities.

Typical field distributions for the modes in an eight-cavity magnetron are illustrated in Fig. 13.17. The modes are designated by a mode number *n*, which is the number of times the RF field pattern is repeated in going around

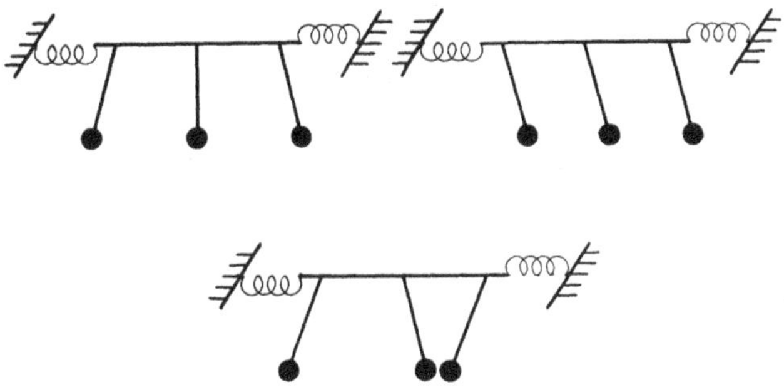

Figure 13.16 Modes of Oscillation of Three Pendulums Supported by a Movable Bar

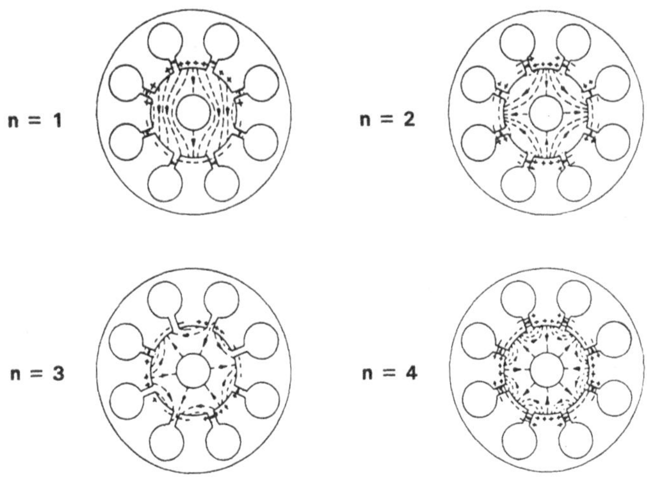

Figure 13.17 Typical Field Distribution for the Modes in an Eight-Cavity Magnetron

(From H.A.H. Boot and J. T. Randall, *IEEE Trans. Electron Devices*, July 1976.)

the anode once. The largest value of n is $N/2$, where N is the number of cavities. This mode is called the π mode and is always the desired mode of operation.

In addition to there being several modes of oscillation for a magnetron, there is, quite literally, an infinite number of electron velocities that can excite any given mode. As an example, the situation existing in the π mode is illustrated in Fig. 13.18. The eight anode segments of an eight-cavity magnetron are across the top of the figure. The potential distribution along the anode at various time intervals separated by one-half of a period are shown below the anode segments. An electron, such as one at point A and time t, is in a decelerating region and gives up energy to the RF field. An electron in the region between A and B is considered to be essentially shielded from the RF field by the anode segment.

The dashed lines represent the progress of electrons traveling in favorable phases (continually giving up energy) with the RF field. For example, electrons represented by line no.1 have a transit angle from cavity to cavity of π radians and travel in synchronism with the oscillating field. Electrons represented by line no.3 travel at one-third of the velocity of those represented by line no.1 and have a transit angle of 3π radians. A complete cycle of oscillation of the RF circuit occurs while these electrons pass by each anode segment in traveling from one cavity to the next. Still, these electrons are synchronous with the RF field and give up energy on every third cycle of oscillation of the RF field. This situation is analogous to keeping the pendulums in Fig. 13.16 moving by giving them a push on every third swing.

If the individual trapezoidal potential profiles in Fig. 13.18 are broken into their harmonic components, odd-numbered harmonics(1,3,5,7,et cetera) result, and it is found that the electrons represented by line no.3 are in synchronism with the third harmonic component of the field. Similarly, electrons represented by lines no.5 no.7, and an infinite set not shown are synchronous with still higher order harmonic components of the field. In principle, any of these harmonic interactions can produce oscillation in the π mode. Each of the other magnetron modes can be broken into an infinite set of harmonic components and can, therefore, in principle, be driven by electrons having an infinite range of velocities.

At this point, it should be remembered that there is an electron cloud circling about the cathode prior to magnetron oscillation. As an increasing anode voltage is applied to a magnetron when it is turned on, the radius of this space-charge cloud grows. If there are no fluctuations in this space-charge cloud, its radius will increase until it impinges on the anode, at which point dc current will flow and the magnetron will not oscillate.

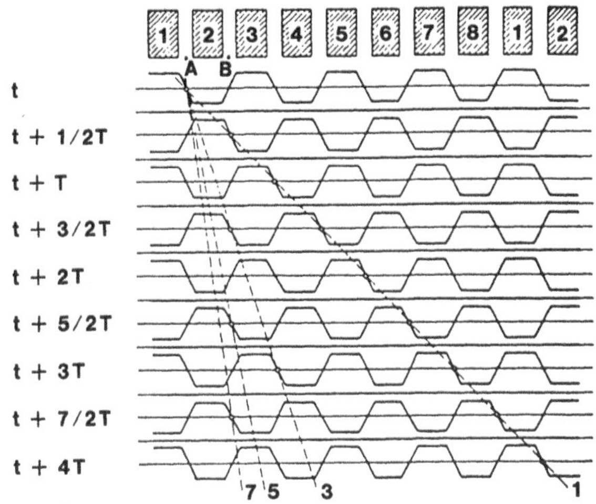

Figure 13.18 Potential Distribution in an Eight-Cavity Magnetron Operating in the π Mode (Dashed Lines Represent the Progress of Electrons Traveling in a Favorable Phase)

(From *Theory and Application of Microwaves* by Bronwell and Beam, Copyright 1947 by McGraw-Hill, Inc., Reproduced with permission.)

Because of fluctuations in the emission process (see CH. 5) and noise amplifying effects in the space-charge cloud, electron velocities and the charge density in the rotating space-charge cloud fluctuate randomly with time. The frequency of the fluctuations varies over such a wide range that the noise is referred to as "white noise," which implies fluctuations over an infinite frequency range. While the frequency range of fluctuations is not infinite, it is rich in components in the microwave range, and so there are components present to initiate any of the frequencies that may be supported by the cavity structure of a magnetron.

The resulting possible array of magnetron operating conditions is bewildering and this is, perhaps, what led Boot and Randall in their 1976 article to comment: "If this point had been appreciated in 1939 it is doubtful that the magnetron would ever have been made." Fortunately, there are a number of selection processes, both natural and by design, which very significantly limit the range of operating conditions of a magnetron. For example, the amplitudes of the harmonic fields decrease rapidly as harmonic number increases so that the probability of interaction at harmonic frequencies decreases.

As an increasing anode voltage is applied to a magnetron, increases occur in both the radius and angular velocity of the space charge cloud surrounding the cathode. When the angular velocity is such that the cloud is rotating synchronously with one of the RF modes, or one of the harmonic components of an RF mode, noise currents on the space charge cloud may cause oscillations to start in that mode. The voltages at which the oscillations are initiated in various modes and various harmonics are called the Hartree threshold voltages. Some of these voltages are plotted as a function of magnetic flux density for an eight-cavity magnetron in Fig. 13.19. The dashed line represents a harmonic of the $n = 3$ mode, which has a threshold voltage nearly identical to the $n = 4$ which is the (π) mode. As a result, unless precautions are taken, operation may occur in the $n = 3$ mode, rather than the π mode.

In considering precautions that can be taken to force a magnetron to operate in the desired mode, it is important to understand that the frequencies of the various modes are very close together. This is indicated by the solid

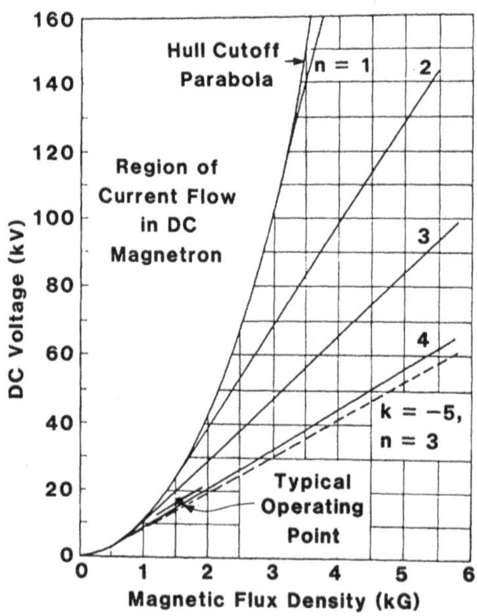

Figure 13.19 Hartree Threshold Voltage Diagram for an Eight-Cavity Magnetron

(From H.A.H. Boot and J. T. Randall, *IEEE Trans. Electron Devices*, July 1976.)

line in Fig. 13.20 for an eight-cavity tube. The separation between the $n =$ 3 and $n = 4$ modes is on the order of 1 to 2%. For structures having more than eight cavities (in practice, the number of cavities usually ranges from eight to eighteen), the separation in frequency between the π mode and its nearest neighbor becomes even smaller than 1 to 2%. Because of the frequency instabilities encountered when the mode separation is small, structural changes are made to the resonator to increase this separation.

Two techniques, the use of straps and the rising-sun structure, for increasing the mode separation are discussed at this point. The third, the use of a coaxial cavity, is discussed in a section on coaxial magnetrons (Sec. 13.5).

In the desired π mode of operation, alternate anode segments are at the same RF potential. Thus, if alternate segments are connected together by wires (straps) no additional inductance will be introduced because the ends of the straps are at the same potentials. The straps will add capacitance to the circuit, and so the π-mode frequency will be altered.

In modes other than the π mode, the voltage differences between alternate anode segments is not zero, and so the straps introduce inductance as well as capacitance resulting in different frequency shifts than occur for the π

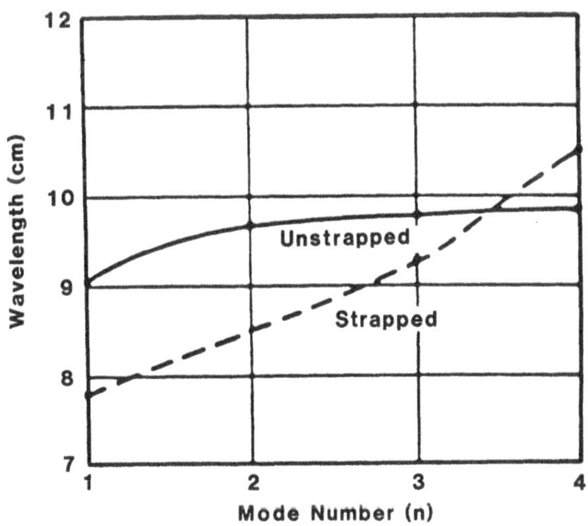

Figure 13.20 Mode Spectrum for an Eight-Cavity Magnetron

(From *The Magnetron Oscillator* by E. Kettlewell, Copyright 1971 by Mills and Boon, Ltd., Reproduced with permission.)

mode. The effect of strapping on mode separation in an eight-cavity magnetron is shown in Fig. 13.20. In low-power tubes, mode separations of 25 to 35% are achieved with strapping. In high-power tubes, the mode separation is normally about 10 to 14%. Thus, the undesired modes are at frequencies far enough removed from the π mode so that the tube can be prevented from operating in these modes.

Shown in Fig. 13.21 are examples of strapped structures and an unstrapped resonator known as a rising-sun structure. The way in which the rising-sun structure causes mode separation to occur is not as easy to understand as the strapping technique. If the field patterns associated with the various modes are examined, it is found that for the low-numbered modes the fields are such that the large cavities control the frequency of operation, and so the frequencies associated with these modes are relatively low. For the higher numbered modes (expect the π mode), the field patterns are such that the small cavities control the frequency of operation, and so the frequencies associated with these modes are relatively high. In the π mode, the frequency is midway between the resonant frequencies of the two cavity configurations.

The rising sun resonator system is generally used in the millimeter wave region where dimensions are small and it is difficult, if not impossible, to fabricate strapped resonators.

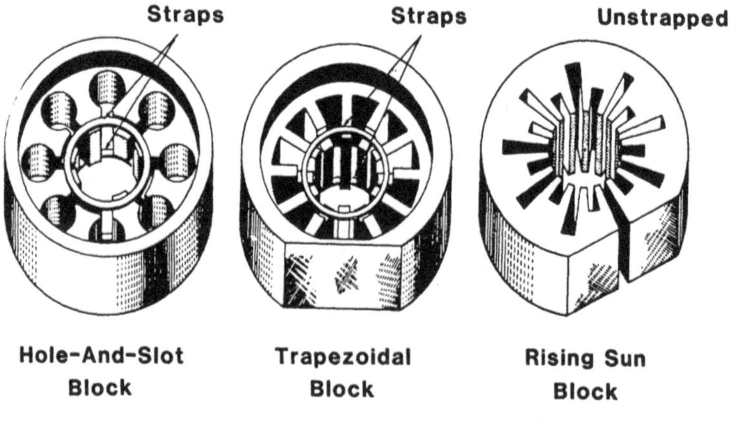

Figure 13.21 Techniques for Achieving Mode Separation

(From *Microwave Devices* by James T. Coleman, Copyright 1982 by Reston Publishing Company, Inc., Reproduced with permission.)

13.2.5 Coaxial Magnetrons [13.5]

An indication of the significance of the development of the coaxial magnetron is that it is listed by Skolnik [13.6], along with the cavity magnetron and the high-power klystron as significant events in the development of radar technology. The coaxial magnetron was developed as an alternative to the strapped cavity magnetron.

In the conventional magnetron, the strap-vane structure results in anode segments that are sometimes small and operate at an excessively high power density. Close interelectrode spacing leads to occasional voltage breakdown and the cathode loading is often excessive. As a result, in early radars, the conventional magnetron was often the least reliable component of the system.

As Fig. 13.22 shows, the coaxial magnetron contains an anode resonator structure very much like that in a conventional magnetron. This structure is surrounded by a single high-Q cavity operating in the TE_{011} mode. Slots in the back walls of alternate cavities of the anode resonator structure tightly couple the fields in these resonators to the surrounding cavity. In the π mode, the fields in every other cavity are in phase, and so they couple in the same direction into the surrounding cavity. Thus, the surrounding coaxial cavity helps stabilize the magnetron in the desired π mode of operation.

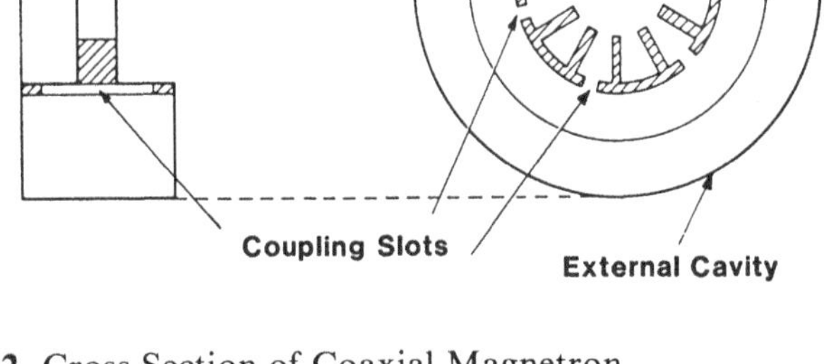

Figure 13.22 Cross Section of Coaxial Magnetron

(From *The Magnetron Oscillator* by E. Kettlewell, Copyright 1971 by Mills and Boon, Ltd, Reproduced with permission.)

Because straps are not required, the anode resonator for the coaxial magnetron can be larger and less complex than for the conventional strapped structure. A comparison of anode sizes for the same operating frequency and power is shown in Fig. 13.23. Because the anode is larger, the cathode and interelectrode spaces can also be larger. As a result, cathode loading is lower and voltage gradients are reduced. These factors all combine in the coaxial magnetron to produce relatively conservative magnetron designs having enhanced reliability compared with conventional magnetrons.

In the coaxial magnetron, the multiple-cavity anode structure has a series of modes like those of the conventional magnetron. In addition, there are various modes of oscillation of the coaxial cavity. Fortunately, the use of the TE_{011} mode makes possible the use of attenuators to damp unwanted modes.

In the desired TE_{011} mode, the RF electric field exists in circular form within the cavity as indicated in Figs. 13.22 and 13.24, and decreases to zero at the walls of the cavity. Current flow in the TE_{011} mode is in the walls of the cavity in circular paths about the axis of the tube. An attenuator placed as shown in Fig. 13.24 at the junction of the inner cylinder to the fixed end of the cavity couples very little to TE_{011} currents, but couples heavily to other modes, thereby suppressing these unwanted modes of oscillation.

Similarly, when the fields near the coupling slots from the anode resonator to the coaxial cavity are examined, it is found that the ends of the slots are free of TE_{011} mode current. Thus, an attenuator within the inner slotted

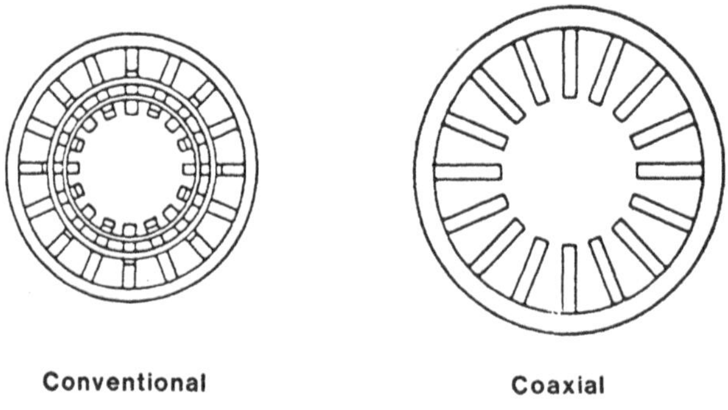

Figure 13.23 Comparison of Anodes of Conventional Strapped Magnetron and Coaxial Magnetrons

(From *Microwave Tube Manual* by Varian Associates, Air Force Publication Number T.O. 00-25-251, October 1979.)

cylinder near the ends of the coupling slots does not couple to the TE_{011} mode currents, but it does dampen undesired modes of the vane structure.

Another advantage of the TE_{011} mode is that one of the cavity end plates can be made movable (tuning piston in Fig. 13.24) without interrupting the flow of current. As a result, tuner reliability is high for coaxial magnetrons.

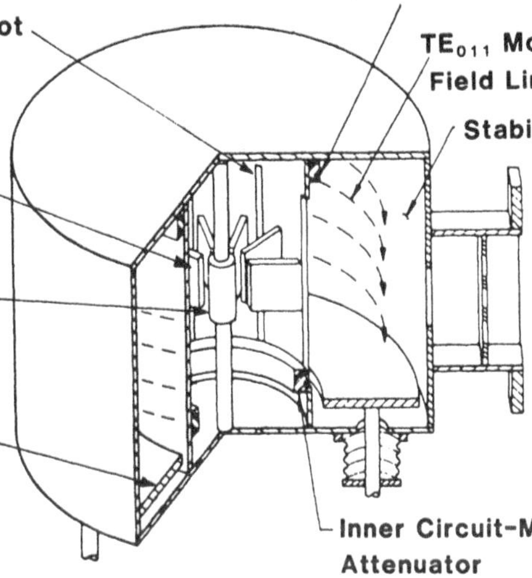

Figure 13.24 Cutaway View of Coaxial Magnetron

(From *Microwave Tube Manual* by Varian Associates, Air Force Publication Number T.O. 00-25-251, October 1979.)

13.2.6 Inverted Magnetrons

A further advantage of the coaxial-magnetron configuration is that it can be turned inside out, or inverted, as shown in Fig. 13.25. In this figure, an inverted structure is compared with a conventional rising-sun magnetron designed to operate at the same frequency. In comparing the two structures, it should be noted that the cathode size in the inverted configuration is, perhaps, an order of magnitude larger than the cathode in the conventional magnetron. Also, the number of anode vanes and cavities can be large in the inverted structure because moding is controlled by the center coaxial cavity.

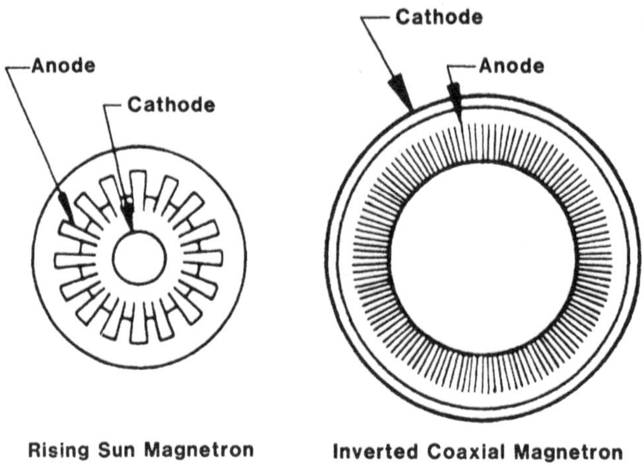

Figure 13.25 Comparison of Rising-Sun Magnetron Structure with Inverted Coaxial Magnetron Designed for Operation at the Same Frequency

(From *Microwave Tube Manual* by Varian Associates, Air Force Publication Number T.O. 00-25-251, October 1979.)

The inverted coaxial configuration has application in the millimeter-wave region where dimensions are extremely small. In particular, the power of the conventional rising-sun magnetron is limited by the high cathode loading. This problem is alleviated in the inverted structure. In addition, the center TE_{011} cavity lends itself to coupling to an output waveguide operating in the circular electric mode which has extremely low loss.

It should also be noted that experiments are being carried out with inverted magnetrons at S band to determine their ability to generate extremely high power levels [13.7].

13.2.7 Magnetron Tuning

Many techniques are used for changing the frequency (tuning) of magnetrons. In one way or another, these tuning techniques employ changes in the capacitance or the inductance of the resonant cavities. For example, the conventional hole-and-slot structure may be tuned as indicated by the single-cavity segment in Fig. 13.26. The capacitance of the cavity may be varied by the movable metallic element (or dielectric element) in the vicinity of the slot portion of the cavity. The inductance of the cavity may be decreased

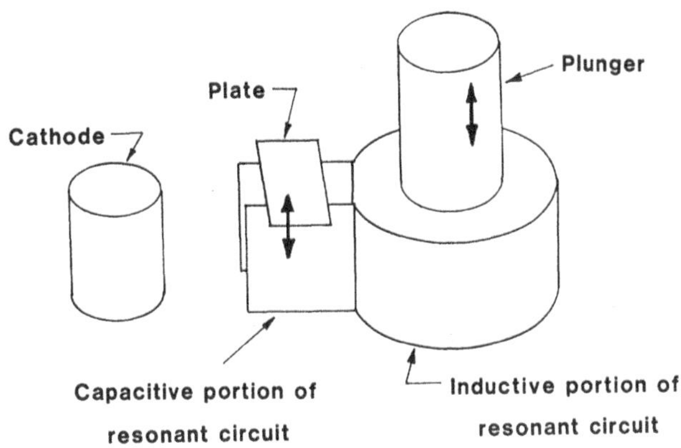

Figure 13.26 Basic Tuning Techniques for a Hole-and-Slot Resonator

by inserting a metallic plunger into the hole portion of the cavity. Configurations of capacitive and inductive tuning elements are shown in Fig. 13.27.

In coaxial magnetrons, tuning may be accomplished by changing the volume of the cavity with a movable end plate as indicated in Fig. 13.28. A fine-tuning technique using a compressible ring in the wall of the coaxial cavity is also shown in Fig. 13.28.

In the coaxial magnetron, dielectric tuning may be accomplished as indicated in Fig. 13.29 by inserting rotatable dielectric paddles into the high-electric-field region of the coaxial cavity. The lowest frequency occurs when the paddles are aligned with the electric field. The highest frequency occurs when the paddles are perpendicular to the electric field lines.

There are numerous techniques by which the tuning elements may be positioned in a magnetron. Examples are the mechanical gear arrangements on early magnetrons in which the position of the tuning element was adjusted through a bellows using a gearing arrangement. Tuning speed, backlash, and drive torque are limiting factors. Early frequency-agile magnetrons were tuned by a hydraulic system that replaced the mechanical gear system. Tuning was slow and maintenance difficult, so these systems are becoming obsolete.

There are various electromagnetic techniques for positioning tuning elements. The electromagnetically driven tuner shown in Fig. 13.30(a) operates much like the voice coil of a speaker. Current through a coil produces a magnetic field that acts with the field of a permanent magnet for axial positioning of a shaft to which the tuning element is attached. An electric motor can be coupled through a gear-and-cam arrangement to position a tuning plunger as indicated in Fig. 13.30(b). Accurate tuning may be accomplished through a resolver.

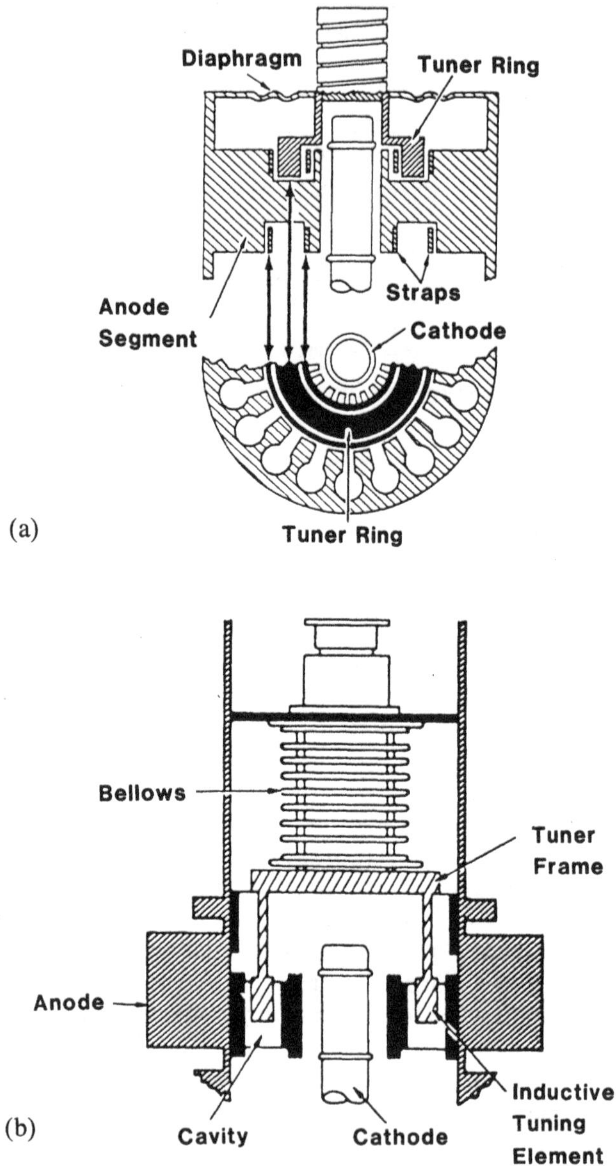

Figure 13.27 Configurations of Capacitive (a) and Inductive (b) Tuning Elements

(From *Microwave Devices* by James T. Coleman, Copyright 1982 by Reston Publishing Company, Inc., Reproduced with permission.)

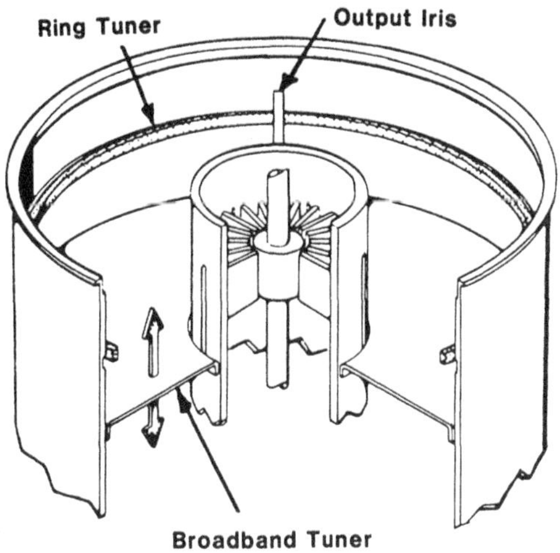

Figure 13.28 End Plate (Broadband) Tuning and Ring Tuning of Coaxial Magnetron

(From *Microwave Tube Manual* by Varian Associates, Air Force Publication Number T.O. 00-25-251, October 1979.)

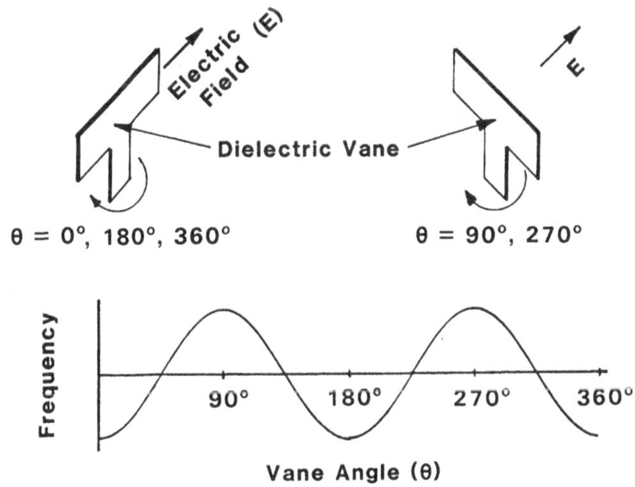

Figure 13.29 Dielectric Tuning of Coaxial Magnetron

(From *Microwave Tube Manual* by Varian Associates, Air Force Publication Number T.O. 00-25-251, October 1979.)

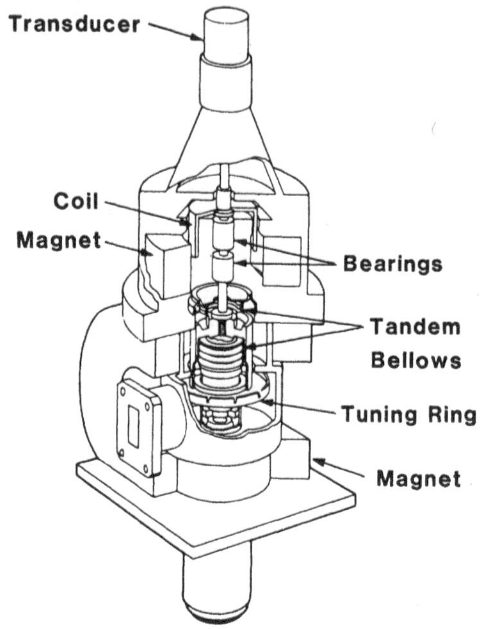

(a)

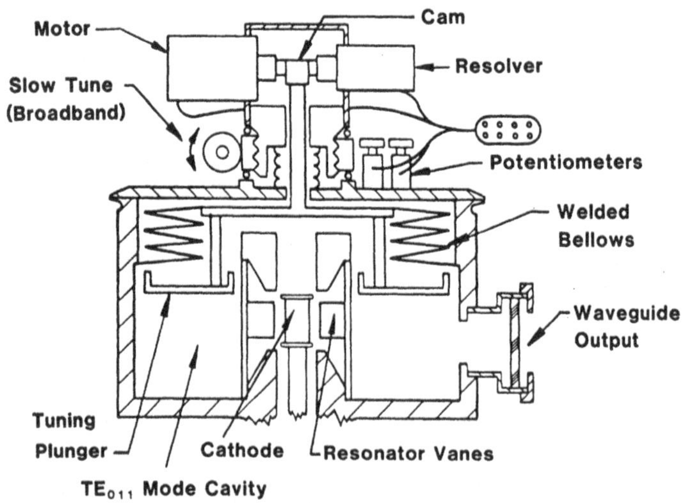

(b)

Figure 13.30 Electromagnetic Tuning Techniques: (a) Electromagnetic Driven Tuner; (b) Motor-Driven End-Plate Tuner

(From *Microwave Tube Manual* by Varian Associates, Air Force Publication Number T.O. 00-25-251, October 1979.)

One of the most sophisticated motor-driven tuning systems is in the spin-tuned magnetron. As shown in Fig. 13.31, a rotary tuner mounted on a ball-bearing system in vacuum is used for very fast inductive tuning. The tuner is driven by an electric motor through the vacuum envelope by a magnetic coupling system. The tuner is a copper cylinder that forms part of the inductive portions of the cavity resonator. As the tuner is rotated, holes in the cylinder move into and out of alignment with the cavities, and thereby inductively tune the cavities.

In X-band versions of the magnetron, there are 16 cavities and a rotor with 16 holes. With a nominal rotor speed of 4000 rpm, the frequency of the magnetron is changed over 1000 complete tuning cycles per second. The frequency excursion of the magnetron is approximately 5% of the carrier frequency.

Because the ball bearings supporting the rotary tuner must operate in vacuum, bearing life would appear to be a matter of considerable concern. In the Philips versions of the spin-tuned magnetrons, bearing life is over 4000 hours and, in most cases, exceeds cathode life.

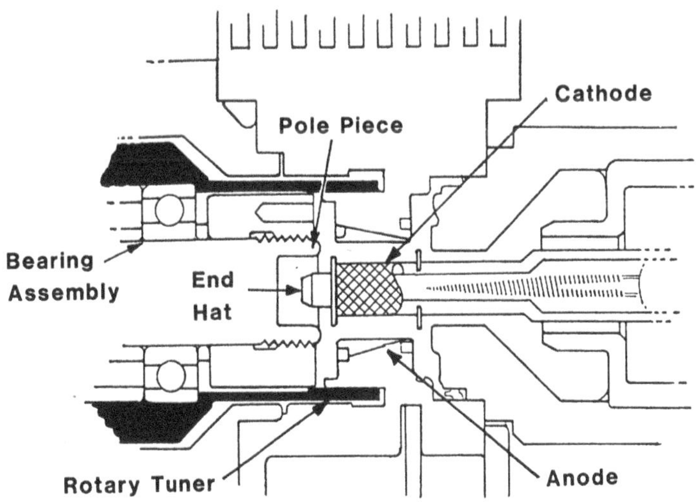

Figure 13.31 Cross Section of Spin-Tuned Magnetron

13.2.8 Cathode and Heater Operation

In crossed-field devices, electrons that are out of phase with the RF fields are driven back into the surface of the cathode (see Figs. 13.13 and 13.14, for examples). These electrons produce secondary emission and heating of the cathode. The secondary emission process was discussed in the chapter on cathodes (Ch.5). In particular, its significance in the operation of some magnetrons was illustrated by the example of the Philips YJ 1183 spin-tuned magnetron in which it is estimated that the thermionic emission is 3 A/cm^2 and the secondary emission is 23 A/cm^2 for a total cathode emission of 26 A/cm^2.

In the microwave oven magnetron, the cathode used is carburized thoriated tungsten for which the secondary emission coefficient is low. As a result, thermionic emission is the main emission mechanism. Carburized thoriated tungsten is used to achieve long tube life.

The importance of cathode heating by out-of-phase electrons cannot be overstated. Typically, on the order of 5 to 10% of the anode power input may go into cathode heating. In some magnetrons, this effect is such that the filament voltage can be removed once the tube is operating and the cathode temperature is maintained by bomdarding electrons. A more typical case is illustrated by Fig. 13.32, which shows how magnetron filament voltage should be reduced as average anode current is increased if essentially constant cathode temperature is to be maintained.

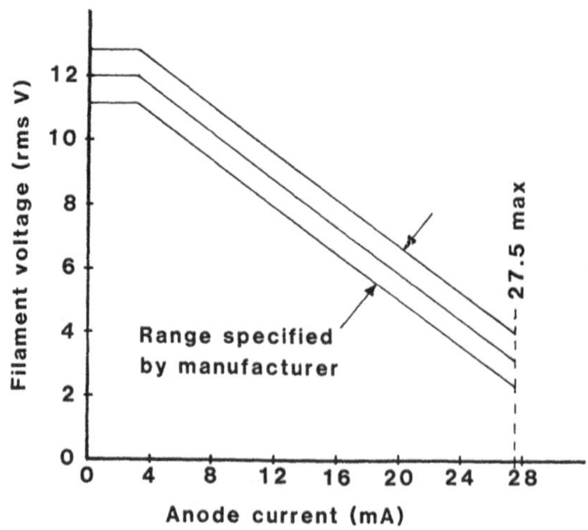

Figure 13.32 Magnetron Filament Voltage as a Function of Average Anode Current

The primary effect of cathode overheating, which occurs if filament voltage is not adequately reduced, is excessive evaporation of cathode material, resulting in early depletion and low emission. The effect of underheating is that, since the emissive surface of the cathode is renewed by migration of material from below the surface and, since migration rate is a function of temperature, depletion of emissive material on the cathode surface occurs.

Either overheating or underheating eventually result in low emission and, in turn, low output power. Low emission may also lead to moding, which is evidenced by higher voltages (resulting from improper loading of the modulator), lower currents, much lower output power, and a shift in frequency. Finally, low emission may also lead to arcing because of the higher voltages or because of the increased electric field at the cathode surface in the absence of adequate space charge. Arcing results in gas evolution and the destruction of the cathode surface and, along with overheating, leads to the deposition of emissive material on other critical areas within the magnetron.

An example shown in Fig. 13.33 of a particularly critical area is the end hat on the cathode in a spin-tuned magnetron. If precautions are not taken, the deposition of emissive material on the end hat leads to electron emission from it. The resulting electron current impinges on the magnetic pole piece and may heat it to the Curie temperature at which point it becomes nonmagnetic. As a result, the magnetic field collapses and anode current may increase to the point where the anode melts.

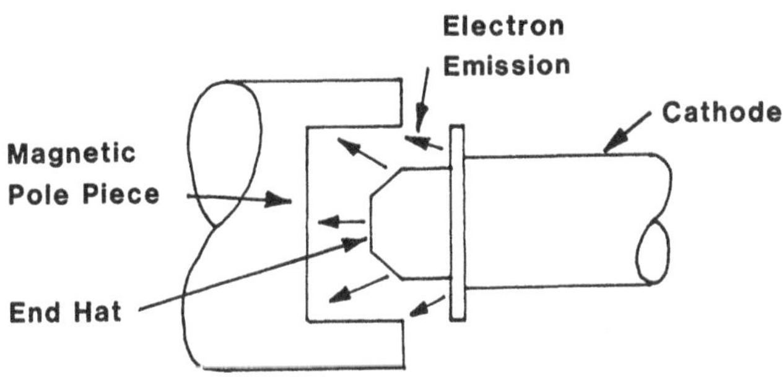

Figure 13.33 End-Hat Emission Resulting from Migration of Cathode Material

Finally, the disasterous effect of overheating a magnetron cathode might possibly be best illustrated by the actual example in which the filament voltage schedule given in Fig. 13.25 was not followed. In this case, the manufacturer of the equipment utilizing the magnetron did not properly understand the effect of cathode overheating and failed to reduce filament voltage adequately at high average anode currents. The result was a mean time between failures (MTBF) of approximatley four hours for a magnetron costing over $30,000 (1980 dollars).

13.2.9 Performance

The voltage-current characteristics of a magnetron are usually displayed on a performance chart like that shown in Fig. 13.34. Magnetic flux density is fixed by the manufacturer, and so operation is along one of the constant flux density lines. Normally, the optimum operating condition is indicated, for example, by the small circle at 26 kV and 38 A. As anode voltage is varied, current changes, and so power output and, to some extent, efficiency change.

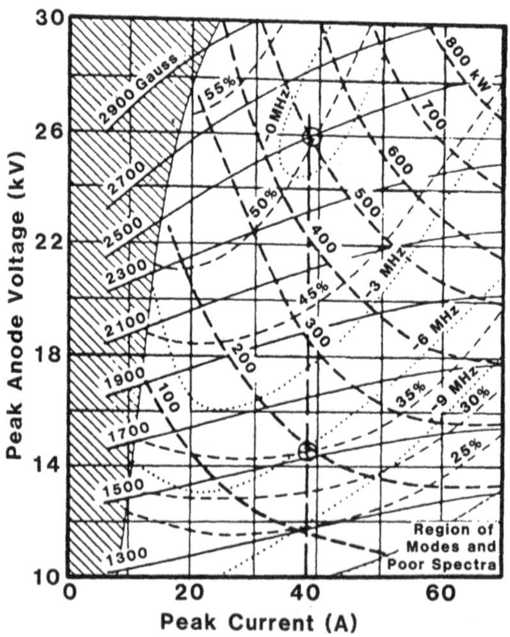

Figure 13.34 Typical Magnetron Operating Characteristics

(From *Microwave Tube Manual* by Varian Associates, Air Force Publication Number T.O. 00-25-251, October 1979.)

Finally, as current changes, the electronic impedance varies, which, in turn, affects the resonant frequency of the resonator, and so the frequency of oscillation changes. The change in frequency with current is called frequency pushing and the frequency shift per ampere is known as the pushing figure.

Varying load impedances reflected into a magnetron affect the resonance condition and with it the operating frequency and power output. A diagram, known as a Rieke diagram, is normally used to show the effects of load impedance on performance. As shown in Fig. 13.35, a Rieke diagram is a Smith chart on which are plotted contours of constant power and frequency. It should be noted that, for a given load mismatch and resulting voltage standing wave ratio (VSWR), the power and frequency are highly dependent on the electrical distance between the tube and load. This is because a load may appear to be inductive, resistive, or capacitive depending on its distance from the tube. Thus, in Fig. 13.35, for example, for a load VSWR of 1.5, the frequency may vary from –0.7 to +0.75 MHz relative to the value at the VSWR of 1.0, and the power may vary from 1.07 to 1.44MW.

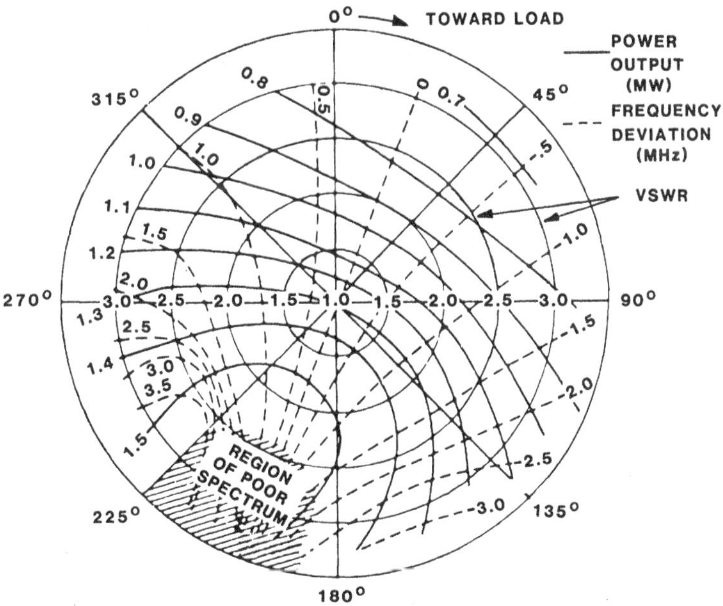

Figure 13.35 Rieke Diagram for Displaying Effect of Load on Tube Operation

(From *Microwave Tube Manual* by Varian Associates, Air Force Publication Number T.O. 00-25-251, October 1979.)

If a magnetron operates into a load, the impedance of which varies with time, then frequency and power will vary with time. As an example, a rotating antenna will usually present a varying impedance to a magnetron, so frequency and power will vary with antenna position. The use of ferrite isolators significantly reduces this variation. The resulting variation in frequency with changes in load conditions is called frequency pulling. The pulling figure is measured by operating the tube into a load with a specified VSWR (usually 1.5 to 1) and determining the frequency shift as the phase of the mismatch is varied through 360°.

In addition to frequency pulling and frequency pushing, it should be pointed out that the operating frequency of a magnetron is subject to variations in tube temperature. This is because cavity size determines frequency and cavity size is affected by temperature. In a low frequency magnetron, the RF circuit elements are relatively massive and the time required to reach thermal equilibrium may be as much as an hour, as indicated by Fig. 13.36. At X band, cavity dimensions are small and the time to reach equilibrium is a few minutes.

In some devices, compensation techniques are used to offset the temperature effects. For example, bimetallic disc elements can be used in the cavity of a coaxial magnetron to correct for cavity expansion. Similarly, bimetallic elements can be used in the drive linkage to the tuner to compensate for temperature induced frequency drift.

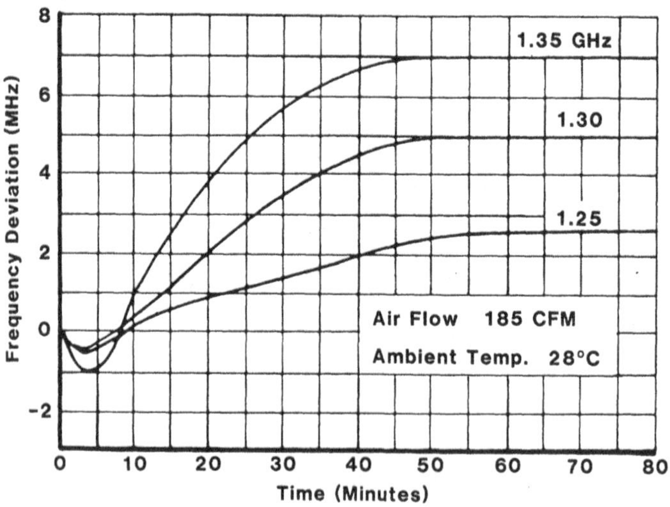

Figure 13.36 Effect on Temperature on 1.3 GHz Magnetron

(From *Microwave Tube Manual* by Varian Associates, Air Force Publication Number T.O. 00-25-251, October 1979.)

The overall publishable power characteristics of magnetrons are shown in Fig. 13.37. Peak power levels well in excess of one MW have been available since the early days of magnetron development. In the laboratory, peak powers in the gigawatt range have been generated (not shown).

Average and CW power levels are limited because the beam power must be dissipated by the RF structure. This matter is discussed further in following sections concerning crossed-field amplifiers. Still, average powers approaching the megawatt level are available. Magnetron conversion efficiency from dc to RF is normally in excess of 50% and may be as high as 85%.

Fixed-frequency and tunable magnetrons are available that weigh as little as a few ounces or as much as about 200 lbs. These weights may be reduced in weight-critical applications through the use of samarium cobalt magnets. Possible magnet configurations are shown later in this chapter in discussions concerning crossed-field amplifiers.

It would be irresponsible to complete a discussion of magnetrons without commenting on the microwave oven magnetron. These tubes typically operate at a power level of about 600 W and at a frequency of 2.45 GHz. They are now manufactured in quantities of over five million per year at a selling price of below $15.00. The quantity of these tubes manufactured far exceeds any other microwave tube. In his recent article on magnetrons [13.1], Brown notes that there are an estimated 50 million microwave ovens in the world. If turned on at the same time, they would produce 25 GW of microwave power!

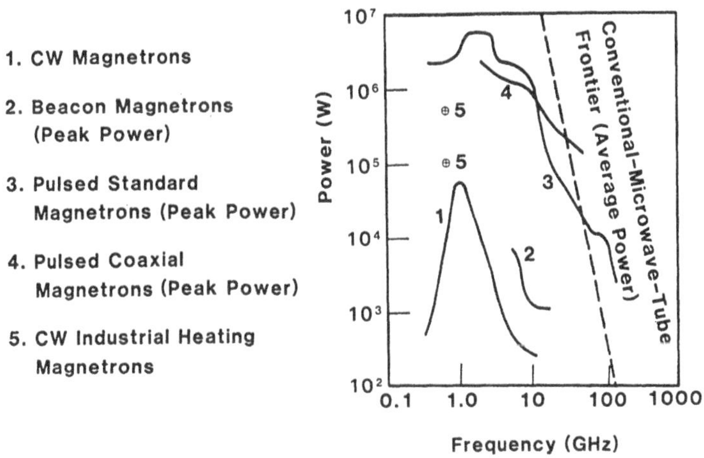

Figure 13.37 Magnetron Power Capabilities

(From J. Skowron, (MPTD), Raytheon Co., Workshop on High-Power, Space-Based Microwave Systems, Los Alamos, NM, March 1985.)

13.3 Crossed-Field Amplifiers [13.8]

As is shown in Fig. 13.1, crossed-field amplifiers (CFAs) are classified as either distributed-emission or injected-beam devices. Configurations of these devices are shown in Fig. 13.38. Like a magnetron, the distributed-emission CFA contains a cathode adjacent to the full length of the RF structure. In an injected-beam CFA, an electron gun generates a ribbon-shaped electron beam that passes through the device near an RF circuit. Operation of the injected beam CFA is similar in many respects to an O-type TWT.

The injected-beam CFA was the first approach to the development of a magnetron-interaction type amplifier. Very-high-power pulsed devices were developed in France. In the US, much development work was devoted to CW and low peak power injected-beam CFAs for very wideband use in air borne electronic countermeasures systems. In spite of significant development efforts, injected-beam CFAs are not now being produced because the applications for which they might be suited are better filled by other devices. For this reason, only distributed-emission CFAs are discussed in this book and henceforth they are referred to simply as CFAs.

13.3.1 Distributed-Emission CFAs

The first practically successful CFA was introduced by Brown in 1953 [13.1]. The device was very similar to a magnetron, except that, as shown in Fig. 13.39, the RF circuit was interrupted to provide input and output connections. The rf structure was a strapped-vane network similar to that in a magnetron. The tube had a gain of 13 dB over a frequency range of 3780 to 3850 MHz. The peak power output was 24 kW and the efficiency was 40%.

Since the first demonstration of the CFA, every conceivable configuration has been examined, as indicated in Fig. 13.40. The circular format is the one shown in Figs. 13.38 and 13.39. In the linear format, the cathode and RF circuit are parallel.

In the re-entrant circular format, the electron cloud circles around the cathode as it does in a magnetron until electrons give up sufficient energy to reach the anode. This contributes to high efficiency and, as it is believed, to improved turn on and starting. In some devices, a drift region is used between the output and input ports (see Fig. 13.38(b)) so that space charge forces cause electron bunches to disperse. This removes the modulation from the re-entering cloud of electrons and, thus, minimizes feedback from the output to the input.

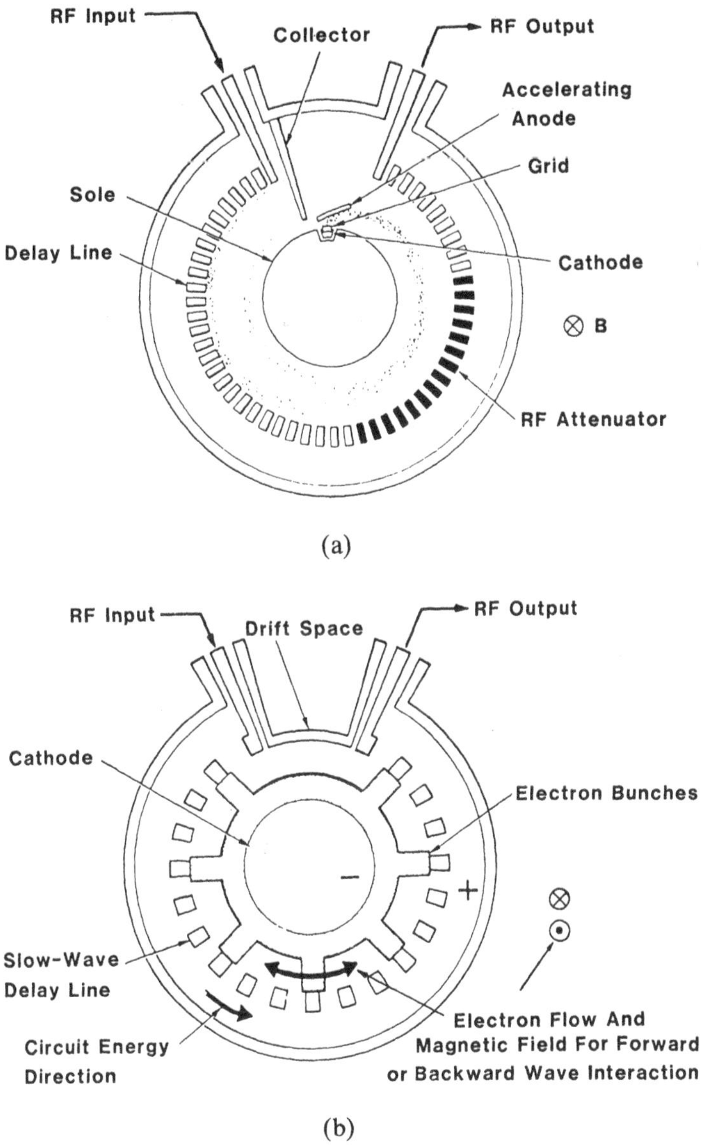

Figure 13.38 Basic Crossed-Field Amplifier Types: (a) Injected Beam; (b) Distributed Emission

(From J. Skowron, *Proc. IEEE*, March 1973.)

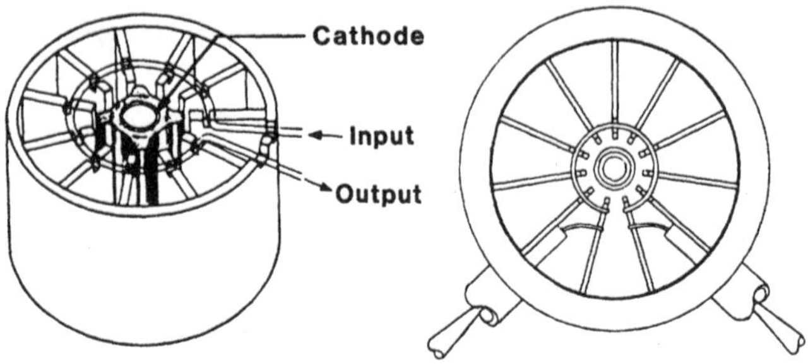

Figure 13.39 Schematic of the First Device to Demonstrate the Principles of Converting a Magnetron into an Amplifier

(From William C. Brown, *IEEE Trans. on Electron Devices*, Nov. 1984.)

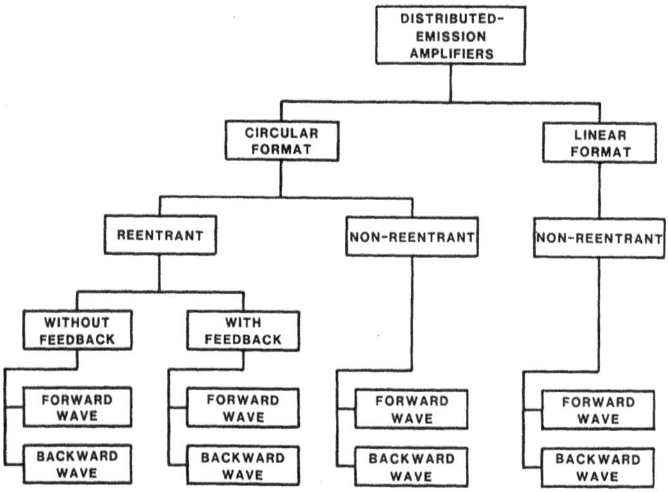

Figure 13.40 Family Tree of Distributed Emission CFA's

(From *Microwave Tube Manual* by Varian Associates, Air Force Publication Number T.O. 00-25-251, October 1979.)

As Fig. 13.40 shows, either forward-wave or backward-wave operation is possible with any of the CFA configurations. The characteristics of both forward-wave and backward-wave devices will be discussed in sections that follow.

A cross-sectional view of a typical CFA is given in Fig. 13.41. Magnets arc not shown, but the iron pole pieces to which the magnets attach are shown. The slow-wave circuit will be discussed in following sections. Helix-coupled-vane circuits are often used in devices of this type.

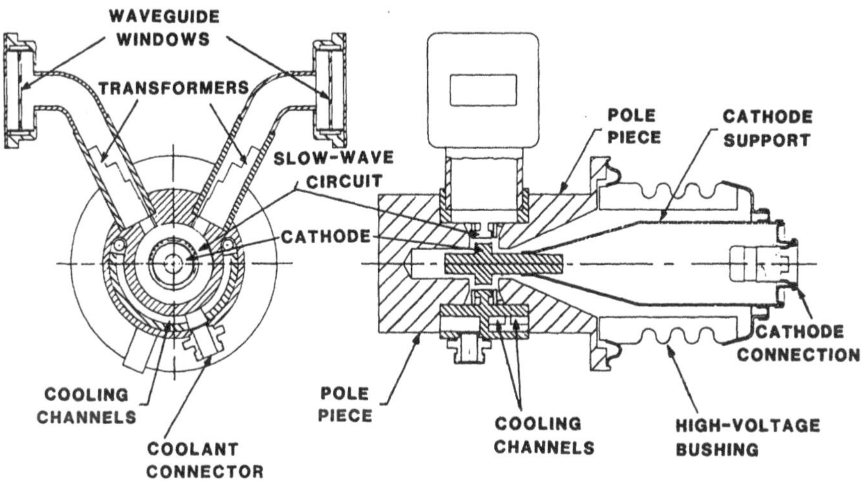

Figure 13.41 Cross Sectional View of Typical CFA

(From *Microwave Tube Manual* by Varian Associates, Air Force Publication Number T.O. 00-25-251, October 1979.)

13.3.2 CFA Emission

Indirectly heated cathodes and cold cathodes are used in CFAs. The cold-cathode technique is applied primarily in high-power amplifiers ($\geq$ 10kW output), although lower power levels are possible. At lower power levels, the RF electric fields are less likely to impart sufficient energy to electrons to produce the required secondary emission.

With the cold-cathode technique, the initiation of emission build-up results from the presence of a relatively small density of free electrons near the cathode surface. These electrons result from various emission processes incuding the very low level of emission that occurs even at, or near, room temperature. For emission to build up properly, the cathode-to-anode voltage must be correct and the RF drive level must be adequate. Surprisingly, the resulting emission build-up is very rapid ($<$ 10ns) and there is very little jitter.

13.3.3 DC Operation

In a CFA with a cold secondary emitting cathode, it is possible to apply anode voltage without drawing current. Upon application of the RF input pulse, out-of-phase electrons are driven into the cathode surface and the secondary emission process is initiated. As shown in Fig. 13.42, the emission process builds up rapidly as is indicated by the fact that the RF output power is at its maximum value within about 10 ns.

At the end of a pulse, space charge in the cathode-anode region must be cleared, otherwise the device can oscillate or generate noise. In non-reentrant CFAs, like the dematron, this occurs automatically, but efficiency is low, so these devices have not found application. In re-entrant CFAs, a control or cut-off electrode is pulsed positive with respect to the cathode to collect the space charge. The configuration of a CFA containing a cut-off electrode is shown in Fig. 13.43.

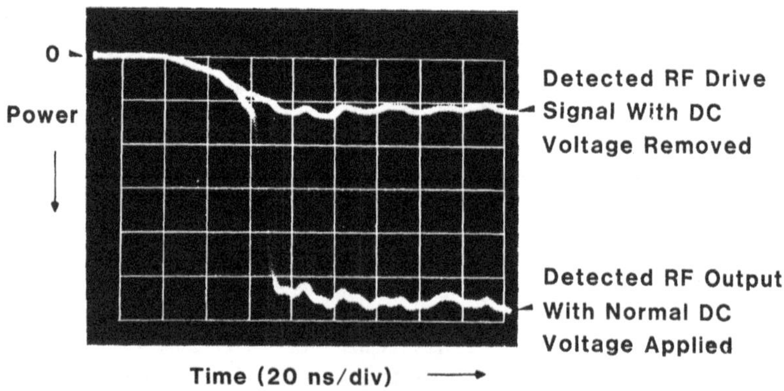

Figure 13.42 Leading Edge of RF Drive and Output Pulse on a dc-Operated CFA

(From J. Skowron, *Proc. IEEE*, March 1973.)

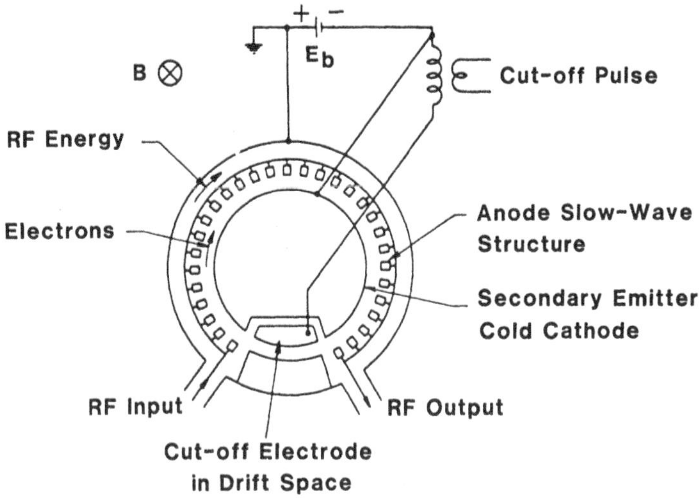

Figure 13.43 CFA with Cutoff Electrode

(From J. Skowron, *Proc. IEEE,* March 1973.)

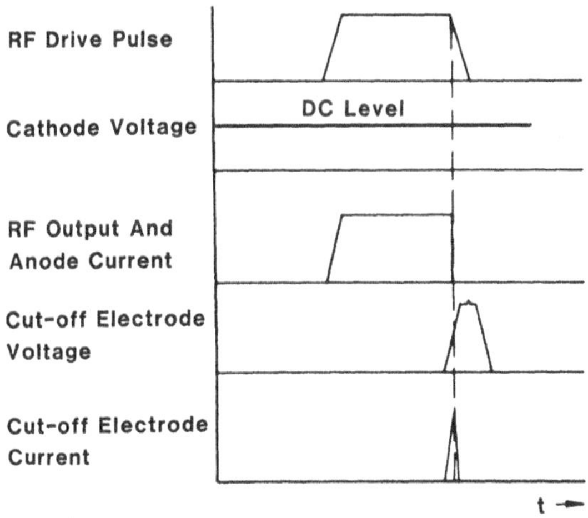

Figure 13.44 Timing of Pulses in a dc-Operated CFA

(From *Microwave Tube Manual* by Varian Associates, Air Force Publication Number T.O. 00-25-251, October 1979.)

The sequence of pulses applied to a CFA during dc operation is shown in Fig. 13.44. The applied RF drive pulse produces anode current and an RF output pulse. At the end of the RF pulse, the voltage pulse to the cut-off electrode produces a short pulse of current corresponding to the charge stored in the interelectrode space.

13.3.4 Slow-Wave Circuits

As is shown in Fig. 13.40, CFAs may be forward-wave or backward-wave devices. The questions of why there are forward-wave and backward-wave CFAs and of when one should be selected over the other do not have straightforward answers. The first practical demonstration of a CFA by Brown (Section 13.3.1) was with a backward-wave circuit similar to a magnetron circuit. Early development of the CFA for radar applications also centered on backward-wave operation.

A problem with the backward-wave amplifier (as was pointed out in the discussion of dispersion characteristics of the TWT in Ch.10) is that it is a voltage-tuned amplifier. As shown in Fig. 13.45, the frequency of operation increases with modulator voltage. Operation over about a 10% bandwidth is possible with corresponding adjustments in anode voltage. In a forward-wave CFA, on the other hand, operation is more like that of a TWT in that a reasonable bandwidth can be realized at a fixed modulator voltage. In high-power forward-wave CFAs, bandwidths on the order of 10 to 20% are typical.

In the final analysis, the selection of a forward-wave or backward-wave circuit is based mostly on the requirements of the intended applications. Perhaps, the most important requirement is that the circuit must have the

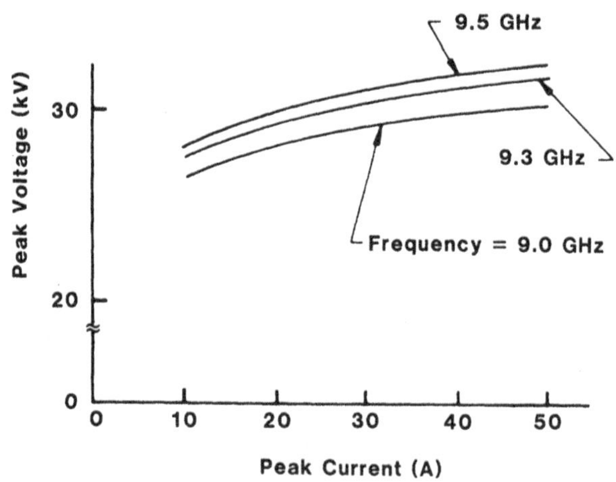

Figure 13.45 Variation of Modulator Voltage with Frequency of Operation of a Backward-Wave CFA

(From *Microwave Tube Manual* by Varian Associates, Air Force Publication Number T.O. 00-25-251, October 1979.)

thermal capability to handle the required power dissipation. This problem of thermal capacity will be addressed at the end of this chapter. At this point, it should be remembered that in the CFA, as in the magnetron, the slow-wave circuit also acts as the collector for the electron beam. In addition to the thermal requirement, bandwidth must be sufficient for the application and high efficiency is always important.

To meet these requirements, the tube designer may sometimes have to select a backward-wave circuit, even though the design and operation of the modulator are more difficult than for the forward-wave circuit.

Innumerable configurations of slow-wave circuits can be visualized and have been examined for use in CFAs. Most have been eliminated from consideration for use because of thermal, mechanical, or electrical deficiencies. Some circuits that have survived are shown in Fig. 13.46.

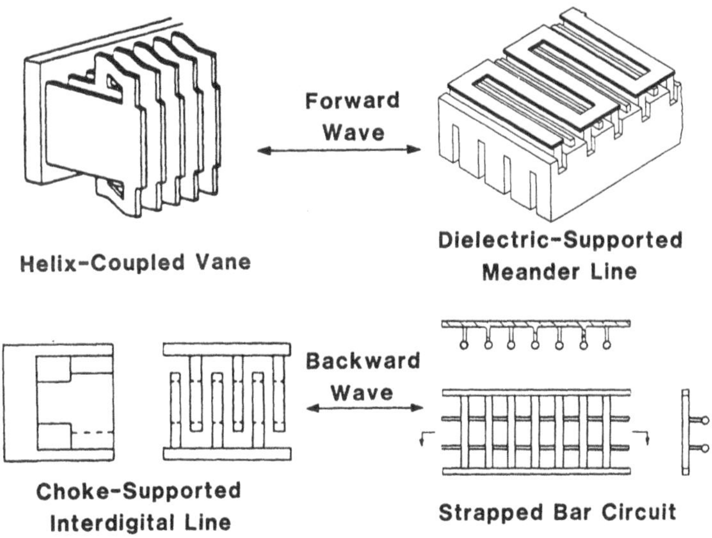

Figure 13.46 Examples of CFA Slow-Wave Circuits

(From *Microwave Tube Manual* by Varian Associates, Air Force Publication Number T.O. 00-25-251, October 1979.)

13.3.5 Voltage-Current Characteristics

Like a magnetron, a CFA is a diode with a magnetic field applied. A typical voltage current *(V-I)* characteristic is shown in Fig. 13.47. Because of the magnetic field, the tube is essentially cut off until a relatively high voltage level is reached. Factors which affect the *V-I* characteristic are frequency (and the dispersion characteristic of the slow-wave circuits), magnetic field, and RF drive power.

The impedance characteristics of CFA's are of importance when considering their operation with a modulator. As Fig. 13.47 shows, there is a significant difference between static impedance, which is simply the operating voltage divided by the operating current, and the dynamic impedance, which is the rate of change of voltage with current. The dynamic impedance may be an order of magnitude smaller than the static impedance.

The magnetic field applied to a CFA is not normally adjustable in a system. The magnetic field, however, affects the *V-I* characteristics in the manner shown in Fig. 13.48.

In connection with comments concerning the magnetic field, it should be noted that the use of samarium cobalt can significantly reduce magnet and

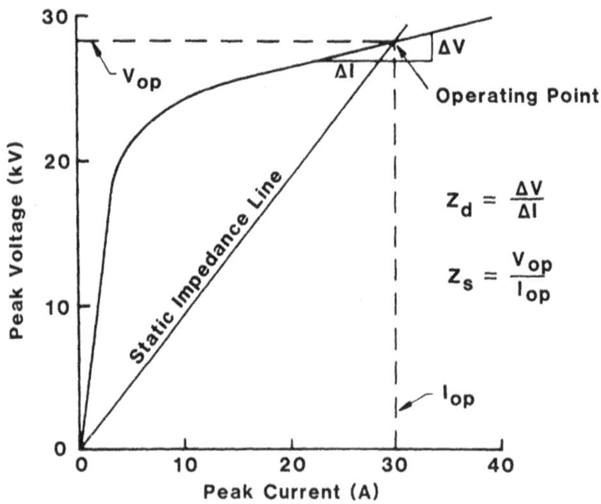

Figure 13.47 Voltage-Current Characteristics of a CFA

(From *Microwave Tube Manual* by Varian Associates, Air Force Publication Number T.O. 00-25-251, October 1979.)

CFA weight. For example, a size comparison between Alnico V and samarium cobalt magnets is illustrated in Fig. 13.49. Still another possibility in cold-cathode CFAs is the placement of the magnet inside the cathode as illustrated in Fig. 13.50. The use of samarium colbalt in this manner has been demonstrated in low frequency CFAs.

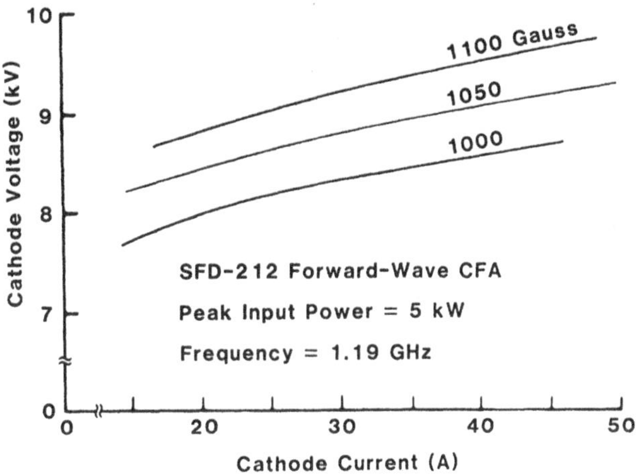

Figure 13.48 Affect of Magnetic Field on V-I Characteristic

(From *Microwave Tube Manual* by Varian Associates, Air Force Publication Number T.O. 00-25-251, October 1979.)

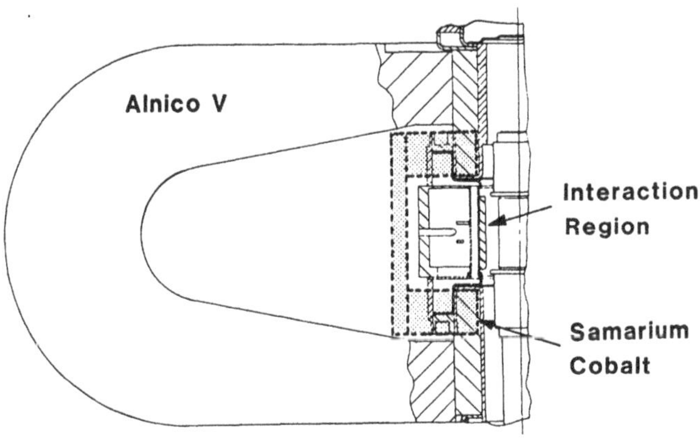

Figure 13.49 Size Comparison Between Alnico V and Samarium Cobalt Magnets for a CFA

(From J. Skowron, (MPTD), Raytheon Co., Workshop on High-Power, Space-Based Microwave Systems, Los Alamos, NM, March 1985.)

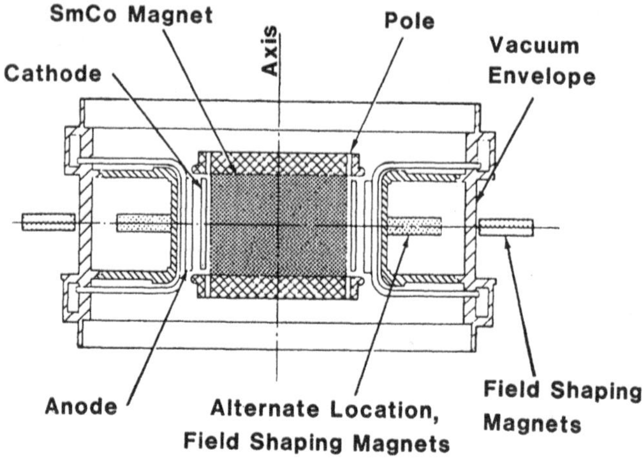

Figure 13.50 Placement of Samarium Cobalt Magnet Inside Cathode in Cold-Cathode CFA

(From J. Skowron, *Proc. IEEE,* March 1973.)

13.3.6 CFA Performance

A typical chart of voltage-current characteristics for a CFA is shown in Fig. 13.51(a). This chart is similar to a portion of a magnetron performance chart (see Fig. 13.34, for example). Of course, magnetic flux density is not normally a variable to the user of a CFA, and so the user is likely to receive a "one-line" version of this performance chart (as shown in Fig. 13.51(b)). When voltage and current are dependent on frequency of operation, as in a backward-wave CFA, the performance chart will show the frequency dependence as illustrated in Fig. 13.52. With this chart, the user can determine the variations in power, current, and voltage that will result from the modulation technique used in tuning the amplifier across the required frequency range.

The gain of a CFA is often shown on a plot of output power *versus* input power (referred to as a compression curve) like that shown in Fig. 13.53. Values of gain shown in Fig. 13.53 are typical, and they are low compared with those for linear-beam tubes. It should also be noted that the CFA does not exhibit any region in which operation is linear (constant gain).

The gain-limiting factors in CFAs are varied and complex and lead to the chart in Fig. 13.54, showing upper and lower boundaries of output power

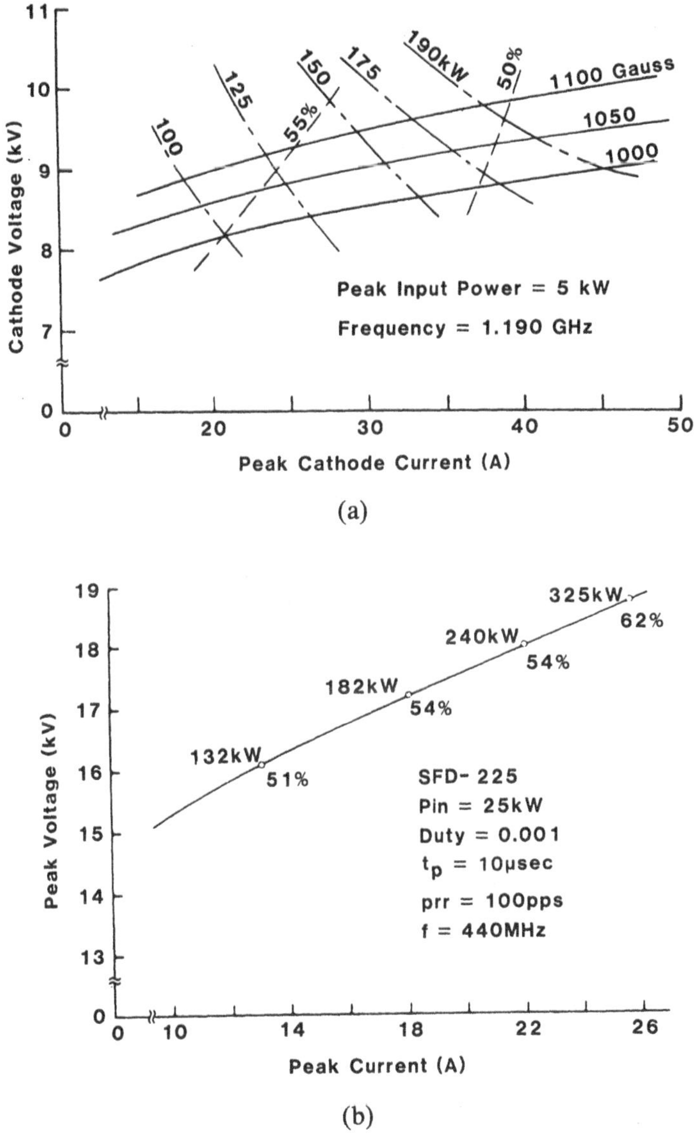

Figure 13.51 CFA Performance Chart (a) and "One-Line" Performance Chart (b)

(From *Microwave Tube Manual* by Varian Associates, Air Force Publication Number T.O. 00-25-251, October 1979.)

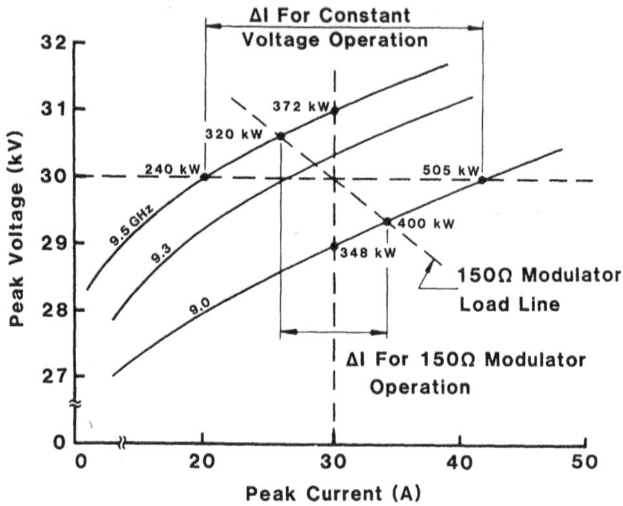

Figure 13.52 Frequency Dependence of Voltage and Current

(From *Microwave Tube Manual* by Varian Associates, Air Force Publication Number T.O. 00-25-251, October 1979.)

for given values of input power. The upper boundary can be caused by any one of the following three factors:

1. a limitation in cathode current;
2. the onset of oscillation because the synchronous voltage of a competing mode has been reached;
3. sufficiently large output powers so that the drive level is insufficient to retain lock in the main amplifying mode.

The first two of these limitations may be eased with appropriate design, but the third is an intrinsic limit resulting only from the basic interaction process.

The lower boundary of the working region may also result from oscillation because the synchronous voltage of another mode has been reached. As with the upper voltage competing mode, this lower voltage limitation may be eased with proper design procedures.

As the frequency of operation is varied, changes occur in the various factors leading to the upper and lower boundaries of operation of a CFA. As a result, the boundaries change with frequency. In the example shown

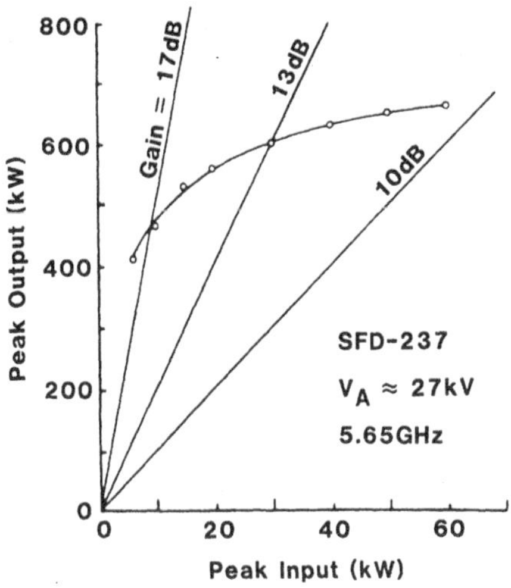

Figure 13.53 Compression Curve for CFA Showing Lines of Constant Gain

(From *Microwave Tube Manual* by Varian Associates, Air Force Publication Number T.O. 00-25-251, October 1979.)

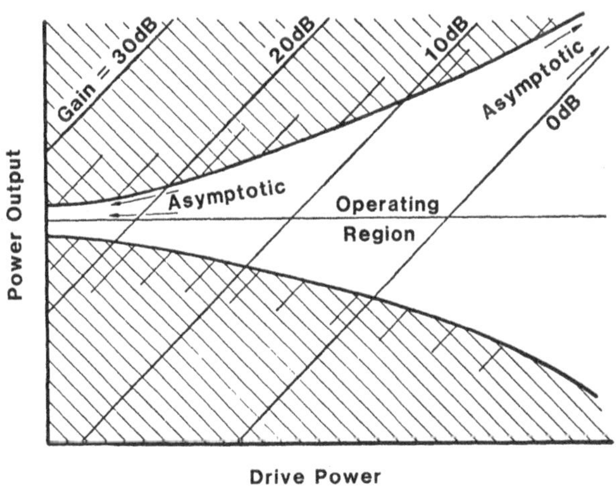

Figure 13.54 Upper and Lower Boundaries of Output Power of a CFA

(From J. Skowron, *Proc. IEEE*, March 1973.)

in Fig. 13.55, the relationship of the actual operating current to the currents that would lead to upper- or lower-mode instabilities is shown. In this case, a limiting factor in bandwidth is the intersection of the operating current line with the upper-mode characteristic.

At low drive levels, the gain may be 20 to 30 dB or more and operation is limited by feedback (for example, by imperfect matching or through the electron stream). At low currents, if the RF circuit is designed to avoid resonances, the tube will produce a broad noisy output. In a backward-wave device, the center of the frequency band is determined by the anode voltage, and so a voltage-tunable noisy output results. In the forward-wave case, the output is broadband noise and crossed-field noise generators are built on this concept.

At this point, it is appropriate to comment on cathode-modulated operation of CFAs. As indicated in Fig. 13.56, the RF drive pulse is applied before high voltage is applied to the cathode. There are two primary reasons for this. First, if the amplifier employs a cold cathode when RF drive is absent, the emission process may fail to build up. As a result, the modulator

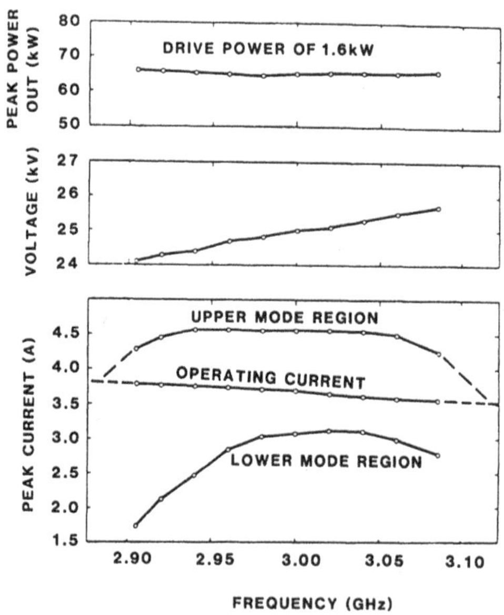

Figure 13.55 Performance Characteristics of a QKS 1267 Backward-Wave CFA and Relation of Operating Current to Currents that Would Lead to Upper- or Lower-Mode Instabilities

(From J. Skowron, *Proc. IEEE*, March 1973.)

providing the cathode voltage will be essentially unloaded, and so excessive voltage may be applied to the tube. This can result in internal arcing and damage to the tube.

The second reason for applying RF drive before cathode voltage is of importance in CFAs using cathodes heated by a heater or by back bombardment so that thermionic emission occurs. In these devices, RF drive is required to help control the space charge and to help prevent oscillations at band edge or the generation of broadband noise.

Regardless of whether the cathode is hot or cold, the RF drive pulse should continue throughout the cathode pulse. This is to ensure proper control of the space charge so that band-edge oscillations or the generation of broadband noise will not occur.

In practice, mode interference is often the limiting factor in determining the useful range of CFA operation. Mode interference is an oscillation that takes time to build up. In pulsed operation, the rate of cathode voltage rise has a significant effect on mode interference. Care must be taken in modulator design, however, because a fast rate of rise, which prevents a lower mode from starting, may initially result in insufficient tube current to load the modulator. As a result, voltage can swing through the values appropriate for operation in the amplifying mode and cause oscillation in an upper mode.

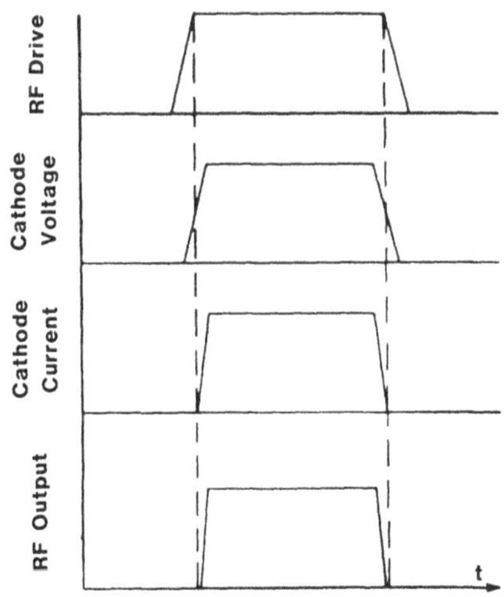

Figure 13.56 Typical Pulse Nesting

(From *Microwave Tube Manual* by Varian Associates, Air Force Publication Number T.O. 00-25-251, October 1979.)

In connection with a discussion of mode interference, it is instructive to examine the impedance characteristics of a CFA. The input VSWR as a function of frequency is typically as shown in Fig. 13.57. Within the operating band, the VSWR is low, typically between 1.2:1 and 1.5:1. The corresponding return loss (magnitude of reflected power) is between 20 and 14 dB below the input power.

Outside of the operating band of the amplifier, the frequency regions where the VSWR increases to very high values are referred to as the band edges of the circuit. It is at one of these band edges (the low-frequency edge in a backward-wave amplifier) that oscillations may occur during the rise and fall of the voltage pulse. The resulting spurious outputs of the tube are frequently referred to as "rabbit ears," because of the way they appear on a time display of RF output.

In addition to the input VSWR being very low in the operating band, it should be pointed out that the transmission loss of the signal through a CFA in the absence of beam current may be very low. This is indicated in Fig. 13.58 for a low-frequency device. As a result, the amplifier can be considered essentially transparent, and this is of use in some system applications. Transmission losses are referred to as insertion losses and are related to electrical characteristics like resistivity as well as mechanical characteristics such as roughness and size of the circuit.

In practice, CFAs are capable of impressive performance. For example, Fig. 13.59 shows an efficiency of approximately 50% over a 10% bandwidth for a pulsed C-band amplifier having a power output of about one megawatt.

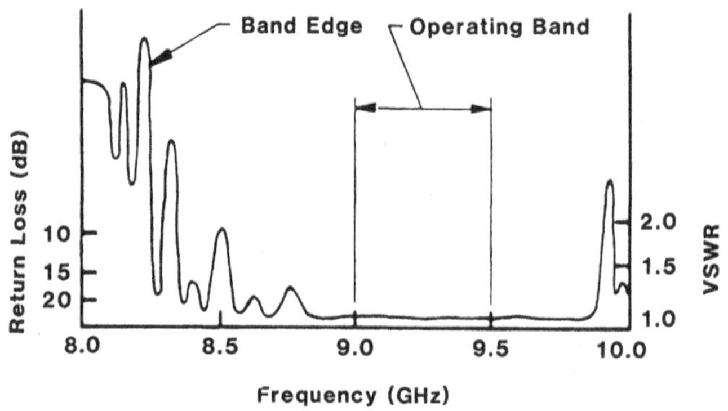

Figure 13.57 Input VSWR for CFA Circuit with Output Terminated

(From *Microwave Tube Manual* by Varian Associates, Air Force Publication Number T.O. 00-25-251, October 1979.)

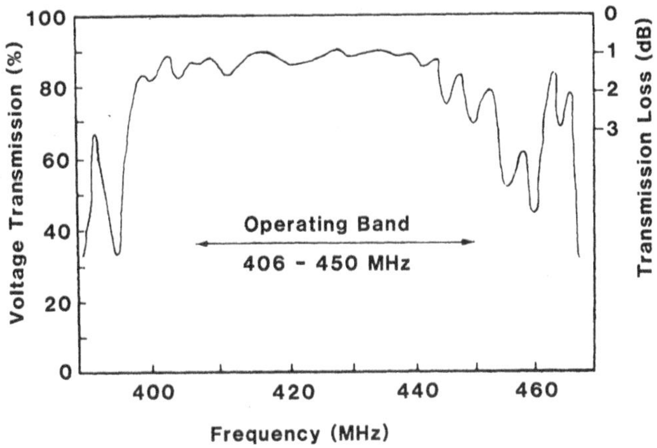

Figure 13.58 Transmission Characteristic of a CFA

(From *Microwave Tube Manual* by Varian Associates, Air Force Publication Number T.O. 00-25-251, October 1979.)

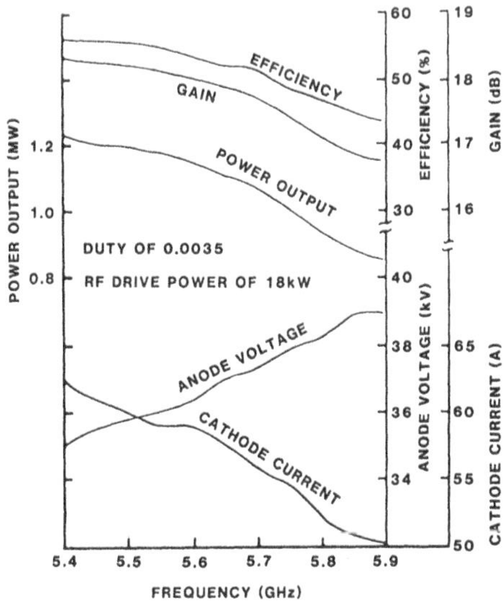

Figure 13.59 CFA Performance Over 500 MHz Band

(From *Microwave Tube Manual* by Varian Associates, Air Force Publication Number T.O. 00-25-251, October 1979.)

A particularly important feature of CFAs is the relatively low voltage at which they operate. The voltage of less than 40 kV for the megawatt device in Fig. 13.59 is a factor of from two to three below the voltage that is required for a linear-beam device at the same power level.

The performance characteristics of an extremely high-power ("super power") CW CFA operating at 3 GHz with a bandwidth of over 5% are shown in Fig. 13.60. Output power was as high as 425 kW and efficiency exceeded 70%. The performance characteristics of a pulsed "super power" CFA are shown in Fig. 13.61. These high-power tubes were developed for use in very high-power radar systems, which were replaced by the development of active phased array technology using large numbers of lower powered tubes.

The overall state of the art of the publishable power capabilities of CFAs are shown in Fig. 13.61. Note that megawatt power levels are available in pulsed devices at frequencies in excess of 10 GHz.

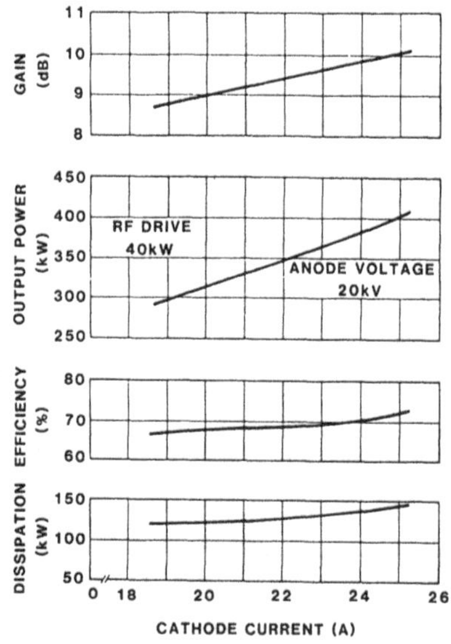

Figure 13.60 Performance Characteristics of "Super Power" CW CFA

(From J. Skowron, (MPTD), Raytheon Co., Workshop on High-Power, Space-Based Microwave Systems, Los Alamos, NM, March 1985.)

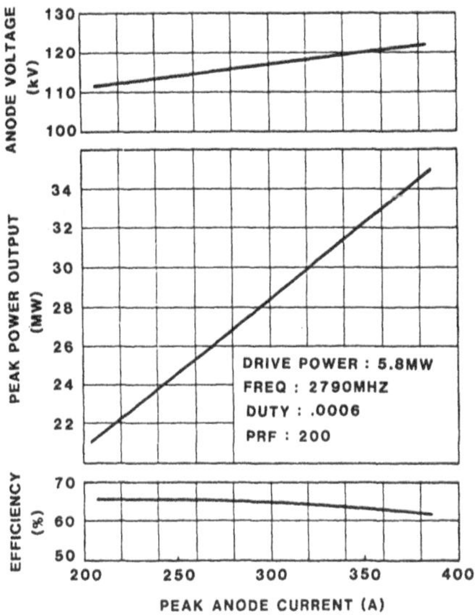

Figure 13.61 Performance Characteristics of "Super Power" Pulsed CFA

(From J. Skowron, (MPTD), Raytheon Co., Workshop on High-Power, Space-Based Microwave Systems, Los Alamos, NM, March 1985.)

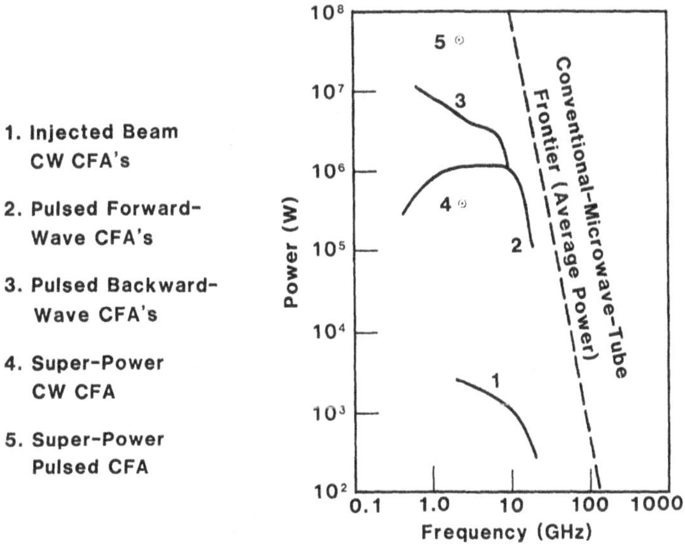

Figure 13.62 State of the Art of CFAs

(From J. Skowron, (MPTD), Raytheon Co., Workshop on High-Power, Space-Based Microwave Systems, Los Alamos, NM, March 1985.)

13.3.7 CFA Phase Characteristics

Three types of phase sensitivities are normally of interest in applications of microwave tubes. These are:

- Phase sensitivity to voltage or current
- Phase sensitivity to RF drive
- Phase linearity with frequency

In a linear-beam tube, phase sensitivity to changes in beam voltage are readily estimated. Total phase shift through a tube is proportional to the transit time of the beam which, in turn, is inversely proportional to beam velocity. Beam velocity, in turn, varies with the square root of beam voltage. For a TWT, phase changes about 30° for a one-percent change in beam voltage. Because current varies with voltage to the three-halves power, the phase changes about 45° for a one-percent change in beam current.

In a CFA, the rotational rate of the spokes in the space-charge cloud is locked to the velocity of the signal on the RF circuit. The speed of the spokes tends to be controlled by the propagation characteristics of the circuit and the signal level in the circuit. In addition, the electrical length of the RF circuit in a CFA is short. As a result, phase sensitivity to changes in anode voltage would be expected to be small.

In a linear-beam tube, because of the three-halves power relationship between voltage and current, the phase shift for a one-percent change in voltage or current is of the same order of magnitude. In a crossed-field device, as can be seen in Figs. 13.51 and 13.52, for example, a small change in voltage is accompanied by a large change in current. Thus, the phase shift for a one-percent change in voltage would be expected to be much larger than the phase shift for a one-percent change in current.

The phase variations with voltage for three RF drive levels in an X-band CFA are shown in Fig. 13.63. The phase changes indicated are for a one-percent change in anode voltage. Thus, at low drive levels and high gain, the phase shift ($15° / 1\% \, \Delta V$) is much larger than at high drive levels and low gain ($3.8° / 1\% \, \Delta V$).

In Fig. 13.64, the phase shift per one percent change in current is shown for a low-power CFA with a thermionic cathode and for a medium-power, cold-cathode device. In both cases, the phase shift at all drive levels is small. While the comparison is not strictly valid because of the large difference in devices, the phase sensitivity to voltage in Fig. 13.63 in much larger than the phase sensitivity to current in Fig. 13.64, as would be expected. In all cases, the phase sensitivity is much smaller than for linear-beam devices.

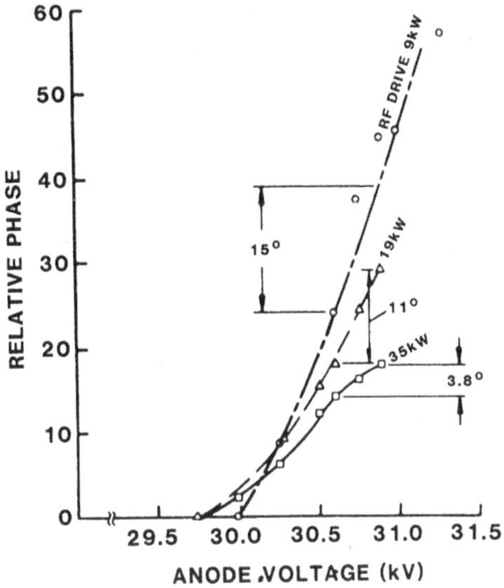

Figure 13.63 Phase Variation with Voltage for Three RF Drive Levels (X-Band CFA)

(From *Microwave Tube Manual* by Varian Associates, Air Force Publication Number T.O. 00-25-251, October 1979.)

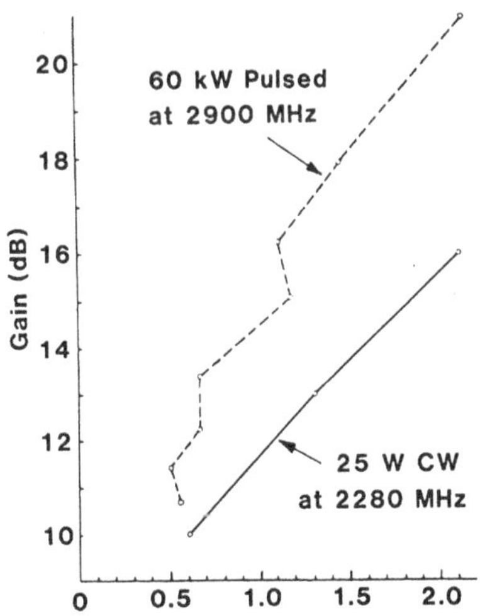

Figure 13.64 Effect of Gain on Phase Sensitivity for Low- and Medium-Power Backward-Wave CFAs

(From J. Skowron, *Proc. IEEE*, March 1973.)

The phase linearity of a CFA with frequency is largely determined by the circuit reflections and the gain of the device. The effect of circuit reflection depends on electrical length and gain. Because the CFA is electrically short (compared with linear-beam devices) and gain is low, phase deviations from linearity of less than ±5° can be achieved.

The favorable phase characteristics of CFAs lead to their use in systems where arrays of devices are used and phasing is important. This is indicated in Fig. 13.65. Also indicated are other applications where the transparent characteristic is used (for interpulse feed-through) or where tubes are readilty connected in parallel to upgrade power capability.

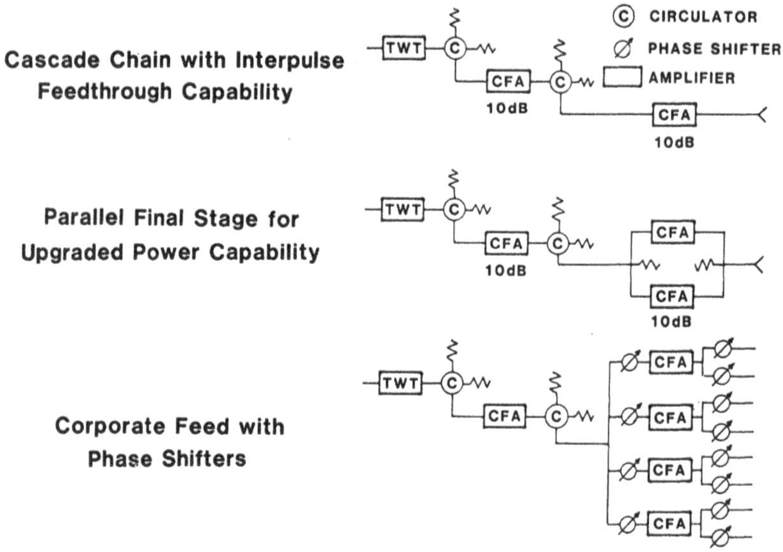

Figure 13.65 Typical Configurations for Applications of CFAs (From J. Skowron, *Proc. IEEE,* March 1973.)

13.3.8 Thermal Considerations

As has been mentioned several times, the electron beam in an efficient crossed-field device is collected on the RF circuit. As a result, of course, the circuit must be capable of dissipating the spent-beam power. In addition, there may be a significant amount of heating of the cathode by back bombardment so that the cathode may have to be cooled.

As an example of the way in which power may be divided in a CFA, the power distribution is given in Table 13.1 for the CW CFA, the performance characteristics for which were shown in Fig. 13.60. Note that the anode dissipation is over 100 kW and the cathode dissipation is nearly 25 kW.

TABLE 13.1
Distribution of Power in "Super Power" CW CFA
(From J. Skowron, (MPTD), Raytheon Co., Workshop on High-Power, Space-Based Microwave Systems, Los Alamos, NM, March 1985.)

RF Power Generated	71.7%	350 kW*
Anode Dissipation	21.4	104.5
Cathode Dissipation	5.0	24.6
Transmission Line	1.9	9.3
Totals	100.0%	488.4 kW

*Note: The RF drive power is added to the RF power generated for a total RF output power of 405 kW.

The severity of the cooling problem becomes apparent when it is realized that the total cross-sectional area of the anode segments ranges from a fraction of a square centimeter in high-frequency devices to a few tens of square centimeters in low-frequency devices. These relatively small areas may result in power densities in excess of a kW/cm^2 that must be dissipated by the anodes.

Table 13.2 contains a listing of the cooling capabilities possible with various techniques. Over 10 kW/cm^2 can be removed by high-velocity water when nucleate boiling occurs on the surfaces being cooled. Thus, this is the cooling technique used in extremely high-power applications.

TABLE 13.2
Cooling Capability of Various Media ($\Delta T = 200°$ C)
(From J. Skowron, (MPTD), Raytheon Co., Workshop on High-Power, Space-Based Microwave Systems, Los Alamos, NM, March 1985.)

Black body radiation (T ~ 200° C)	0.25 W/cm^2
Air, natural convection to finned forced convection	0.5–110
Organic oils, natural convection to forced convection, boiling	5–90
Water, pool boiling	125
Water, pressurized, non-boiling	155
Liquid metals, forced convection non-boiling	450
Water, high velocity, nucleate boiling	
Operational	10,000
Maximum achieved	20,000

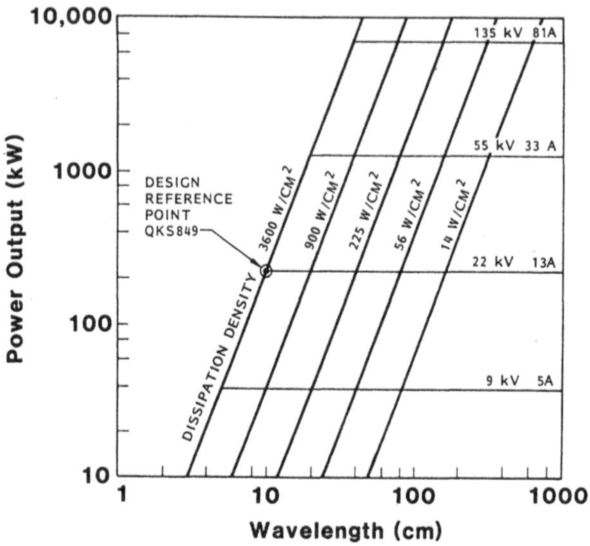

Figure 13.66 Predicted Power Capabilities of CW CFAs Based on Anode Dissipation Power Densities

(From J. Skowron, (MPTD), Raytheon Co., Workshop on High-Power, Space-Based Microwave Systems, Los Alamos, NM, March 1985.)

Shown in Fig. 13.66 are predicted power capabilities of CW CFAs based on anode dissipation capability. The Raytheon QKS 849, which uses nucleate boiling to achieve an anode dissipation of 3600 W/cm^2, is used as a design reference point. Based on this power density, power outputs in excess of one megawatt should be possible at frequencies below about 1500 MHz.

In pulsed CFAs, electron bombardment of the anode results in a transient temperature rise of the surface during the pulse. The amount of energy that can safely be imparted to the anode is directly related to the heat capacity of the anode material and to the temperature to which the surface may rise.

The calculation of the surface temperature and temperature distribution within an anode segment during a pulse is relatively straightforward. Because the heat is concentrated in the surface of the anode, a one-dimensional model like that illustrated in Fig. 13.67 may be used.

Shown in Figs. 13.68 and 13.69 are temperatures calculated for the case of the 200 μs pulse length, 100 pulses per second, and a peak power density of 67,850 W/cm^2. Notice that the temperature of the face of the anode vane reaches nearly 500°C during the pulse and that the temperature only .010 inch into the vane is just 200°C. During the interpulse period, the temperature

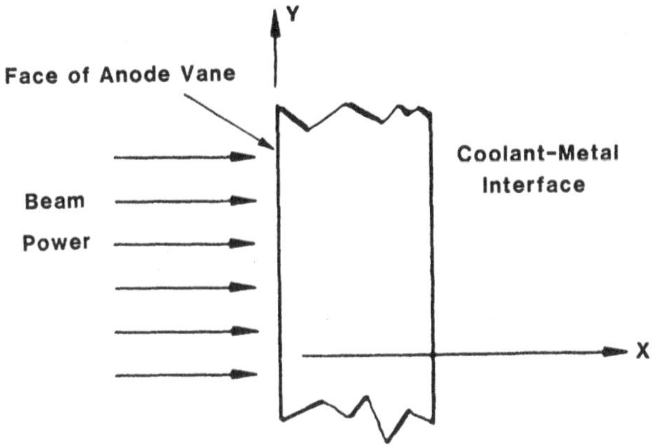

Figure 13.67 Model Used for Solution of One-Dimensional Heat Flow Equation

(From J. Skowron, *Proc. IEEE*, March 1973.)

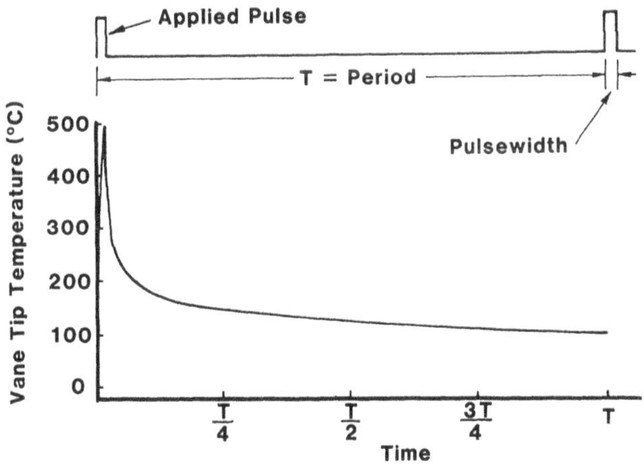

Figure 13.68 Vane Tip Temperature as a Function of Time

(From J. Skowron, *Proc. IEEE*, March 1973.)

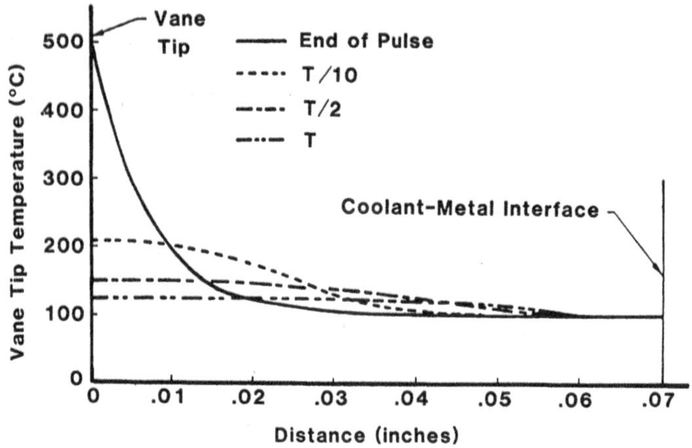

Figure 13.69 Temperature as a Function of Distance Into Vane Tip

(From J. Skowron, *Proc. IEEE*, March 1973.)

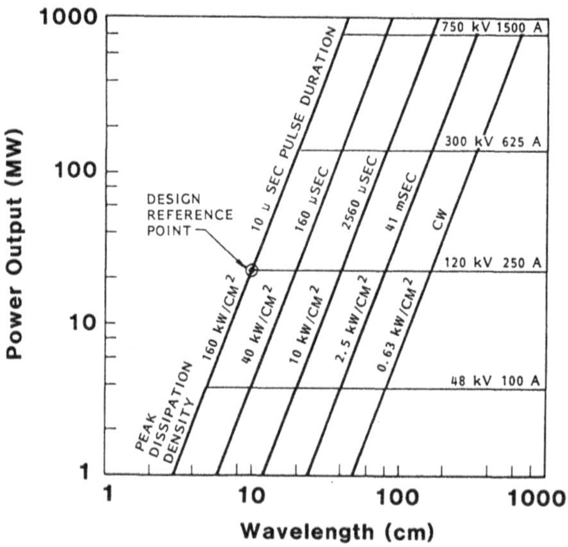

Figure 13.70 Predicted Power Capabilities of Pulsed CFAs Based on Anode Surface Temperature Limitations

(From J. Skowron, (MPTD), Raytheon Co., Workshop on High-Power, Space-Based Microwave Systems, Los Alamos, NM, March 1985.)

of the entire vane nearly drops to that of the coolant-metal interface. It is because of the high vane surface temperature reached during a pulse that the vane tips of CFAs are sometimes fabricated from a high temperature metal such as molybdenum.

Shown in Fig. 13.70 are predicted power capabilities of pulsed CFAs based on vane-tip surface temperature. For a given surface temperature, as pulse duration is reduced, the peak-power density can be increased before the limiting energy input is reached. At 3000 MHz, 20 MW has been achieved and is used as the design reference point.

13.4 M-TYPE BACKWARD-WAVE OSCILLATORS

The M-type backward-wave oscillator (MBWO), which is sometimes called a carcinotron, is a voltage-tunable oscillator usually capable of supplying more than 100 W of CW power with an efficiency as high as 50%. MBWOs resemble magnetrons in appearance, but have the advantage over magnetrons of being able to tune nearly instantaneously over a wide frequency range. They are useful as ECM noise transmitters, swept signal sources, communications transmitters, or as driver oscillators for high-power transmitters.

As shown in Fig. 13.71, the format for the MBWO is similar to that for the injected-beam CFA shown in Fig. 13.38. The MBWO format is circular so as to minimize size and weight. A ribbon-shaped electron beam is injected into the region between the sole and the RF circuit (anode). The slow-wave structure is a folded waveguide (interdigital delay line) as shown in Fig. 13.72. This is a backward-wave circuit, and so the signal grows as it travels from the collector end to the electron-gun end of the device where the RF output connector is located. The MBWO requires a relatively large attenuator, well bonded to the heat sink of the tube, to absorb reflections from the output or along the anode.

The electron-beam velocity may be controlled by either the anode or the sole voltage in a linear manner. As a result, frequency may be controlled by either electrode or by simultaneous control of voltages as shown in Fig. 13.73. As voltages are varied, beam power and efficiency vary, and so RF output power varies as shown in Fig. 13.74.

Shown in Fig. 13.75 are tuning curves for MBWOs for operation from L band through X band. The power output is about 200 W and the weight is 16 lbs, except for the lowest frequency device which weighs 27 lbs.

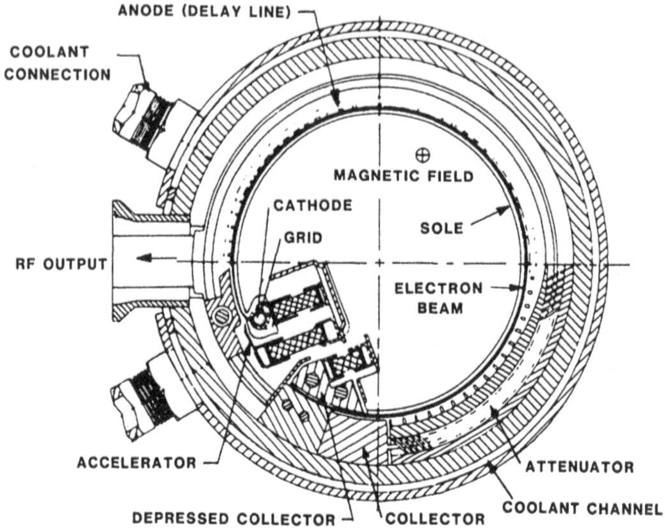

Figure 13.71 MBWO Cross Section

(From *Microwave Tube Manual* by Varian Associates, Air Force Publication Number T.O. 00-25-251, October 1979.)

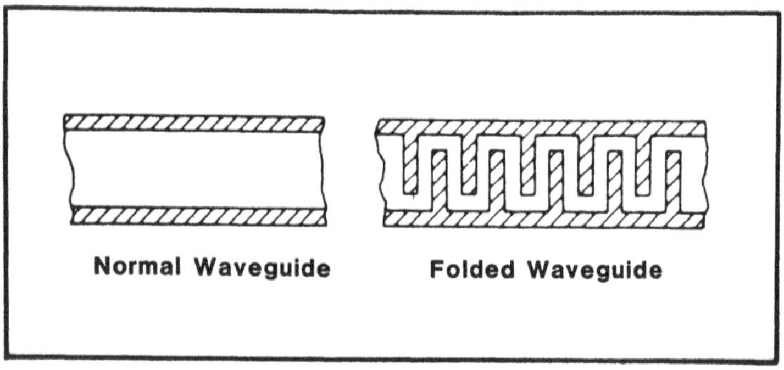

Figure 13.72 Folded Waveguide (Interdigital Delay Line)

(From *Microwave Tube Manual* by Varian Associates, Air Force Publication Number T.O. 00-25-251, October 1979.)

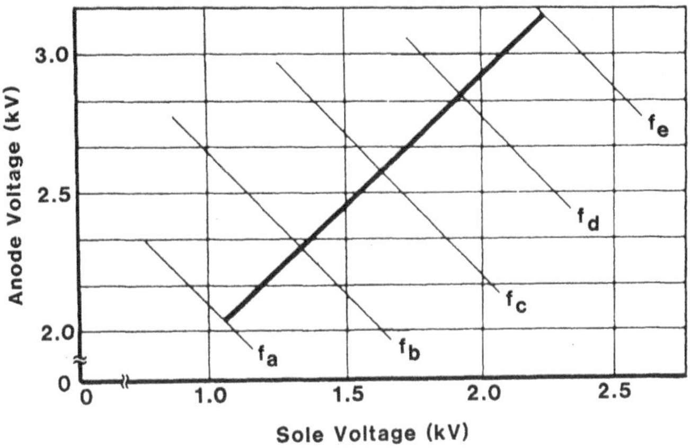

Figure 13.73 Anode *versus* Sole Voltage for Simultaneous Tuning of an MBWO

(From *Microwave Tube Manual* by Varian Associates, Air Force Publication Number T.O. 00-25-251, October 1979.)

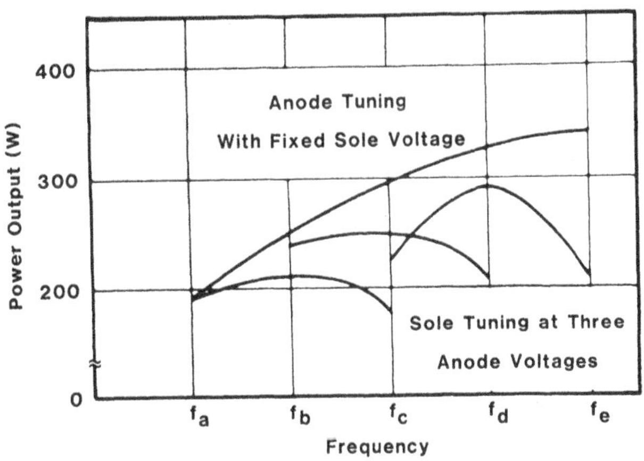

Figure 13.74 Variation of Output Power with Anode and Sole Voltages

(From *Microwave Tube Manual* by Varian Associates, Air Force Publication Number T.O. 00-25-251, October 1979.)

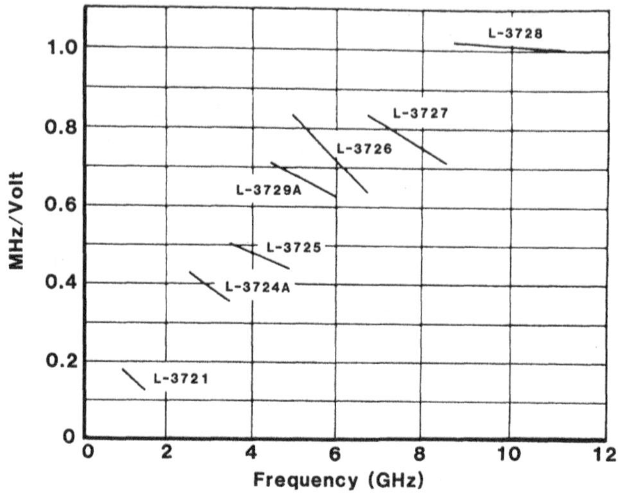

Figure 13.75 Tuning Sensitivity for Medium Power MBWOs

(From *Microwave Tube Manual* by Varian Associates, Air Force Publication Number T.O. 00-25-251, October 1979.)

13.5 Voltage-Tunable Magnetrons

Voltage-tunable magnetrons (VTMs) are low-power (watts to hundreds of watts) CW oscillators that use a hollow injected electron beam in a circular format. As indicated in Fig. 13.76, The electron beam originates from a short cylindrical cathode at one end of the device. Electrons are accelerated radially outward from the cathode and are then formed into a hollow beam by the electric and magnetic fields near the cathode. The beam is injected into the region between the sole and the anode, and rotates about the sole at the rate governed by the axial magnetic field and the dc voltage applied between the sole and anode.

The RF circuit used is a low-Q interdigital line operating in the π mode. Electron interaction with the circuit is similar to that in a conventional magnetron. Bunching of the hollow beam occurs and the frequency of oscillation is determined by the rate at which bunches pass by the circuit, which, in turn, is determined by the rotational velocity of the beam. The primary difference between a VTM and a conventional magnetron is that the high Q of the resonator system in the conventional magnetron governs the frequency of operation and only very minor changes in frequency occur when voltage is changed. In the VTM, the low Q of the resonator system makes possible its operation over bandwidths as high as 50% or more. As a result, frequency can be controlled by varying the velocity of the electrons, and this is done by varying the applied voltage between the anode and sole.

A typical tuning characteristic for an S-band VTM is given in Fig. 13.77. Notice that the variation of frequency with voltage is reasonably linear at about 1.5 MHz/V.

Figure 13.78 shows a composite curve exhibiting the general bandwidth capabilities of VTMs for various frequencies and power levels of operation. At high power levels and high frequencies, the bandwidth is limited to a few percent. At low power levels, the bandwidth may exceed 50%.

In VTMs, as in conventional magnetrons, much of the beam current is eventually collected on the RF circuit as the dc energy of the electrons is converted to RF energy. As a result, efficiency is high and may approach 70% at low frequencies. As anode voltage is varied, the power into the beam and the efficiency vary. As a result, output power varies over a wide range as indicated by Fig. 13.79. Power can be adjusted to some extent through the use of the control electrode in the electron gun.

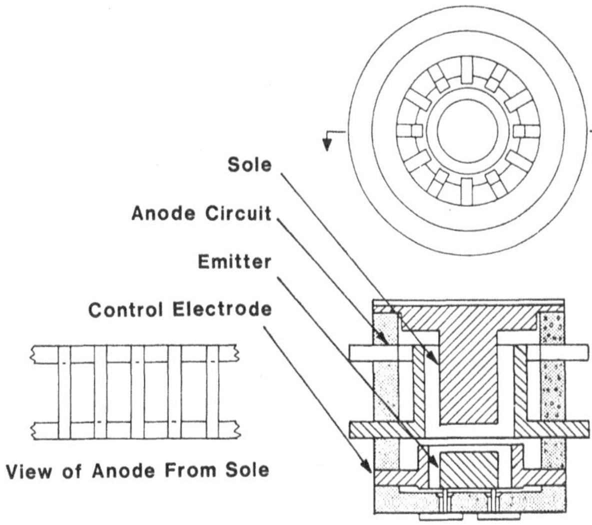

Figure 13.76 Cross Section of Voltage-Tunable Magnetron

(From *Microwave Tube Manual* by Varian Associates, Air Force Publication Number T.O. 00-25-251, October 1979.)

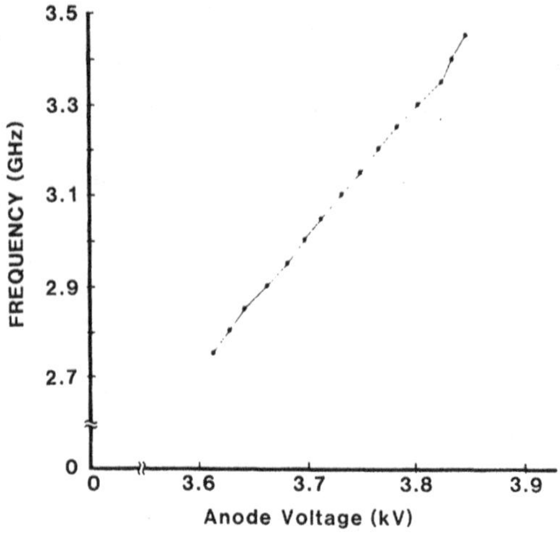

Figure 13.77 Typical Tuning Characteristic for S-Band VTM

(From *Microwave Tube Manual* by Varian Associates, Air Force Publication Number T.O. 00-25-251, October 1979.)

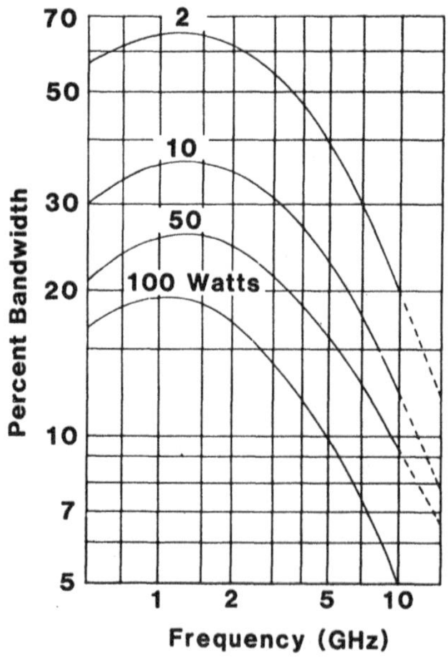

Figure 13.78 Bandwidth Capabilities of VTMs Operating at Various Frequencies and Power Levels

(From *Microwave Tube Manual* by Varian Associates, Air Force Publication Number T.O. 00-25-251, October 1979.)

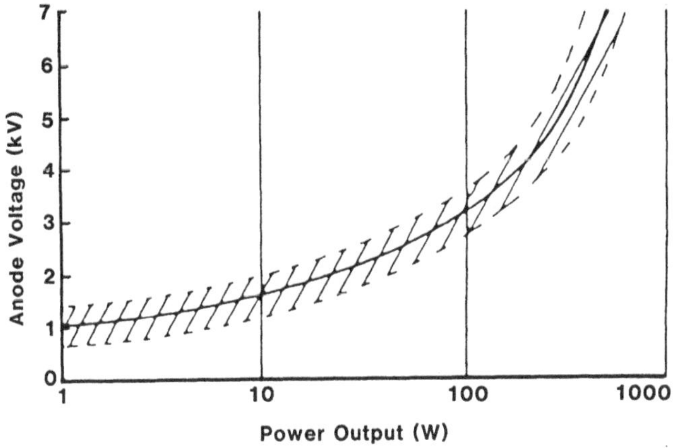

Figure 13.79 Relation between VTM Anode Voltage and Output Power

(From *Microwave Tube Manual* by Varian Associates, Air Force Publication Number T.O. 00-25-251, October 1979.)

REFERENCES

[13.1] Recommended for further reading:

H.A.H. Boot and J.T. Randall, "Historical Notes on the Cavity Magnetron," *IEEE Trans. Electron Devices*, Vol. ED-23, July 1976, pp. 724–729.

G. Ross Kilgore, "Recollections of Pre-World War II Magnetrons and Their Applications," *IEEE Trans. Electron Devices*, Vol. ED-31, November 1984, pp. 1593–1595.

William C. Brown, "The Microwave Magnetron and Its Derivatives," *IEEE Trans. Electron Devices*, Vol. ED-31, November 1984, pp. 1595-1605.

Joseph F. Hull, "A Young Engineer Enters the Microwave Tubc Field During World War II," *IEEE Trans. Electron Devices*, Vol. ED-31, November 1984, pp. 1611–1613.

[13.2] Article by J.J. Coupling, A Pseudonym of J.R. Pierce, which appeared in the February, 1948 edition of *Astounding Science Fiction*, a Street and Smith publication.

[13.3] Recommended for further reading:

George B. Collins, *Microwave Magnetrons.* New York: McGraw-Hill, 1984.

E. Kettlewell, *The Magnetron Oscillator.* London: Mills and Boon, Ltd., 1971.

[13.4] Recommended for further reading:

M. Chodorow and C. Susskind, *Fundamentals of Microwave Electronics.* New York: McGraw-Hill, 1964.

A.S. Gilmour, Jr., "Basic Principles and Some Aspects of the Electromagnetic Compatibility of Microwave Tubes," *Inter-System Electromagnetic Compatibility.* New York: Holt, Reinhart and Winston, 1972.

[13.5] For additional information, see Varian brochure entitled "Introduction to Coaxial Magnetrons."

[13.6] Merrill I. Skolnik, "The Future of Microwave Engineers: From the Technical Perspective," *Microwave Journal*, Vol. 18, October 1985, pp. 24–36.

[13.7] W.M. Black, *et al.*, "A High Power Magnetron For Air Breakdown Studies," *Technical Digest*, 1980 International Electron Devices Meeting.

[13.8] Recommended for further reading:

John F. Skowron, "The Continuous-Cathode (Emitting-Sole) Crossed-Field Amplifier," *Proc. IEEE*, Vol. 61, March 1973, pp. 330–336.

Varian Brochure, "Introduction to Pulsed Crossed-Field Amplifiers."

Chapter 14

Fast-Wave Tubes

Conventional microwave tubes are commonly referred to as slow-wave tubes. In some way, the RF structure is configured so that the phase velocity of the electromagnetic field is slowed to the electron beam velocity. This is, perhaps, most evident in the helix-type TWT. The electromagnetic wave can be thought of as propagating in a helical path along the helix at the velocity of light, but advancing much more slowly along the axis where the electron beam is placed.

In fast-wave tubes, the RF structure is usually a smooth waveguide or a large resonator in which no attempt is made to reduce the velocity of the wave. Instead, the electron beam is injected into the electromagnetic field in a manner such that interaction can take place.

There is intense interest in these devices at the present time. This stems from the simplicity of the RF structure and the fact that the electron beam is normally placed well away from the RF structure. The result is that the size limitations found in conventional microwave tubes (where RF circuit dimensions are typically a fraction of a wavelength) are significantly relaxed. With larger dimensions, the power-handling capability is increased. As an example, peak powers of a megawatt or more at 100 GHz have been generated in one type of fast-wave device (a gyrotron), whereas conventional devices are capable of only a few kilowatts at this frequency.

Many schemes have been devised for achieving fast-wave interactions. While none have been developed to the point where they are widely used in systems, some hold promise for future applications and are described in the following sections. These are the ubitron (and the related free-electron laser), the peniotron, and various gyrotron devices. The ubitron is discussed first because the interaction of the electrons with the electromagnetic wave, conceptually, is the easiest to describe and understand. The interaction in the peniotron is, in some ways, similar to that in the ubitron, but it also leads naturally to the gyrotron family of devices.

14.1 UBITRONS [14.1]

In a ubitron, an electron beam, which has been caused to undulate by a periodic magnetic field, is passed through a waveguide containing a TE-mode electromagnetic wave as shown in Fig. 14.1. If the beam velocity is adjusted so that the beam is synchronous with the wave in the manner indicated in Fig. 14.2, the beam can give up energy to the wave and produce amplification.

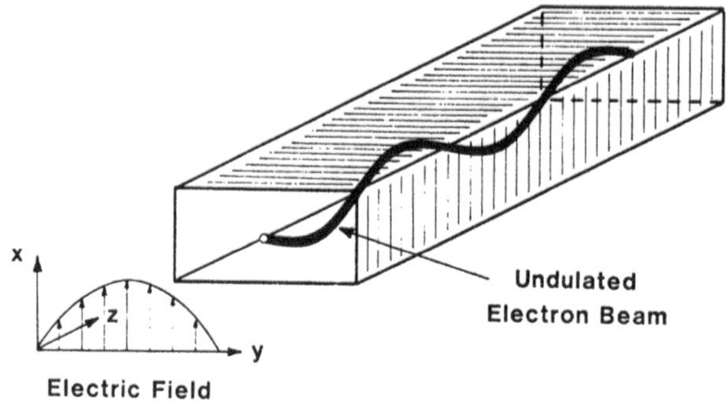

Figure 14.1 Undulated Electron Beam in TE-Mode Waveguide

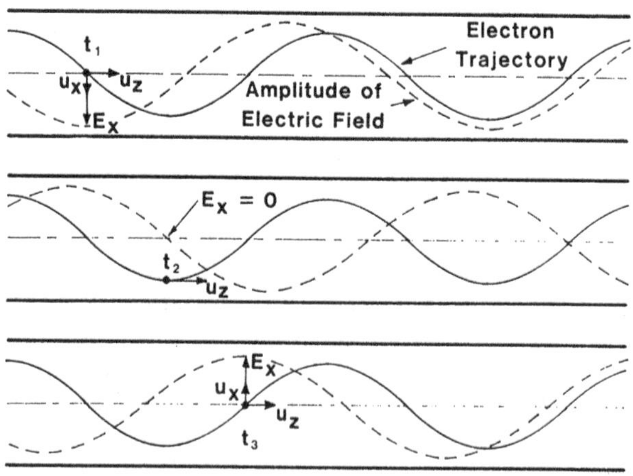

Figure 14.2 Position of Electron with Respect to Traveling Electric Wave

To understand how amplification can occur, consider first a single electron having the most favorable phase with respect to the electromagnetic wave. The position of this electron with respect to the electric field is indicated at three times: t_1, t_2 and t_3 in Fig. 14.2. At time t_1, the electron has a velocity component, u_x, in the minus-x direction and the electric field, E_x, is also in the minus-x direction. As a result, at this time, the electron is being decelerated by the field.

At time t_2, the electron is moving in the z direction with no x-directed velocity component. The electromagnetic wave has advanced so that the electric field is zero at the position of the electron. At time t_3, the electron has a velocity component in the plus-x direction and the electromagnetic wave has advanced so that the electric field is in the plus-x direction. As a result, the electron is again decelerated.

There will be another electron in the least favorable phase with respect to the electromagnetic wave that will be accelerated by the field. Those electrons at other positions will receive transverse accelerations or decelerations in varying degrees. Up to this point, the transverse velocity of the beam has been modulated, but there has been no net exchange of energy between the beam and the wave.

The transverse velocity modulation, in conjunction with the periodic magnetic field, results in bunching of the electron beam. By causing the electron bunches to move into the decelerating field regions, beam energy is converted to wave energy, and so amplification occurs.

Ubitron amplifiers have been operated at 1.6 MW and 16 GHz, and at 150 kW and 54 GHz [14.2]. A drawing of the 54-GHz device is shown in Fig. 14.3.

In this device, a circular waveguide and a hollow electron beam were used. The waveguide was formed from alternate steel and copper disks, which, when placed in the field of a solenoid, produced both a periodic and a dc field to undulate and focus the electron beam. The output of the tube was a circular waveguide through a circular ceramic window.

The concept of causing electrons to undulate and interact with an electromagnetic field has application over a wide range of frequencies. For example, during the past few years, operation at very high power levels at light-wave frequencies has been achieved in the free-electron laser, which, in many respects, is similar to the ubitron. It appears that the full range of application of this technology is only in its early stages of development.

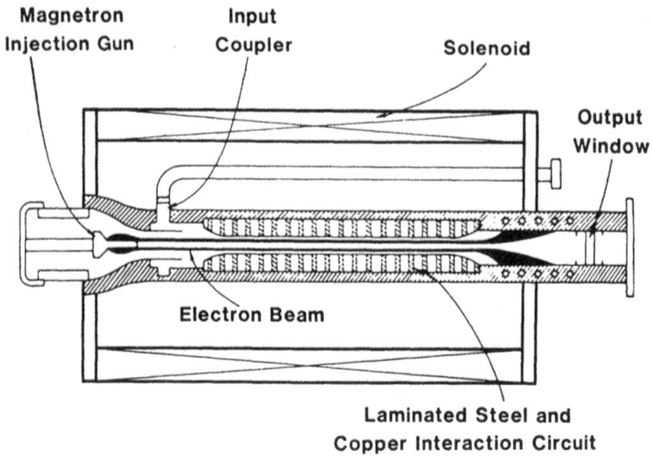

Figure 14.3 Cross Section of 54-GHz Ubitron

(From C.E. Enderby and R.M. Phillips, *Proc. IEEE*, October 1965.)

14.2 PENIOTRONS

The peniotron [14.3, 14.4, 14.5], from "*penio*" meaning spool in Greek, is a member of the family of gyro-devices to be discussed in the following section. It is singled out for discussion in this section for two reasons. The first is the relatively straightforward way in which the interaction between the electrons and the electromagnetic wave occurs. The second is that all of the devices in the following section depend upon a relativistic change in the electron mass for their operation, whereas the peniotron does not.

In the peniotron, as indicated in Fig. 14.4, a thin hollow electron beam is immersed in an axial magnetic field in a waveguide. The waveguide is usually ridged to help concentrate the electric fields near the electron orbits. The TE_{10} mode of propagation is used.

The axial velocity of the electrons is such that the phasing of the electrons with respect to the RF field is like that in the ubitron (see Fig. 14.2). In fact, an electron with the most favorable phase follows a helical path, which, from the side, looks the same as the electron trajectory shown in Fig. 14.2. When the frequency of rotation of the electrons (cyclotron frequency) is nearly the same as the frequency of the electromagnetic wave, these favorably phased electrons move up and down in the direction of the electric field as shown in Fig. 14.5, and thereby give up energy to the field.

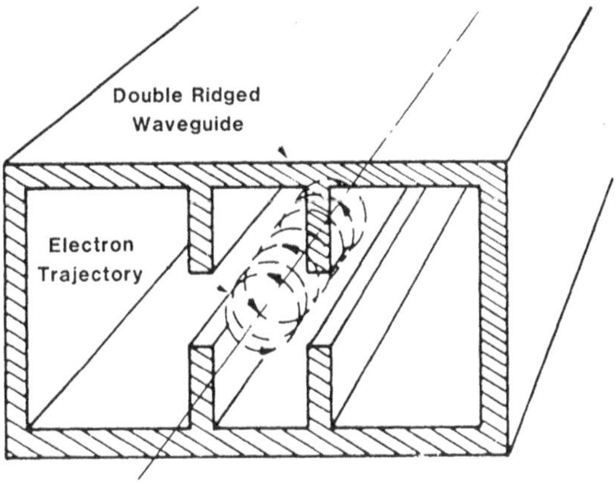

Figure 14.4 Electron Motion in a Peniotron

(From G. Dohler, *et al.*, *Technical Digest*, 1978 IEDM.)

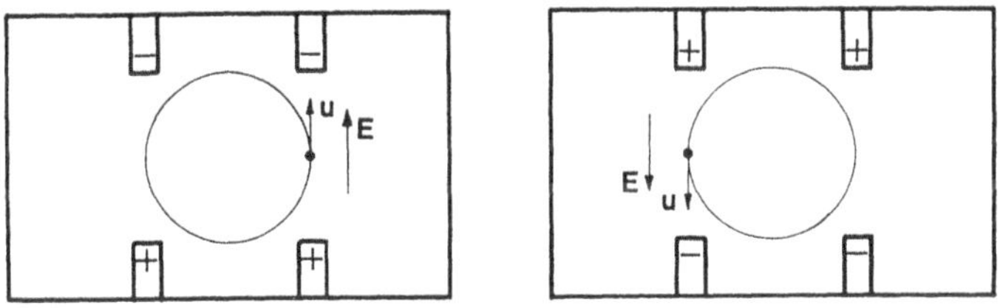

Figure 14.5 Electron Motion in Relation to Direction of Electric Field

The resulting orbit for the electron with a favorable phase, which is, therefore, initially decelerated by the field, is shown in Fig. 14.6(a). For an electron that is initially accelerated by the field, the orbit becomes larger. By the time the electron has moved through one-half of an orbit, the electric field will have advanced one complete cycle and the electron will be decelerated. Because of the enlarged orbit, the electron will be in a stronger field when being decelerated than when it was initially being accelerated. As a result, net deceleration occurs and the electron then follows an orbit similar to that for the initially decelerating electron.

The predicted efficiency of the peniotron is high if an electron beam having the proper velocity characteristics can be generated. Essentially, all of the rotational energy of the beam can be converted to RF energy and then a

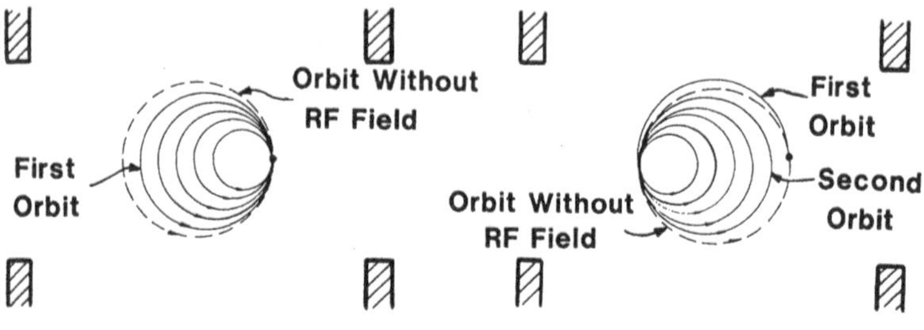

Figure 14.6 Principle of Operation of Double-Ridged Peniotron in the Fundamental Mode

(From G. Dohler, *et al.*, *Technical Digest*, 1978 IEDM.)

depressed collector can be used to recover axial energy. The power capacity is modest because of the small sizes of the components. In a peniotron oscillator, a peak power of 2 kW has been demonstrated at an efficiency of 30 percent and a frequency of 8 GHz. At the same frequency and with the same efficiency, a peak power of 1 kW has been demonstrated in a peniotron amplifier. Because the peniotron is in the very early stages of development, the ultimate capabilities are not at all clear at the present time.

14.3 GYROTRONS [14.6]

The name "*gyrotron*" was originally used by the Russians for a single-cavity oscillator, now often referred to as a gyro-monotron. The name now refers to a class of devices including both oscillators and amplifiers. The most important gyrotrons to be discussed in this section are gyrotron oscillators (gyro-monotrons) and gyro-TWTs. Other devices such as gyro-klystrons, gyro-BWOs, and even gyro-twystrons have received attention, but have not been developed to the same extent as gyro-monotrons and gyro-TWTs.

In the previous section, the peniotron was discussed. In that device, coupling between the electrons and the microwave field was achieved by means of the cyclotron resonance condition. In a gyrotron, as in a peniotron, electrons rotate in a magnetic field in synchronism with the alternating electromagnetic field. The mechanisms by which the electrons are sorted and brought into synchronism with the alternating field are completely different in the two devices. In the peniotron, it is the spatial variation of the electromagnetic field that produces the sorting action. In the gyrotron, it is the relativistic mass change of an electron when its velocity is increased or decreased that results in the sorting action.

In a gyrotron, the electron beam, which is normally in the shape of a thin hollow cylinder, is injected into a strong axial magnetic field and passed through a circular cavity or waveguide containing an electromagnetic wave with a circular electric field pattern. The rotational velocities of the electrons are normally 1.5 to 2 times the axial velocity so that most of the electron energy is rotational. Because the magnetic field is very large, the orbit diameter for the electrons is small. As a result, the thickness of the hollow electron beam may be several times the diameter of the electron orbit as shown in Fig. 14.7, and, in effect, the hollow beam contains a large number of small beams, referred to as beamlets by Dohler *et al.*

At this point, it is informative to again compare the peniotron and the gyrotron. As is shown in Fig. 14.8, the gyrotron contains a hollow electron beam made up of a large number of beamlets, each one of which is like the single beamlet in the peniotron. Whereas the cross-sectional dimensions of the peniotron circuit are fractions of a wavelength, the diameter of the gyrotron circuit can be several wavelengths. Because of this relatively large size and because the beam is located well away from the circuit, the gyrotron is capable of extremely high power operation.

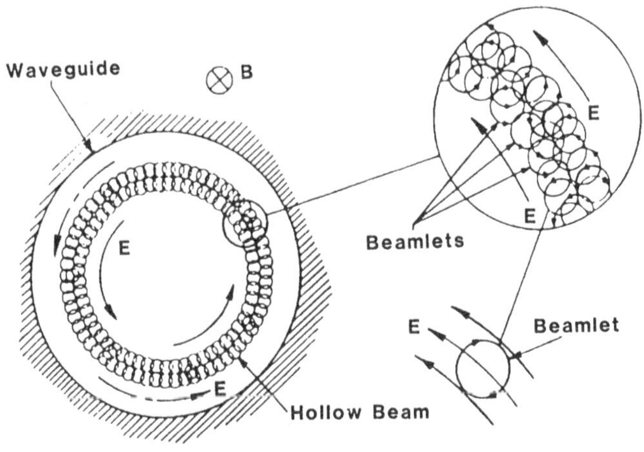

Figure 14.7 Gyrotron Cross Section Showing Electron Trajectories

(From G. Dohler, *Int. J. Electronics* April 1984.)

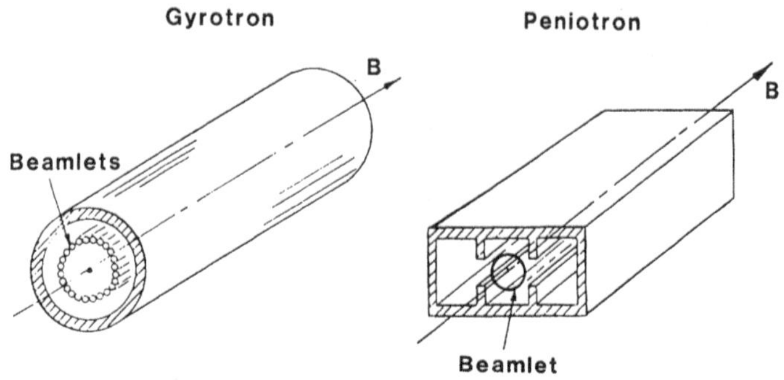

Figure 14.8 Comparison of Peniotron and Gyrotron

(From G. Dohler and F. Voltaggio, 1982 Microwave Power Tube Conference.)

The basic operating mechanism of a gyrotron may be explained by examining the interaction of the electrons in a single beamlet (like that shown in Fig. 14.7) with the electric field. In Fig. 14.9(a), it is assumed that the electrons in a beamlet are uniformly distributed along a single helical path prior to interaction with the RF electric field. The electrons are assumed to rotate in the counterclockwise direction as they move through the RF field. The rotational frequency of the electrons is, of course, the cyclotron frequency which is given by

$$\omega_c = \frac{e}{m} B$$

where B is the applied dc magnetic field, e is the electron charge, and m is the electron mass.

It is very important at this point to realize that, in a gyrotron, the velocity of the electrons is a large enough fraction of the speed of light so that their mass is significantly greater than their mass at rest, that is,

$$m = \gamma m_0$$

where m_0 is the rest mass of an electron and γ is called the relativistic mass factor. The value of γ increases when an electron is placed in an accelerating field.

Now, referring again to Fig. 14.9(a), when the electric field is such that it tends to accelerate electrons (tops of orbits), the electron mass in increased and so the cyclotron frequency, ω_c, decreases for these electrons. Similarly,

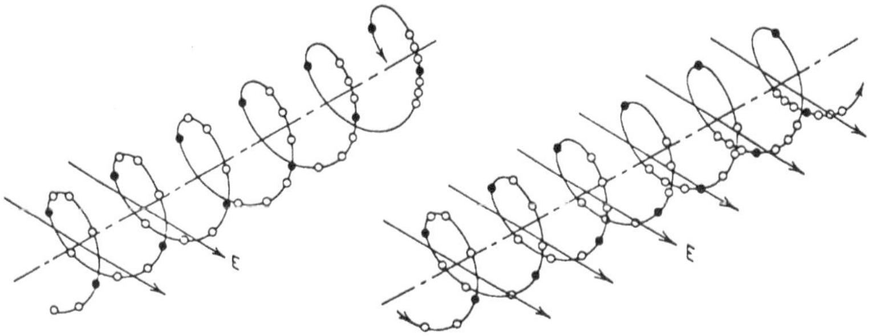

Figure 14.9 Bunching of Electrons in a Gyrotron

(From *Advances in Electronics and Electron Physics*, vol. 55, by R.S. Symons and H.R. Jory, Copyright 1981 by Academic Press., Reproduced with permission.)

when the electric field tends to decelerate electrons (bottoms of orbits), the electron mass is decreased, and so ω_c is increased. Because the rate of rotation is decreased for some electrons and is increased for others, orbital bunching occurs if the electrons are permitted to drift as is indicated at the right-hand side of Fig. 14.9(a).

If the cyclotron frequency is somewhat lower than the frequency of the RF field, then the position of the bunches along the helical orbit is delayed with respect to the phase of the applied field as is indicated in Fig. 14.9(b). At this point, it should be noted that the bunched electrons are all moving in the direction of the electric field, and so they all give up energy to the field. As the electron bunches rotate in near synchronism with the alternating RF field, they continue to give up energy on each half-cycle of rotation.

The interaction that has just been described for the electrons in a single beamlet in a gyrotron also takes place in the other beamlets. Thus, the electron distribution becomes to be as indicated in Fig. 14.10. As the direction of the electric field alternates, the direction of motion of the electrons also alternates, and so electrons throughout the hollow beam give up energy on each half-cycle of operation, and thereby produce amplification of the field.

A problem common to all gyrotrons is the generation of a hollow electron beam in which the division between axial and azimuthal velocities is accurately controlled and where the velocity spreads are extremely small. Accurate control of velocities is necessary for the electrons to maintain the proper phase relationship with the RF field.

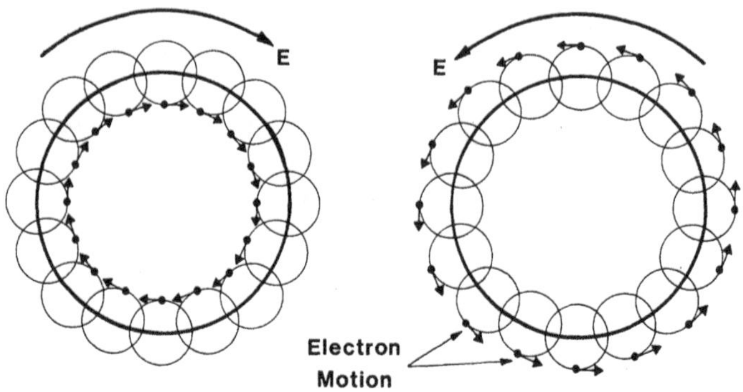

Figure 14.10 Electron Motion in Relation to Direction of Electric Field in a Gyrotron

(From Varian Brochure entitled Introduction to Gyrotrons.)

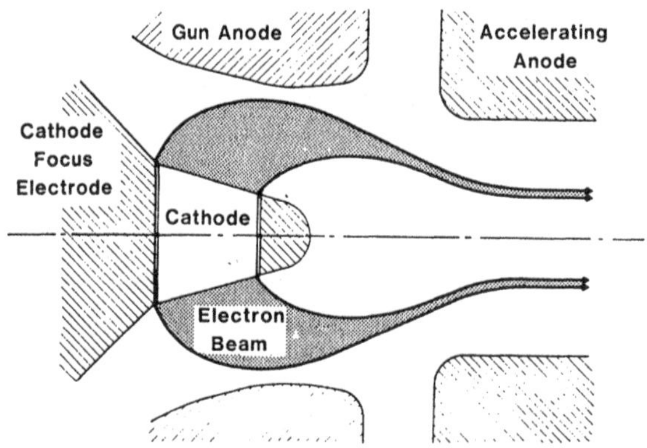

Figure 14.11 Magnetron Injection Gun

Normally, the electron gun used is a magnetron injection gun like that shown in Fig. 14.11. The gun is so-named because of the magnetron-like appearance of the cathode assembly. Electrons are drawn in a generally radial direction from the cathode surface and then formed into a hollow beam by electric and magnetic fields. The resulting electron trajectories are as indicated in Fig. 14.12. The initiation of the helical trajectory followed by each electron is clearly shown.

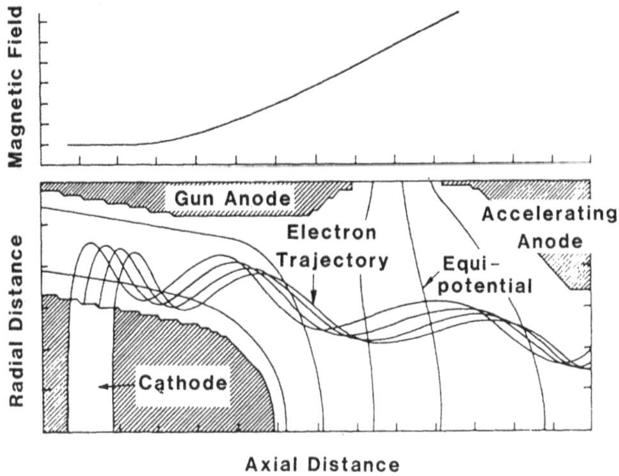

Figure 14.12 Electron Trajectories in Gun Region of Gyrotron

(From Robert W. McMillan, *Proc. IEEE*, Jan. 1985.)

The cathodes used in magnetron injection guns for gyrotrons are operated temperature-limited rather than space-charge-limited to minimize the velocity spread in the beam. Some work has been done on the use of Pierce guns and a very carefully designed "corkscrew" shaped magnetic field for generating low-energy-spread beams. Early results obtained in a gyro-TWT have been promising [14.7].

14.3.1 Gyrotron Oscillator

The hollow beam from the gyrotron gun in injected into a very strong magnetic field. Flux densities of several Tesla are normally required and this usually necessitates the use of cryogenic magnets. The beam is then passed through an interaction cavity. The location of the cavity with respect to the gun and the magnetic-field structure is shown in Figs. 14.13 and 14.15. The shape of the cavity is dependent on the mode of the electromagnetic field with which the beam is intended to interact. The shape is also dependent on whether or not operation with a harmonic of the cyclotron frequency is required. Harmonic operation and mode configurations are discussed in following paragraphs. Some representative cavity shapes are shown in Fig. 14.14.

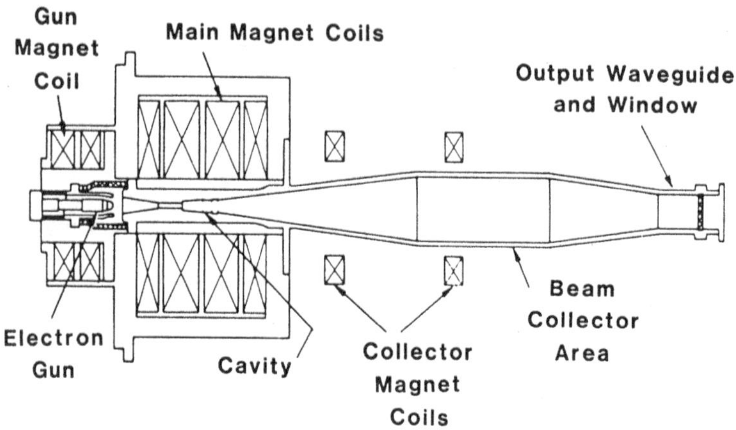

Figure 14.13 CW Gyrotron Oscillator

(From Varian Brochure entitled Introduction to Gyrotrons.)

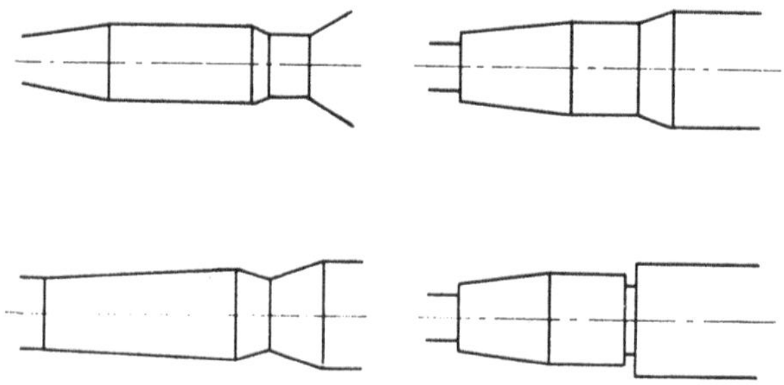

Figure 14.14 Gyrotron Oscillator Cavity Shapes

With the proper shape for the RF field, it is possible for harmonic interactions with the electrons to occur. An example is given in Fig. 14.15 in which the field varies at twice the cyclotron frequency. The direction of the field reverses in the center of the electron orbit. Thus, an electron that is initially decelerated by the field (Fig. 14.15(a)), is moving transverse to the field when the field reverses (Fig. 14.15(b)), and so does not have its orbital energy changed. By the time the field reverses again, the electron has moved another 90° around its orbit and is again in a decelerating field. Thus, during each orbit of the electron, the RF field goes through two complete cycles. For a given operating frequency, the cyclotron frequency is half the value used in the fundamental mode of operation. As a result, the magnetic

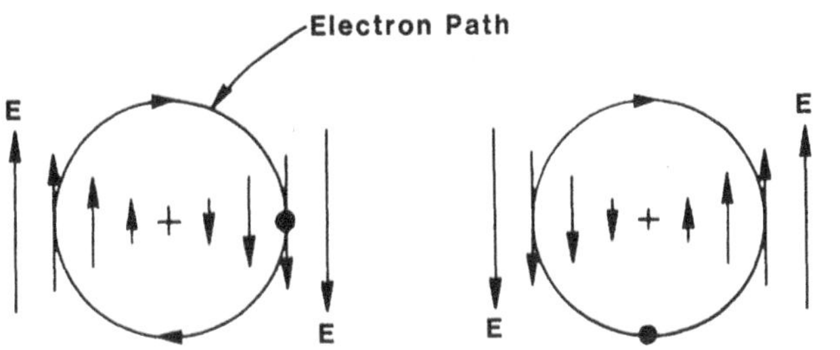

Figure 14.15 Harmonic Interaction of an Electron and a Field Varying at Twice the Cyclotron Frequency

(From *Advances in Electronics and Electron Physics*, vol. 55, by R.S. Symons and H.R. Jory, Copyright 1981 by Academic Press., Reproduced with permission.)

field is reduced by a factor of two. Shown in Fig. 14.16 is the cross section of a section-harmonic gyrotron for which the field level was low enough so that a room-temperature solenoid could be used. This gyrotron operated at a frequency of 34 GHz and a power level of 10 kW CW with an efficiency of 40 percent [14.8].

Operation at frequencies higher than the second harmonic are being examined, but the intensity of the interaction is reduced, making the achievement of significant performance difficult.

Because gyrotron cavities are normally large compared with the wavelength of the electromagnetic wave, many modes of operation are possible. In Fig. 14.17, the amplitude of the electric field as a function of radial position is shown for several modes. Also shown are the positions of the cavity walls and the optimum positions for beam injection for operation at the fundamental frequency in the various modes. The choice of mode and beam position depend on operating characteristics such as frequency and output power required from the device.

The reported CW and pulsed power capabilities achieved with gyrotrons and linear beam-tubes are given in Figs. 14.18 and 14.19. The limit of the capability of conventional microwave tube technology is indicated by the solid line in each figure. It may be noted that gyrotrons are capable of power levels as high as three orders of magnitude above those of conventional tubes.

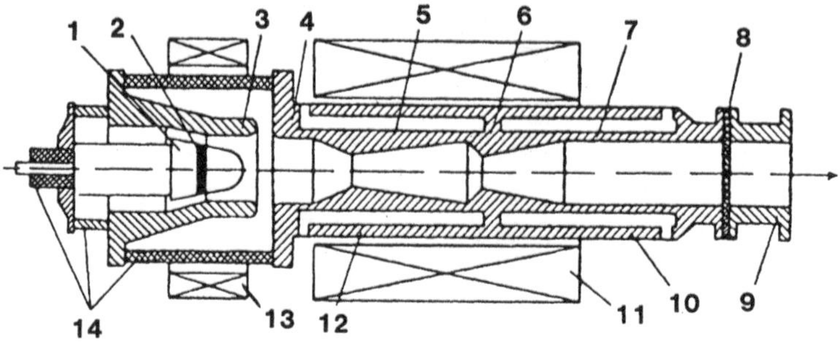

1. Gun Structure	5. Cavity	9. Output Waveguide
2. Emitting Area	6. Output Iris	10, 12. Water Cooling
3. 1st Anode	8. Vacuum Window	11, 13. Solenoids
4. 2nd Anode	7. Collector	14. Ceramic Insulators

Figure 14.16 Gyrotron Employing Second-Harmonic Operation and a Room-Temperature Solenoid

(From D.V. Kisel', *et al., Radio Engineering and Electronic Physics,* 1974., Reproduced with permission from Scripta Technica, Inc., Subsidiary of John Wiley and Sons, Inc.)

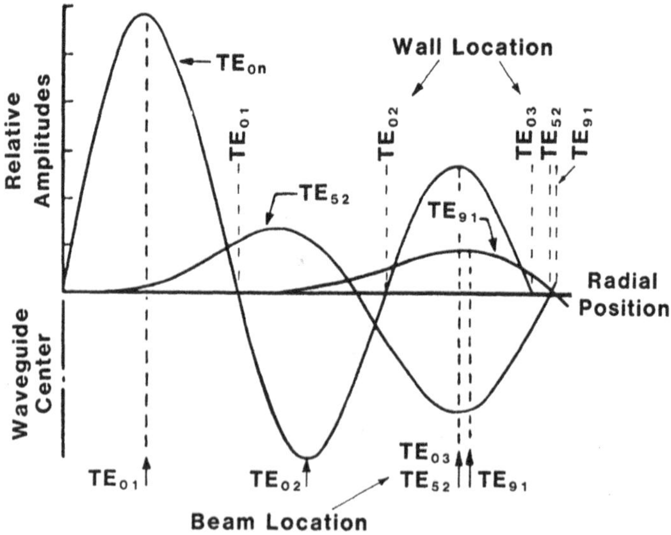

Figure 14.17 Electric Field Distributions and Electron Beam Locations for Operation at the Fundamental Frequency in Several Cavity Modes

(From J. Mark Baird, *Technical Digest*, 1979 IEDM.)

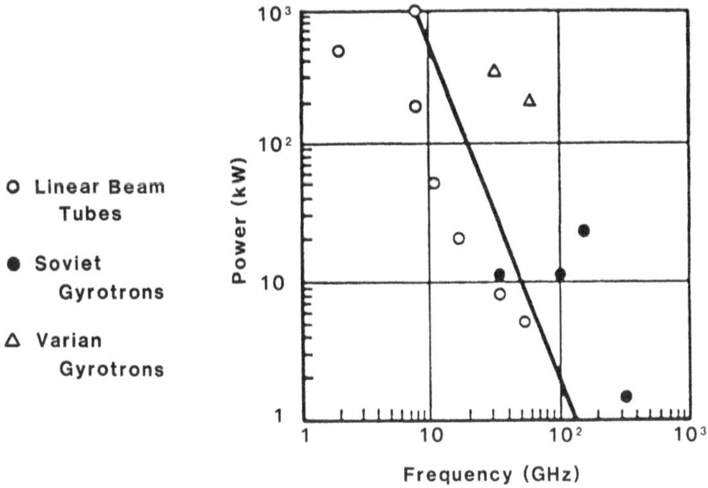

Figure 14.18 Reported CW Capabilities of Gyrotrons and Linear Beam Tubes

(From H.R. Jory, *Technical Digest*, 1981 IEDM; and Varian 200 kW, 60 GHz, CW Result.)

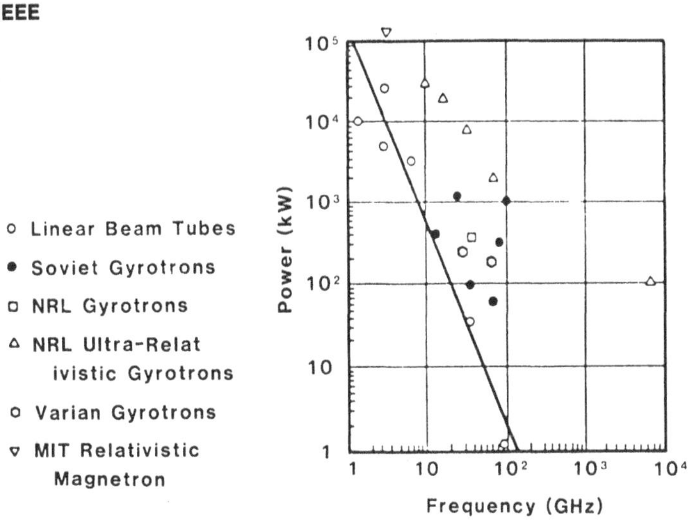

Figure 14.19 Reported Pulsed Capabilities of Gyrotrons and Linear-Beam Tubes

(From H.R. Jory, *Technical Digest*, 1981 IEDM.)

14.3.2 Gyro-klystrons

The operation of a gyro-klystron is similar to a conventional klystron except that electron bunching occurs in the azimuthal direction rather than in the axial direction. In a gyro-klystron, a signal to be amplified is fed into an input cavity where the cyclotron bunching process is initiated. Then, the beam is permitted to drift while bunches form. As the beam passes through a second cavity, the amplified signal may be removed or the bunching process may be enhanced and the signal removed in a subsequent cavity. A cross-sectional view of a two-cavity gyro-klystron is shown in Fig. 14.20. Conceptually, the only difference between the gyro-klystron and the gyrotron oscillator is the addition of the input cavity. The development of gyro-klystrons has not been pursued vigorously, apparently, because of the greater promise of gyro-TWTs.

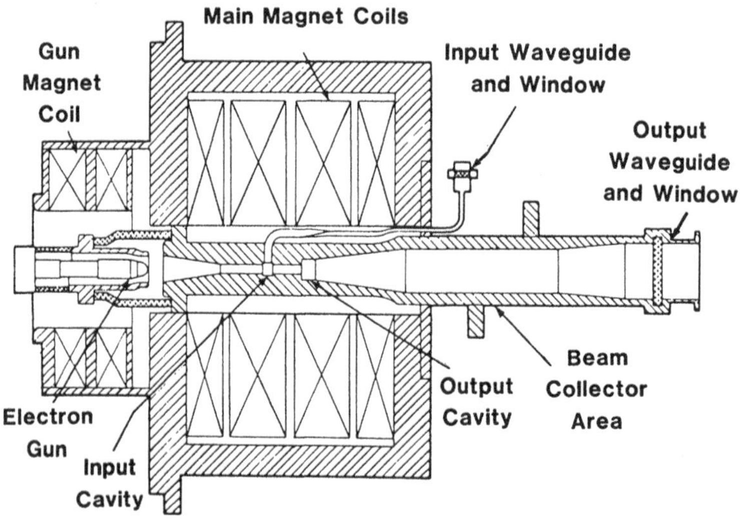

Figure 14.20 Pulsed Gyro-klystron Amplifier

(From Varian Brochure entitled Introduction to Gyrotrons.)

14.3.3 Gyro-TWTs

There has been substantial interest and development effort in gyro-TWTs. In these devices, a nonresonant RF structure is used to produce a traveling wave interaction. Axial phase synchronism is required between the traveling wave and the rotating electrons as indicated in Fig. 14.21. This phase synchronism is similar to that required in the peniotron.

The C-band gyro-TWT developed by Varian [14.9] is illustrated in Fig. 14.22. The interaction waveguide used was a circular waveguide operating in the TE_{11} mode. As indicated by the magnetic flux density plot, optimum results were obtained by using a slightly tapered field. The power output characteristics for this device are shown in Fig. 14.23. At a beam voltage of 65 kV, the maximum power output was 120 kW at an efficiency of 26 percent. The 3-db bandwidth was 7.25%.

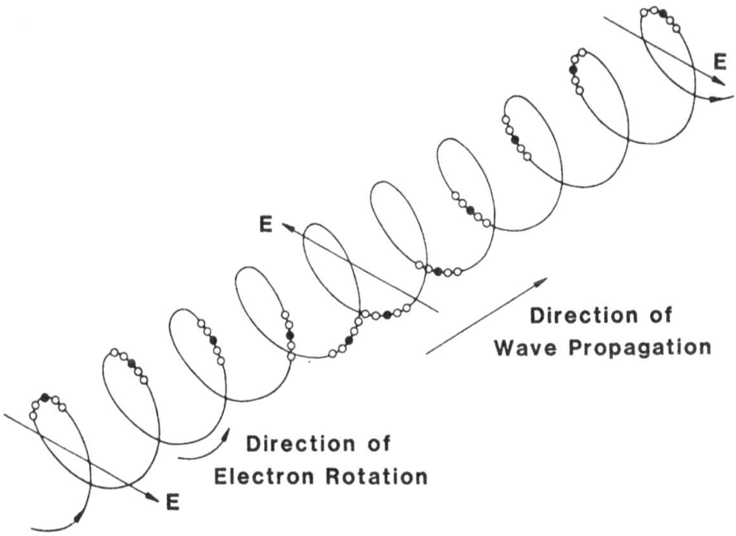

Figure 14.21 Interaction of Beamlet with a Traveling Wave

(From *Advances in Electronics and Electron Physics*, vol. 55, by R.S. Symons and H.R. Jory, Copyright 1981 by Academic Press., Reproduced with permission.)

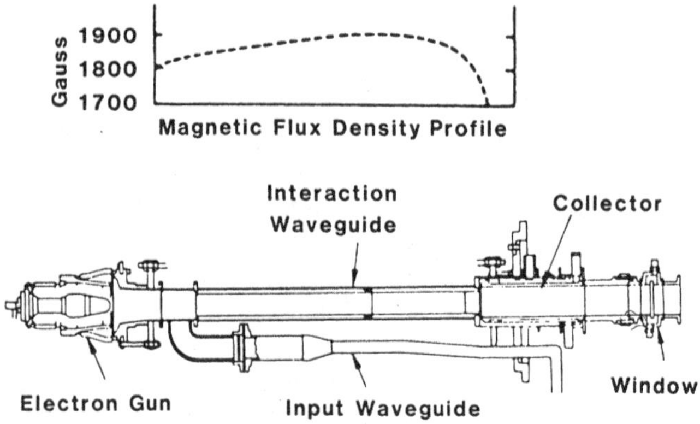

Figure 14.22 C-Band Gyro-TWT

(From Patrick Ferguson and Robert Symons, *Technical Digest*, 1980 IEDM.)

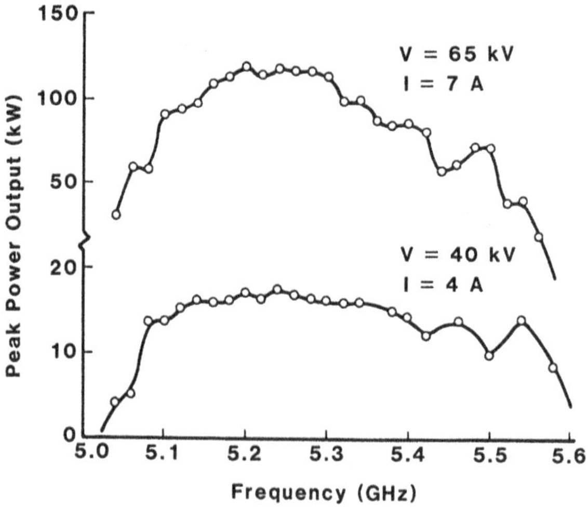

Figure 14.23 Peak Power Output for a Gyro-TWT for Two Operating Conditions

(From Patrick Ferguson and Robert Symons, *Technical Digest*, 1980 IEDM.)

Techniques are being investigated [14.10] for increasing the bandwidth by tapering the RF structure as well as the magnetic field as is indicated in Fig. 14.24. As the input frequency is varied, the region ΔL, over which phase synchronism and amplification occur, moves along the interaction waveguide. The plot of small-signal gain for this device, given in Fig. 14.25, shows that thc bandwidth achieved was about 18 dB.

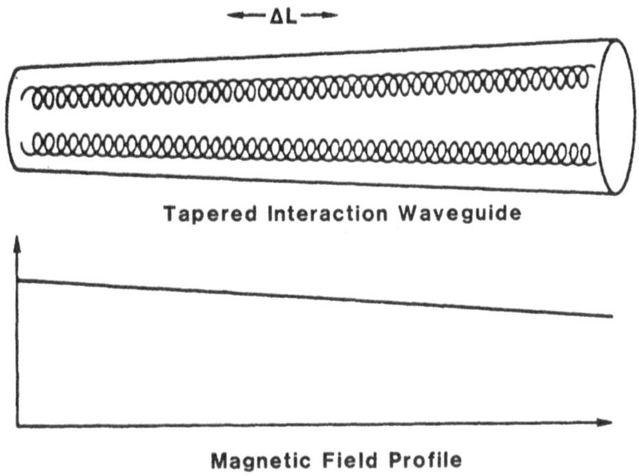

Figure 14.24 Distributed Gyro-TWT

(From K.R. Chu, *et al.*, *IEEE Trans. Electron Devices*, July 1981.)

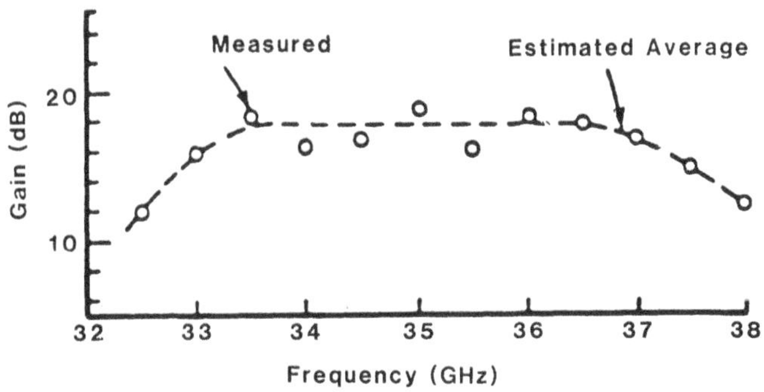

Figure 14.25 Small-Signal Gain for Wide-Band Gyro-TWT

(From L.R. Barnett, *et al.*, *IEEE Trans. Electron Devices*, July 1981.)

In concluding this very brief chapter on fast-wave tubes, it is important to remind the reader that the material for this book grew out of a short course devoted almost exclusively to microwave tubes used in existing systems. As a result, the focus of the book is on conventional microwave tubes. This chapter has been included largely in an effort to complete the overall picture of the scope of microwave tube activity at present.

While conventional microwave tubes are certain to dominate the microwave tube business for years to come, the importance of the development of fast-wave tubes must be recognized. It should be noted that the first report of the invention of the gyrotron was little more than a decade ago. Since that time, there has been intense activity in the fast-wave device area. A strong indication of the intensity of this activity was the devotion to gyrotron-type devices of two full issues and over 650 pages of the *International Journal of Electronics* within a period of 14 months in 1981 and 1982. Fast-wave devices have resulted in the rejuvenation of the vacuum microwave tube area and they comprise an exciting area of activity for the future.

REFERENCES

[14.1] R.M. Phillips, "The ubitron, a high-power traveling-wave tube based on a periodic beam interaction in unloaded waveguide," *IRE Trans. Electron Devices*, vol. ED-7, October 1960, pp. 231–241.

[14.2] C.E. Enderby and R.M. Phillips, "The ubitron amplifier — a high power millimeter-wave TWT," *Proc. IEEE*, vol. 53, October 1965, p. 1648.

[14.3] K. Yamanouchi, S. Ono, and Y. Shibata, "Cyclotron fast wave tube, the double ridges traveling wave peniotron," *Proc. 5th International Conference on Microwave Tubes*, Paris 1964, pp. 96–102.

[14.4] Gunter Dohler, David Gallagher, and Robert Moats, "The peniotron: a fast wave device for efficient high power mm-wave generation," *Technical Digest*, 1978 International Electron Devices Meeting, pp. 400–403.

[14.5] G. Dohler, D. Gallagher, R. Moats, and F. Scafuri, "Peniotron oscillator operating performance," *Technical Digest*, 1981 International Electron Devices Meeting, pp. 328–331.

[14.6] Recommended for further reading:

R.S. Symons and H.R. Jory, "Cyclotron resonance devices," *Advances in Electronics and Electron Physics*, vol. 55, Academic Press, 1981.

Introduction to Gyrotrons, Varian Brochure No. 4368.

J. Mark Baird, "Survey of fast wave tube developments," *Technical Digest*, 1979 International Electron Devices Meeting, pp. 156–163.

H.R. Jory, "Gyro-device developments and applications," *Technical Digest*, 1981 International Electron Devices Meeting, pp. 182–185.

G. Dohler, D. Gallagher and R. Moats, "Research on eigenmode RF generation," Final Report AFWAL-TR-83-1147, November 1983.

International Journal of Electronics, vol. 53, no. 6, 1982, pp. 501–760.

International Journal of Electronics, vol. 51, no. 3, 1981, pp. 275–606.

[14.7] P.E. Ferguson and R.S. Symons, "A gyro-TWT with a space-charge limited gun," *Technical Digest*, 1981 International Electron Devices Meeting, pp. 198–201.

[14.8] D.V. Kisel', G.S. Korablev, V.G. Navel'yev, M.I. Petelin and Sh.E. Tsimring, "An experimental study of a gyrotron, operating at the second harmonic of the cyclotron frequency, with optimized distribution of the high frequency field," *Radio Engineering and Electronic Physics*, vol. 19, no. 4, pp. 95–100 (1974).

[14.9] Patrick Ferguson and Robert Symons, "A C-band gyro-TWT," *Technical Digest*, 1980 International Electron Devices Meeting, pp. 310–313.

[14.10] Larry R. Barnett, *et al.*, "An experimental wide-band gyrotron traveling-wave amplifier," *IEEE Trans. Electron Devices*, vol. ED-28, July 1981, pp. 872–875.

Chapter 15
Breakdown and Protection

There are literally dozens of ways in which microwave tubes can fail. A reasonably thorough discussion of their failure mechanisms is well beyond the scope of this book and, in fact, would require another book of size comparable to this one.

One of the most common and, perhaps, most poorly understood classes of failure mechanisms centers on electrical breakdown in one form or another. In the short course from which this book was developed, electrical breakdown was, at first, only briefly discussed. Because of the strong interest in breakdown phenomena expressed by those taking the course, that section was developed to cover major aspects of breakdown phenomena of concern to those who must use microwave tubes.

Breakdown in microwave tubes may occur inside or outside the vacuum envelope at any of several locations as is indicated below and in Fig. 15.1.

- In the electron gun (between electrodes, between leads or from electrodes or leads to ground);
- In the collector (between electrodes, between leads or from electrodes or leads to ground);
- In high-power portions of the RF structure.

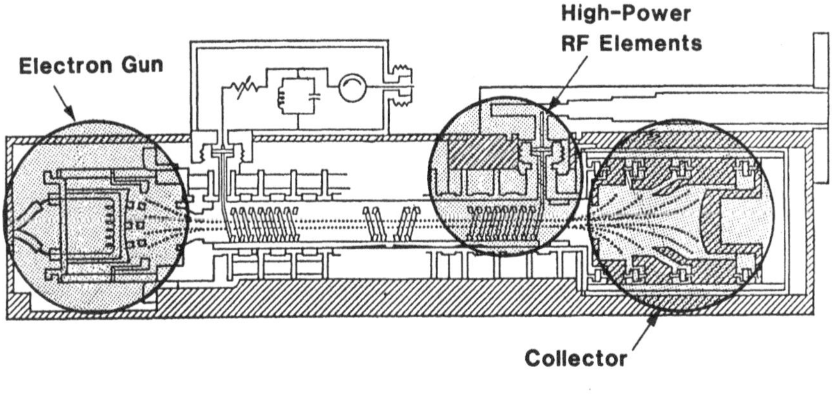

Figure 15.1 Regions in a Microwave Tube Where Electrical Breakdown May Occur

Often, it is not the tube that is initially at fault. For example, there may be a failure at any one of several locations in the RF system attached to a tube. Shown in Fig. 15.2 are some of the common problem areas in these RF systems. If excessive RF power is reflected, the tube may be damaged for any one of several reasons. All too often, the tube is thought to be at fault and is replaced before the original cause of the difficulty is corrected.

Whether or not breakdown will occur depends on two factors:

1. The applied field level and local field enhancement effects;
2. The breakdown field of the medium (gas, vacuum, liquid or solid).

Very approximately, the breakdown fields for various media are as follows:

Gas	– 10's V/cm to 10^5 V/cm depending on pressure and type
Vacuum	$\sim 0.5 - 3 \times 10^5$ V/cm
Liquid	$\sim 0.5 - 1.0 \times 10^6$ V/cm
Solid	$\sim 0.5 - 1.0 \times 10^6$ V/cm

At interfaces between solids, liquids, and vacuum or gas, the breakdown field may be significantly degraded. In addition, in a void in a solid or a liquid, the field is normally enhanced. The amount of enhancement depends on the shape and orientation of the void and the dielectric constants of the media as indicated in Fig. 15.3.

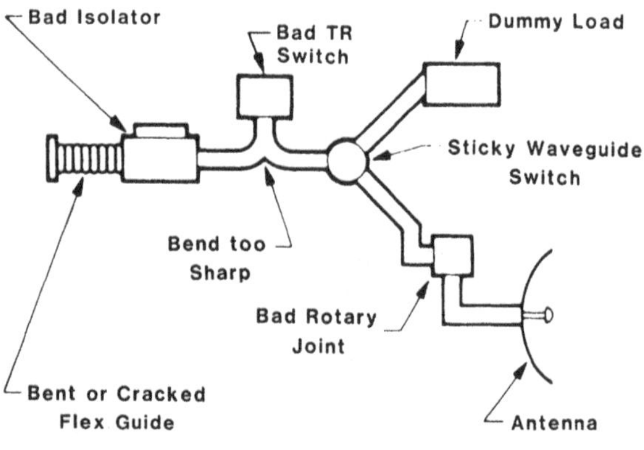

Figure 15.2 Common Problem Areas in RF Systems

(From Microwave Tube Manual, by Varian Associates. Air Force Publication T.O. 00-25-251, October 1979.)

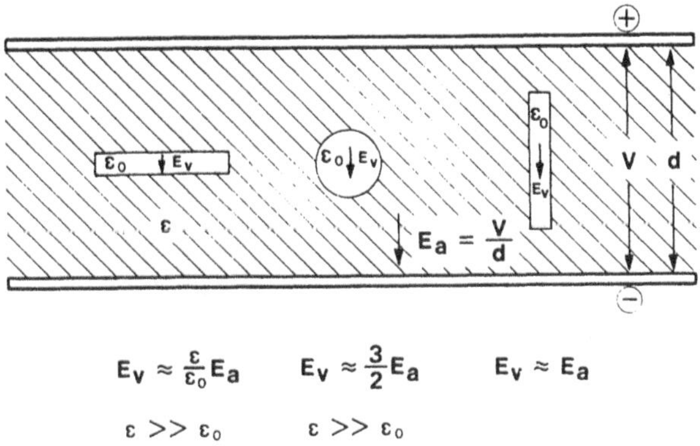

Figure 15.3 Electric Field Enhancement in Voids in Dielectric Media

The low breakdown fields of gas or vacuum relative to those of solids or liquids coupled with various field-enhancement effects results in breakdown occurring, in most cases, in gas or vacuum rather than in solid or liquid. As a result, most of the breakdown discussions in this chapter will concentrate on gas and vacuum. First, a section is devoted to various field-enhancement effects that reduce the value of the applied field at which breakdown occurs. Then, dc and RF breakdown in gas and in vacuum are covered. Finally, the special case of breakdown on the surface of an insulator is discussed.

15.1 FIELD ENHANCEMENT

The electric field between parallel plane electrodes is the voltage applied from one electrode to the other divided by the distance by which the electrodes are separated. In practice, it is common to calculate field levels in all sorts of configurations simply by dividing the applied voltage by the electrode separation distance. Fields calculated in this manner are often in error by a factor of from two to ten or more. Worse yet, because the power varies as the square of the field, the calculated power handling capability is often in error by a factor ranging up to two orders of magnitude or more.

The parallel-plane and coaxial conductors shown in Fig. 15.4 provide a simple case for the comparison of field levels. A potential of 1000 V between electrodes is assumed, and equipotential profiles at 250 V, 500 V, and 750 V are shown. For the parallel-plane electrode case, the potential profiles are equally spaced and the electric field is 1000 V/cm everywhere between the electrodes.

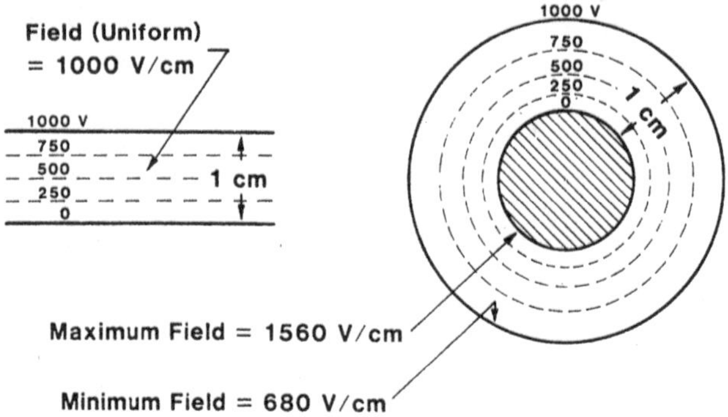

Figure 15.4 Potential Distributions in Parallel-Plane and 50 Ω Coaxial-Electrode Configurations

For the coaxial-electrode case, the equipotential profiles near the center conductor are more closely spaced than in the parallel-plane electrode case. The electric field at the surface of the center conductor is easy to calculate for this configuration and is 1560 V/cm, or over 50 percent above that for parallel-plane electrodes.

If the maximum field were limited to 1000 V/cm because of breakdown considerations, then the maximum voltage that could be applied to the coaxial electrodes would be 1000/1.56, or 641 V. In the TE_{11} mode (for which the potential distributions in Fig. 15.4 are correct), the maximum RF power capability of the coaxial configuration would be $(1/1.56)^2$, or 40 percent of that for the parallel-plane electrodes.

The factor by which the electric field exceeds the field for the parallel-plane electrode case is normally called the field-enhancement factor and is denoted by β. At the center conductor of the 50 Ω coaxial line, the field-enhancement factor is 1.56.

The magnitude of β is highly dependent on the radius of an electrode and its distance to another electrode. These factors are illustrated in Fig. 15.5 for electrode pairs consisting of a sphere and a plane and a hemisphere-on-a-plane and a plane.

Field-enhancement factors have been calculated for many electrode configurations. Some of these are given in Fig. 15.6. In most practical cases of interest, the electrode geometries are close enough to those shown in Fig. 15.6 to permit a reasonable estimate of the value of β to be made. In some cases, values of β for two or more configurations can be combined to obtain an estimate for a configuration in question. For example, the value of β at the

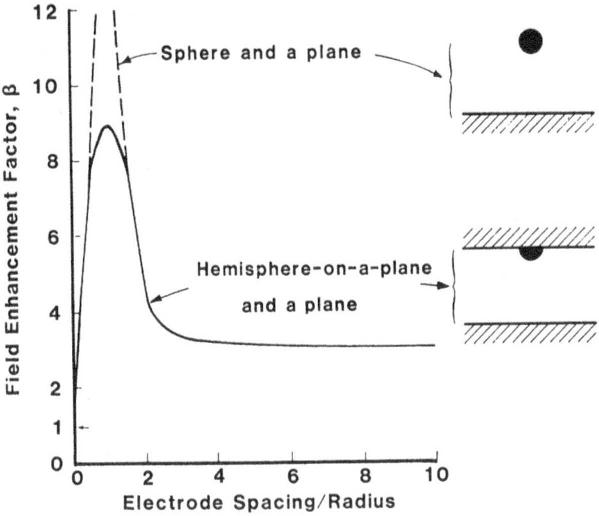

Figure 15.5 Value of β for Two Electrode Configurations

(From V.N. Glazanov, *Elektrichestvo*, vol. 3, 1958.)

Configuration		β	Configuration		β
CONCENTRIC SPHERES	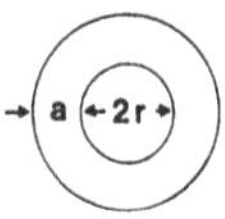	$1 + \frac{a}{r}$	COAXIAL CYLINDERS	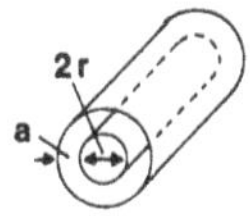	$\frac{\frac{a}{r}}{\ln(1 + \frac{a}{r})}$
SPHERE AND PLANE	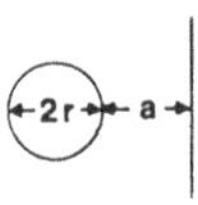	$0.9(1 + \frac{a}{r})$	CYLINDER AND PLANE	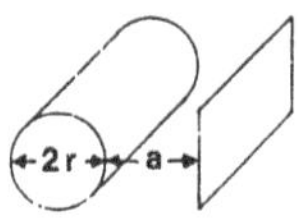	$\frac{0.9\frac{a}{r}}{\ln(1 + \frac{a}{r})}$
TWO SPHERES	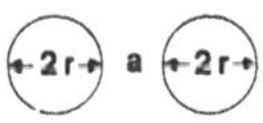	$0.9(1 + \frac{a}{2r})$	PARALLEL OR PERPENDICULAR CYLINDERS	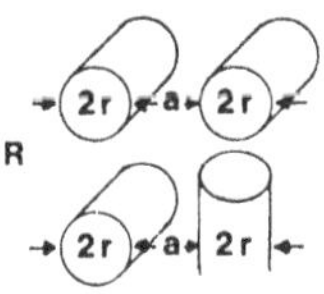	$\frac{0.9\frac{a}{2r}}{\ln(1 + \frac{a}{2r})}$

Figure 15.6 Maximum Field-Enhancement Factor, β, for Various Electrode Configurations

surface of a drop of condensation on the center conductor of a 50 Ω coaxial line is approximately 3 (β for a hemisphere-on-a-plane) times 1.56 (β at the center conductor of a 50 Ω coaxial line), or about 4.7.

Obstacles of various shapes are often purposely placed in waveguides and coaxial lines. This may be for impedance-matching purposes, for mounting of dielectric supports, *et cetera*. Values of β and power-handling degradation are given in Fig. 15.7 for some of these obstacle shapes in rectangular waveguide. Note that the reduction in power capability can easily be an order of magnitude or more.

The importance of field enhancement on electrical breakdown cannot be overemphasized. The most troublesome aspect of field enhancement is that, in many cases, its magnitude is either not known or is badly underestimated. The value of β of 1.56 calculated for the center conductor of a 50 Ω coaxial line is virtually meaningless if there are irregularities such as scratches, machining marks, *et cetera* on the center conductor. These can easily have a β of 10 or more.

Finally, it should be noted here that the breakdown characteristics of electrodes in vacuum may, in some cases, be almost completely controlled by tiny protuberances on the cathode surface having values of β approaching 100. This is discussed in Section 15.4.

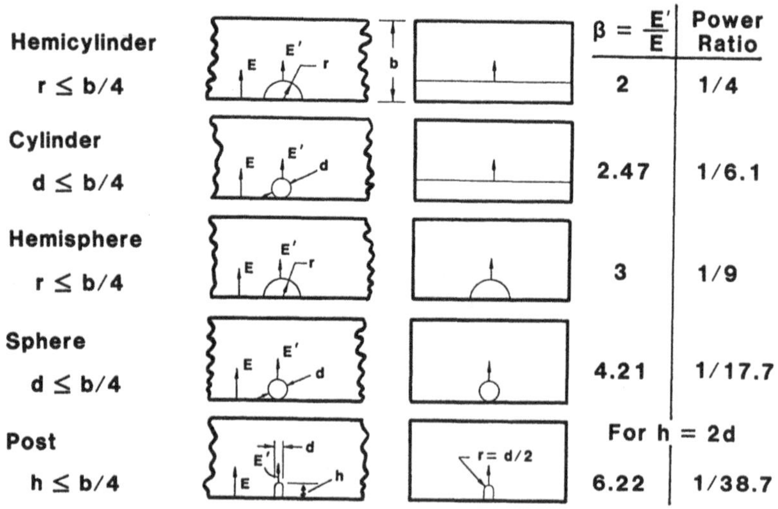

Figure 15.7 Values of β and Power Handling Degradation for Various Obstacles in Rectangular Waveguides

(From R.D. Wegenroth, Wheeler Laboratories Report 953P, June 1960.)

15.2 DC BREAKDOWN IN GAS [15.1, 15.2, 15.3]

Electrical discharges in gases are extremely complex. There are several different types of discharges that can exist, depending on the current and voltage applied to the gas. These various discharge types are indicated in Fig. 15.8. Current and relative voltages shown in Fig. 15.8 are representative values only, and depend on factors such as gas type and area of the electrodes used during testing.

By far, the most damaging discharge is the arc. In the following paragraphs, various phenomena leading up to an arc are discussed and then arc characteristics are described.

The discharges in Fig. 15.8 are roughly categorized as non-self-sustaining and self-sustaining. The non-self-sustained discharges occur at very low current levels. The various phases of the non-self-sustained discharge are indicated in Fig. 15.9, which is an expansion of the left-hand side of Fig. 15.8 with the axes reversed.

The curve in Fig. 15.9 shows the variation of current with voltage between electrodes in a gas at room temperature (the exact temperature is not important as long as it is not absolute zero). It is also assumed that the radiation (cosmic, ultraviolet, *et cetera*) that normally exists is present. Because of the temperature and the radiation, electrons are emitted by the electrodes and are released by ionization of gas atoms and molecules. (At room temperature there are about 10^6 electrons/m^3 in atmosphere air).

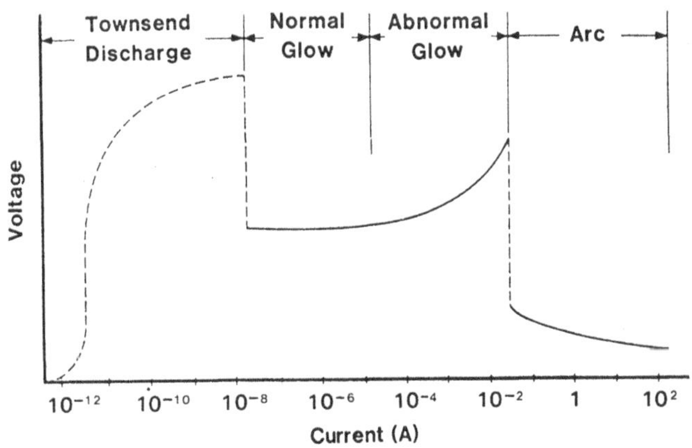

Figure 15.8 Typical Volt-Ampere Characteristics of a Gas Discharge

(From *Principles and Applications of Electron Devices* by Paul D. Ankrum, Copyright 1959 by International Textbook Company, Reproduced with permission.)

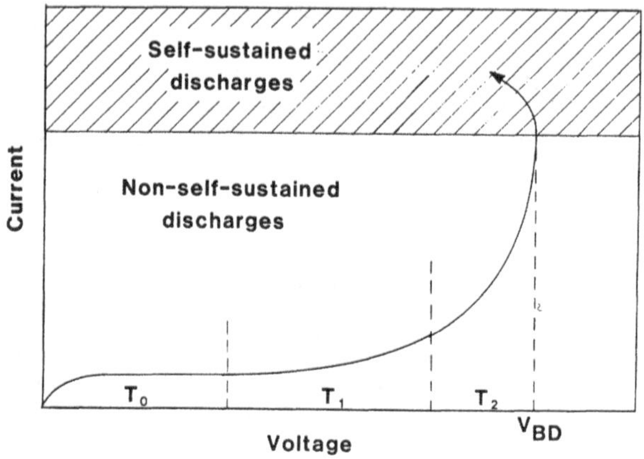

Figure 15.9 Current-Voltage Relations in Pre-Breakdown Regions

(From J.D. Cobine, *Gaseous Electronics*, Published 1958 by Dover Publications. Copyright by J.D. Cobine, Reproduced with permission.)

When voltage is first applied to the electrodes, the current increases slowly and then saturates. That is, all free electrons are drawn to the positive electrode. Further increases in voltage produce no further increase in current (region T_0 in Fig. 15.9). When the voltage is increased to region T_1, the current increases slowly again. In this region, electrons gain sufficient energy from the electric field to produce ionization of the gas and the resulting additional electrons (and ions) contribute to the current level.

In region T_2, the rate of increase of current accelerates because positive ions produced by electron collisions gain sufficient energy from the field to produce secondary emission of electrons from the negative electrode as well as additional ionization.

The discharges in regions T_1 and T_2 are called Townsend discharges, after J.S. Townsend, who conducted early investigations in this field. These discharges (and that in region T_0) are referred to as non-self-sustaining discharges because the current ceases as soon as the external radiation (including the heat to bring the electrodes and gas to room temperature) is removed.

When the voltage reaches a critical level, V_{BD}, current increases very rapidly and is accompanied by a decrease in voltage. The voltage, V_{BD}, is normally called the breakdown voltage, but is sometimes referred to as the firing voltage, ignition voltage, or sparking voltage. The resulting discharge is self-sustaining because each electron leaving the negative electrode establishes secondary processes (ionization, *et cetera*), which, in turn, cause one or more electrons to be emitted. Removal of external radiation does not cause the discharge to extinguish.

The breakdown voltage of a gas depends on the ease with which the gas is ionized and on the effectiveness of the ions in producing electrons when they strike the negative electrode. Thus, the breakdown voltage is a function of gas type and pressure, and electrode spacing and material.

The breakdown voltage is normally plotted in the manner shown in Fig. 15.10. Breakdown voltage plots of this type are referred to as Paschen characteristics because of Paschen's law, which states that, for parallel plane electrodes in a particular gas at a particular temperature, the breakdown voltage is a function of the product of the gas pressure and the electrode spacing.

The shapes of the Paschen characteristics may be explained qualitatively as follows: to the left of the minimum in the Paschen curve, the breakdown voltage increases as pressure and density decrease (with fixed electrode spacing). This is because the gas molecules are farther apart, and more of the electrons travel through the gas and are collected by the positive electrode without having produced ionization. In the limit, under high-vacuum conditions, there is essentially no ionization of the gas between electrodes and breakdown occurs because of effects occurring at the surfaces of the electrodes. Direct current (dc) breakdown in vacuum is discussed in Section 15.4.

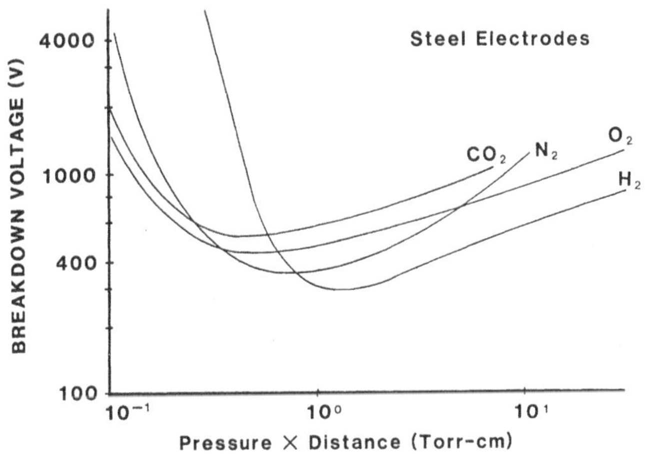

Figure 15.10 Paschen Characteristics for Some Molecular Gases

(From Max J. Schonhuber, *IEEE Trans. P.A.S.* February 1969.)

To the right of the Paschen minimum, the breakdown voltage increases as pressure and gas density increase (with fixed spacing between the electrodes). In this case, gas molecules are closer together, and so a large electric field is required to give the electrons enough energy in the short distances between collisions to produce ionization. Thus, the breakdown voltage is high. At atmospheric pressure and above, the breakdown voltage for a 1 cm gap can be several tens and even hundreds of thousands of volts.

At intermediate pressures and electrode spacings, optimum conditions for ionization occur, and so there is a minimum in the Paschen curve. The minimum breakdown voltage for most gases is a few hundred volts and occurs at a *Pd* (pressure times distance in Torr-cm) near unity.

Breakdown voltages for several commonly used gases are listed in Table 15.1. The values given were obtained with a 1 cm gap at the Paschen minimum (at the critical pressure-spacing product) and at atmospheric pressure. In particular, it should be noted that the freons and sulphur hexafluoride have relatively high breakdown voltages. These gases are often referred to as electronegative gases. Their molecules require one or two electrons to fill the outer shell. As a result, they can reasily capture free electrons and this results in the high breakdown voltages for these gases.

TABLE 15.1
Breakdown Voltages for Several Commonly Used Gases
(From W. G. Dunbar, "High Voltage Design Guide: Aircraft," Vol. IV, AFWAL-TR-82-2057.)

Gas	*Volts (dc)*	*Kilovolts (dc)*
Air	315	33
Ammonia	—	26
Argon	280	4.8
Carbon Dioxide	430	28
Freon 14	480	32
Freon 114	420	90
Freon 115	430	90
Freon 116	500	—
Freon C 138	450	—
Helium	189	1.63
Hydrogen	292	17
Nitrogen	265	32
Oxygen	440	—
Sulfur Hexafluoride	520	89

Electronegative gases are expensive and tend to be somewhat hazardous, and so they are often mixed with more common gases such as air or dry nitrogen. As shown in Fig. 15.11, even a small amount of sulphur hexafluoride (SF_6) in nitrogen significantly increases breakdown voltage.

When breakdown occurs, the discharge increases in intensity very rapidly. Depending on circuit conditions (voltage capability and impedance), either a glow discharge or an arc discharge may develop. The characteristics of these discharges are indicated in Fig. 15.8.

The power supply or modulator for a microwave tube is normally capable of delivering a short-circuit current of at least several amperes and sometimes several kiloamperes. As a result, when breakdown occurs, an arc is established rather than a glow discharge.

Still, glow discharges cannot be ignored because they can occur under special circumstances and go unnoticed until they lead to an arc. For example, a glow discharge can exist in a void at the interface of a potting compound and a conductor. Sputtering and heat from the discharge can enlarge the void and eventually lead to an arc.

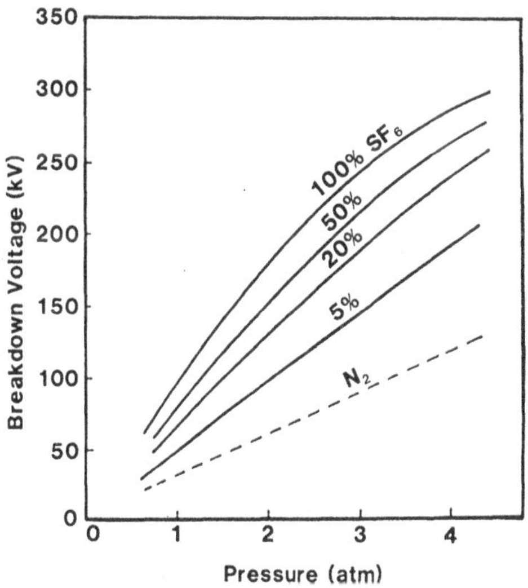

Figure 15.11 Effect on Breakdown Voltage of Mixing SF_6 with Dry Nitrogen

(From N.H. Malik and A.H. Qureshi, *IEEE Trans. on Elec. Ins.*, October 1980.)

The voltage required to maintain a glow discharge is on the order of the minimum breakdown voltage for the gas. That is, from roughly 100 V to a few hundred volts, depending on gas type. The glow discharge is characterized by a glow in the gas and on the cathode surface, and by a low current density on the order of 1 to 10 mA/cm^2. The discharge is maintained by secondary emission of electrons from the cathode as a result of positive-ion bombardment.

The potential distribution between two electrodes is altered significantly when a glow discharge occurs. For parallel-plane electrodes, for example, the potential distribution before breakdown is linear. As shown in Fig. 15.12, after a glow discharge has been established, the potential is close to the anode potential throughout most of the interelectrode region. Near the cathode, the potential drops rapidly to that of the cathode in the cathode fall region.

It is the large number of positive ions formed in the interelectrode space that is the primary reason for the shape of the potential profile. These ions neutralize the negative space charge due to the electrons. Near the cathode a positive-ion space charge cloud forms that is called the cathode sheath. Ions accelerated through the cathode sheath produce secondary-electron emission and sputtering at the cathode surface. The sputtered cathode material may be deposited on nearby insulating surfaces and result in undesirable conduction. Electrons accelerated through the cathode sheath may cause heating and decomposition of materials other than metals.

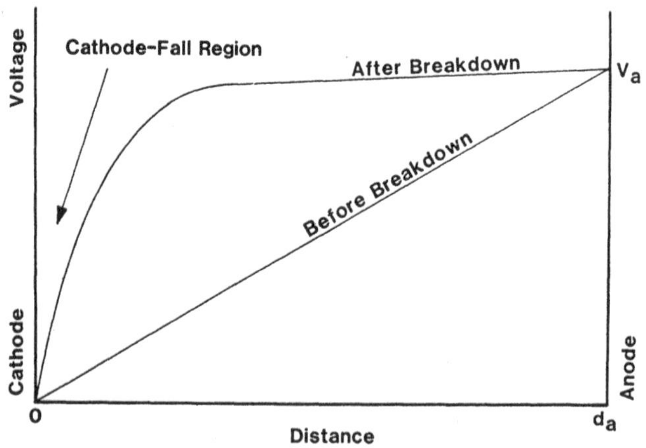

Figure 15.12 Potential Distribution in a Glow Discharge between Parallel-Plane Electrodes

If the current through a glow discharge is increased beyond the point where the cathode is covered with the glow, the discharge voltage will increase to produce the additional ionization required for the current flow. Under these conditions, the discharge is called an abnormal glow.

Continued increases in current result in more intense ion bombardment and heating of the cathode surface. Finally, a sudden transition to an arc discharge will occur. An arc discharge concentrates on a small spot on the cathode and the voltage across the discharge drops to a value comparable to the ionizing voltage of the gas. Because of the concentrated discharge, the current density and electric fields are very high compared with those in the glow discharge. With high-melting-temperature metals, the cathode spot is so hot that the dominant emission mechanism is thermionic emission. In other cases, field emission may dominate, or emission may result from a combination of thermionic and field emission.

At this point, it should be noted that, in spite of intensive research, the cathode and anode regions in an arc are still poorly understood. This is because of the complexities caused by the interactions of electrical, magnetic, thermal, and fluid effects in addition to surface effects.

The voltage profile between electrodes during an arc is somewhat like that of a glow discharge (Fig. 15.12) and is shown in Fig. 15.13. The primary difference in shape is a marked increase in voltage at the anode.

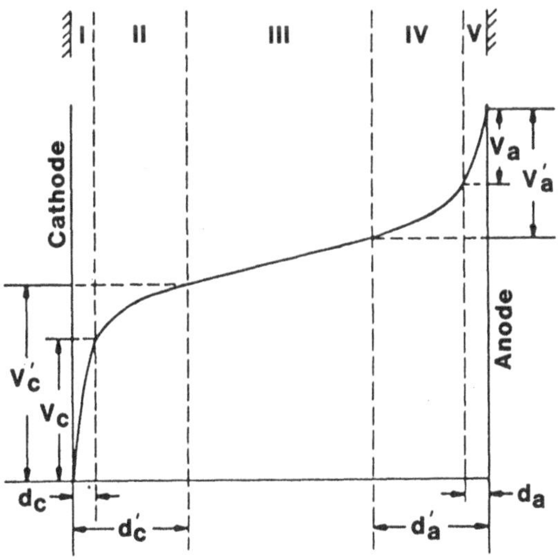

Figure 15.13 The Regions of an Arc

(From *Gaseous Electronics* by Hirsh and Oskam, Copyright 1978 by Academic Press. Reproduced with permission.)

The thickness of the electrode regions is exaggerated in Fig. 15.13. At atmospheric pressure, the thicknesses of the cathode fall (I) and anode fall (V) regions is in the 10^{-4} to 10^{-3} cm range. The thickness of the transition regions (II and IV) is on the order of 10^{-1} cm under the same conditions. In the transition regions, the arc diameter changes from the size at the electrode to the size of the arc column (III) and strong gradients exist in the plasma properties.

The voltage drops in the cathode region (I and II) and in the anode region (IV and V) are highly dependent on gas type, pressure, and electrode material. At atmospheric pressure, the voltage drop in the cathode region is typically on the order of 10 V. The voltage drop in the anode region may be anywhere from 1 or 2 V to 10 V or more.

The voltage gradient in the arc column is largely dependent on gas type and arc current as well as pressure. The arc-column voltage gradients for several common gases at atmospheric pressure are given in Fig. 15.14. The combination of the column gradient of a few tens of volts/cm with the electrode drops results in arc voltages in roughly the 50 to 100 V range for arcs of average length (a few centimeters).

At the end of Section 15.4, estimates are made of the energy and charge levels at which excessive damage may occur during a dc arc inside the vacuum envelope of a tube. Unfortunately, it is not so easy to make comparable

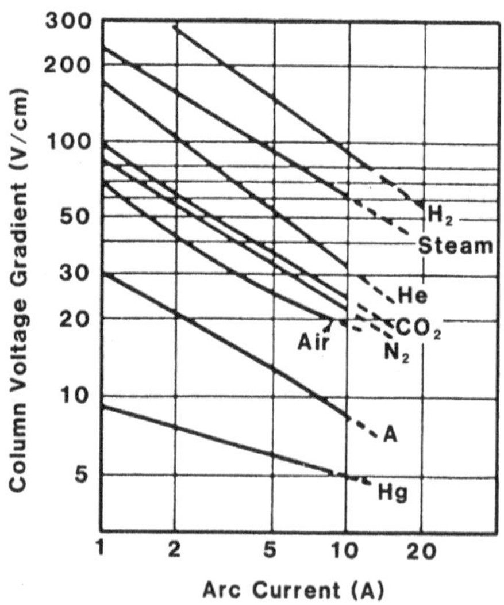

Figure 15.14 Arc-Column Voltage Gradient at Atmospheric Pressure

(From *Gaseous Conductors* by J.D. Cobine, Published 1958 by Dover Publications, Copyright by J.D. Cobine, Reproduced with permission.)

estimates for dc arcs outside the vacuum envelope. If the arc occurs between bare electrodes, it is possible that a substantial amount of energy (hundreds of joules or more) can be dissipated without causing excessive damage.

If, however, the arc occurs in the vicinity of an insulating material, it is possible that the material may be damaged with only a few joules of energy dissipation.

15.3 RF BREAKDOWN IN GAS [15.2]

In the previous section, it was shown that the dc breakdown field for air at atmospheric pressure is about 30 kV/cm. At low frequencies (up to tens of Hz), the breakdown field remains at 30 kV/cm. As frequency is increased to the audio range and then into the RF range, the breakdown field drops, and becomes highly dependent on frequency and electrode spacing (as well as pressure, of course).

The unified plot given in Fig. 15.15 was developed by Woo [15.4] for 50 Ω coaxial transmission lines and gives breakdown power as functions of frequency, pressure, and spacing under laboratory conditions. This plot appears complex at first glance, but it is relatively straightforward once a person becomes familiar with it.

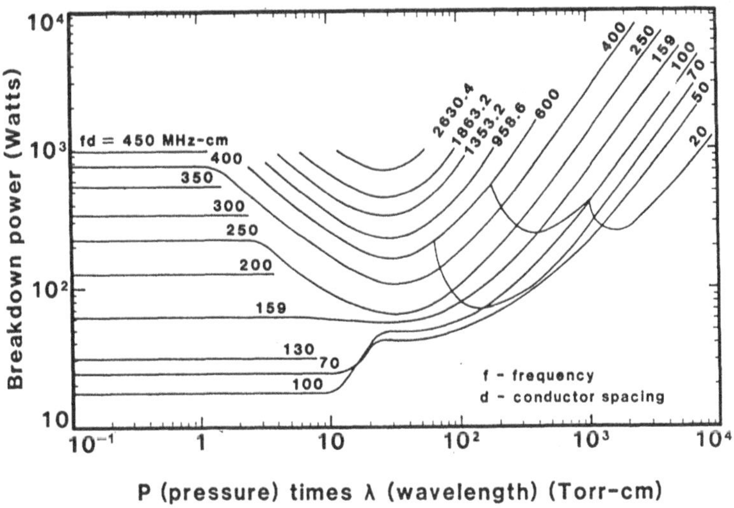

Figure 15.15 Unified Plot for RF Breakdown of Air in 50 Ω Coaxial Transmission Lines

(From R. Woo, *Proc. IEEE*, February 1969.)

Two primary breakdown processes are responsible for the shapes of the plots in Fig. 15.15. These breakdown processes and the ranges of frequency, pressure, and spacing in which they occur are given in Fig. 15.16. Descriptions of the breakdown processes identified in Fig. 15.16 are as follows:

1. Multipacting breakdown occurs when the electron's mean free path length is greater than the electrode spacing. Electrons travel back and forth across a gap between electrodes in synchronism with an oscillating RF field. If the secondary emission coefficient of the electrodes is greater than unity, a discharge develops. Multipacting will be discussed in Section 15.5.

2. In the region identified as the ionization breakdown region, the frequency is high enough and the separation distance large enough so that electrons are not swept out of the discharge by the field. If the field level is high enough, the electrons gain sufficient energy between collisions to cause ionization.

   The minimum breakdown power occurs when the frequency of collision of electrons with neutral molecules equals the applied RF frequency. This minimum occurs at a value of $p\lambda$ of about 30 Torr-cm and is indicated by the dashed line, which is identified as the collision frequency transition.

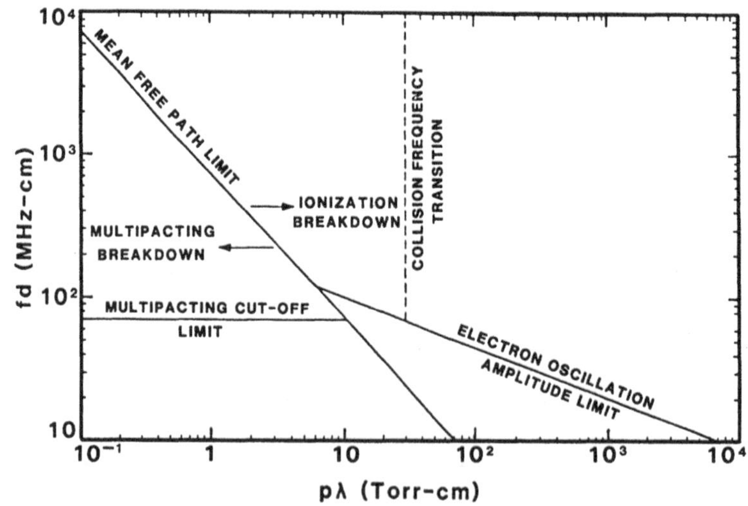

Figure 15.16 The *fd-p*λ Plane Showing Ranges of Breakdown Processes

(From R. Woo, *Proc. IEEE*, February 1969.)

When the frequency-distance (*fd*) product is small, the frequency is low enough and the separation between electrodes is short enough so that the amplitude of the electron oscillations exceeds the electrode separation distance. As a result, at the electron oscillation amplitude limit, electrons are lost to the conductor surfaces and the power required for breakdown increases rapidly.

The actual value of electric field at which breakdown occurs has been determined for selected frequencies in the microwave range. Two such selected frequencies are shown in Figs. 15.17 and 15.18. In both of these cases (992 MHz and 9.4 GHz), extrapolations to atmospheric pressure indicate that the breakdown field is about 20 kV/cm, or comparable to the dc breakdown field level of about 30 kV/cm. The characteristic diffusion length in Figs. 15.17 and 15.18 is a measure of the size and shape of the cavity in which the measurements were made.

The information in Figs. 15.15, 15.17, and 15.18 would lead one to believe that extremely high power levels can be transmitted through waveguides and coaxial lines of relatively modest size. For example, an extrapolation of Woo's unified plots to atmospheric pressure in a 7/8-inch coaxial line and a frequency of 1 GHz indicates that the breakdown power is about 200 kW.

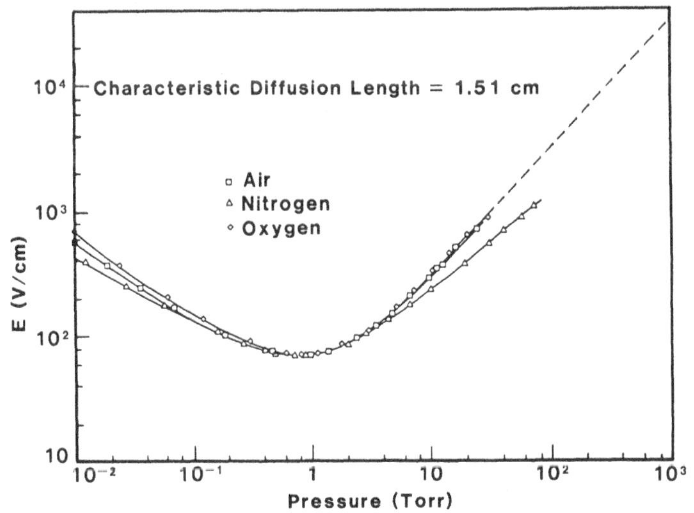

Figure 15.17 Breakdown Field at 992 MHz

(From *Gaseous Electronics,* Volume I, "Electric Discharge," edited by M.H. Hirsh and H.J. Oskam, Copyright 1978 by Academic Press. Reproduced with permission.)

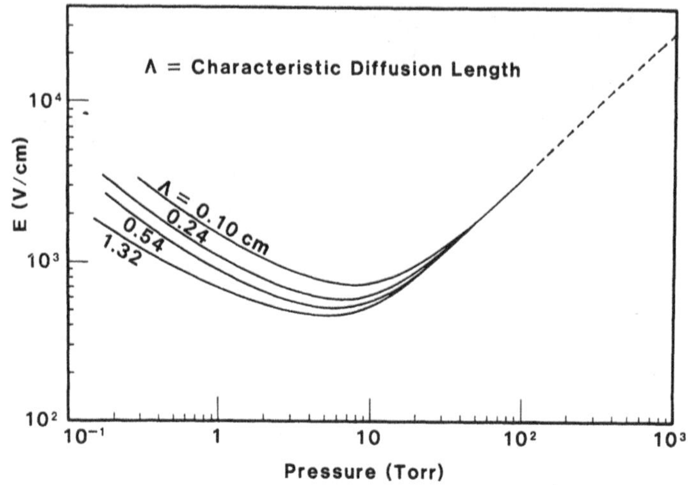

Figure 15.18 Breakdown in Air at 9.4 GHz

(From *Gaseous Electronics,* Volume I, "Electric Discharge," edited by M.H. Hirsh and H.J. Oskam, Copyright 1978 by Academic Press. Reproduced with permission.)

Still, as shown in Fig. 15.19, the power ratings for a standard 7/8-inch coaxial line are 43 kW peak and only about 1250 W average. Some of the degradation results from field enhancement associated with supporting elements for the center conductor. More importantly, for the average power case, heating, especially of the center conductor and gas, is one of the major factors limiting the power rating.

A situation similar to that in coaxial lines exists in waveguides as shown in Fig. 15.20. The ratings in the left-hand plot are for a conductor temperature rise of 42.4°C, and those for the right-hand plot are for a conductor temperature rise of 110°C. Based on a breakdown field of 20 kV/cm, it would be expected that WR 90, which is used at X band and has a small dimension of about 1 cm, should be able to transmit megawatts of power. Instead, the rating is in the kilowatt range.

The power ratings for the waveguides in Fig. 15.20 decrease by a factor of about 3 when the conductor temperature rise increases from 42.4°C to 110°C. Part of this decrease in rating may be attributed to the decrease in gas density in the waveguide as temperature is increased while pressure is held constant. This is equivalent to moving to the left in the high pressure region on the Paschen curve toward lower breakdown voltages. The reduction in pressure at constant temperature in moving to the left on the Paschen curve is equivalent to a reduction in gas density. This reduction in power capability with temperature is illustrated more clearly in Fig. 15.21.

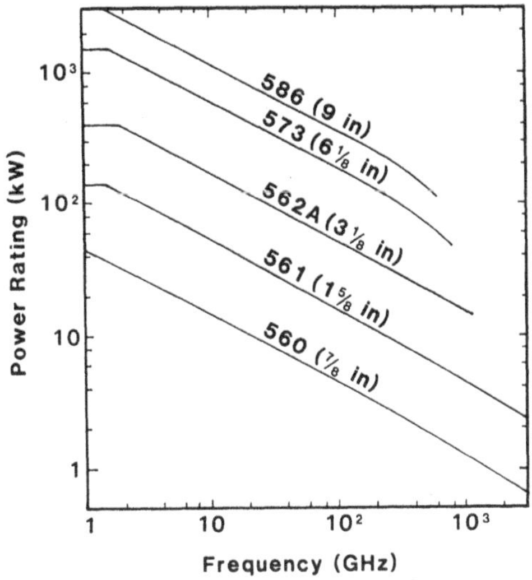

Figure 15.19 Average Power Ratings for Standard Rigid Lines: Peak ratings are constant with frequency and are based on unity VSWR, no modulation, and 1 atmosphere absolute dry nitrogen

(From *Microwave Engineers Handbook*, edited by T. Saad, Copyright 1971 by Artech House. Reproduced with permission.)

As would be expected, a change in pressure in waveguide or a coaxial line has an extremely large effect on power handling capability. As an example, the change in pressure from sea level to about 50,000 ft is given in Fig. 15.22, and the corresponding reduction in waveguide power capacity is shown in Fig. 15.23. Notice that at 50,000 ft, the power capacity drops to about three percent of the sea level capacity.

Band 5/6 of the ALQ-99 electronic countermeasures system is a practical example of a case where the material given in preceding paragraphs was apparently ignored in the system design. The RF power level was high enough so that operation was marginal at sea level. The system was unpressurized and was used by the US Air Force in airborne applications. As a result, frequent failures occurred, which often resulted in damage to or the destruction of a very expensive TWT. Unfortunately, this was one of those situations where the tube manufacturer and the system manufacturer pointed fingers at each other without really getting down to the business of solving the problem. The severity of the situation was compounded by the fact that the young government personnel assigned to solve the problem lacked the knowledge and experience to deal with the situation.

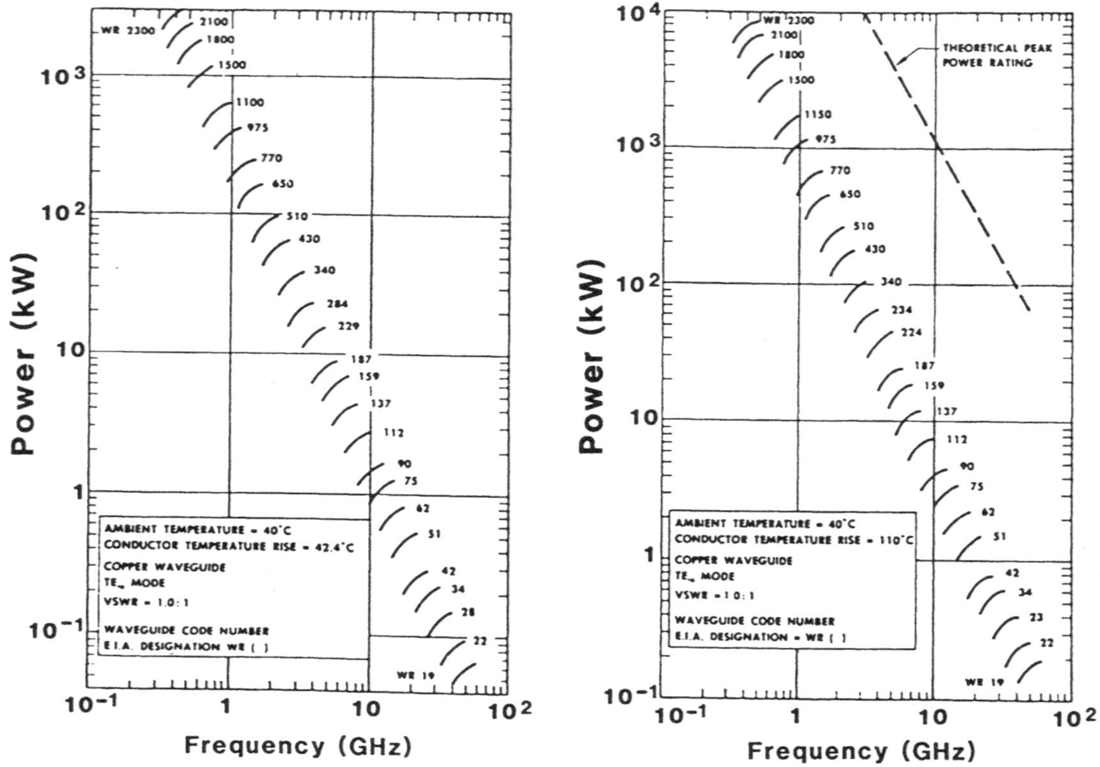

Figure 15.20 Average Power Rating for Copper Rectangular Waveguide

(From *Microwave Engineers Handbook*, edited by T. Saad, Copyright 1971

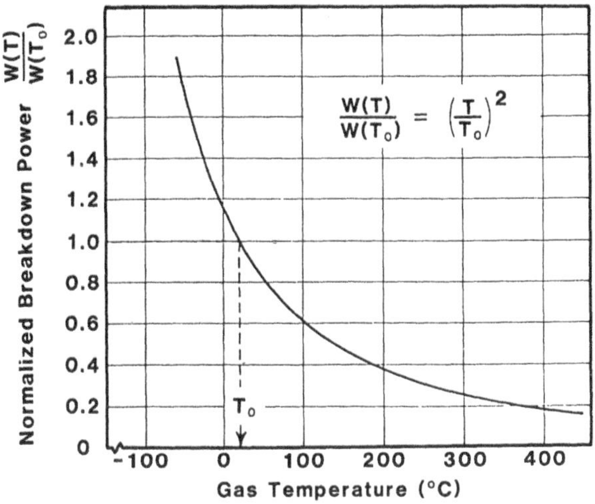

Figure 15.21 Waveguide Breakdown Power as a Function of Temperature at Fixed Pressure

(From R.H. Stone and R.M. White, Final Report TIS R61 ELM 205 General Electric Co., January 1961.)

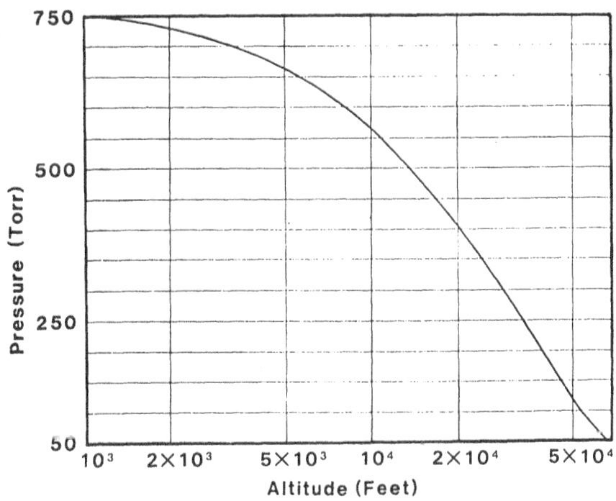

Figure 15.22 Pressure as a Function of Altitude (Normalized to 20° C)

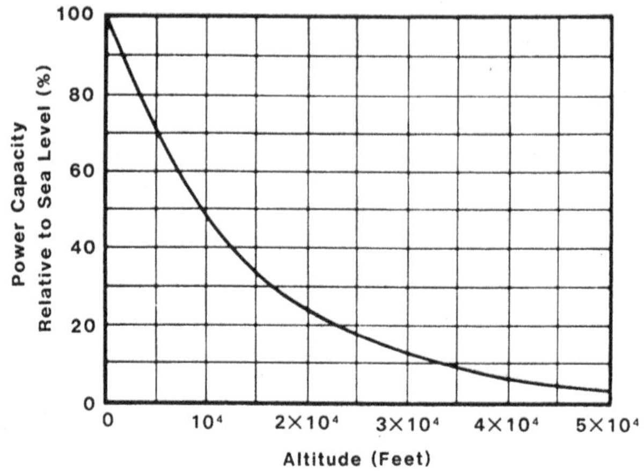

Figure 15.23 Waveguide Power Capacity *versus* Altitude

(From Microwave Tube Manual, by Varian Associates. Air Force Publication T.O. 00-25-251, October 1979.)

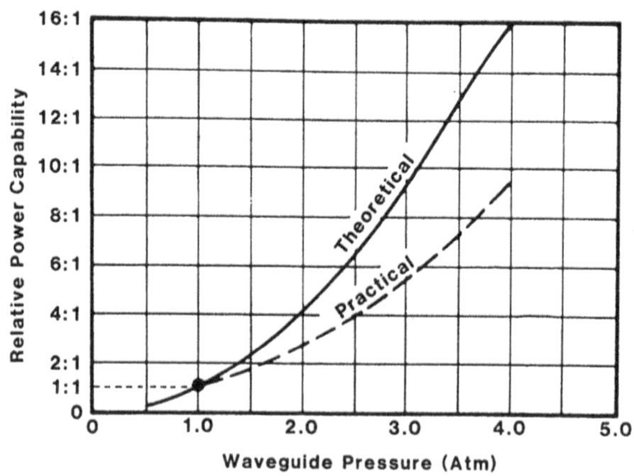

Figure 15.24 Improvement in Power Capability with Pressurization

(From Microwave Tube Manual, by Varian Associates. Air Force Publication T.O. 00-25-251, October 1979.)

If an RF system is pressurized, a significant increase in power handling capability can be realized as shown in Fig. 15.24. Even a modest amount of pressurization of the ALQ-99 components mentioned in the previous paragraph would have eliminated those particular breakdown problems.

In Section 5.2, the use of electronegative gases for increasing breakdown voltage was discussed. As shown in Fig. 15.25, similar improvements are possible in RF components. The improvement in power handling capability with pure sulphur hexafluoride (SF_6) at a gauge pressure of 5 psi is a factor of about 16. Even a small percentage of SF_6 mixed with air produces a significant increase in power capability. Other electronegative gases such as freon 12 and freon 114 produce similar improvements in power handling capability.

In pulsed RF systems, the peak power handling capability increases for at least two reasons. The first has already been pointed out in connection with Fig. 15.19. That is, the heating effect is decreased. The other reason is based on the fact that it takes time for a gas to break down. First of all, there is the time required for the breakdown process to start. This is called the formative time. Secondly, there is the time required for the discharge to develop to the point where appreciable RF power is absorbed or reflected.

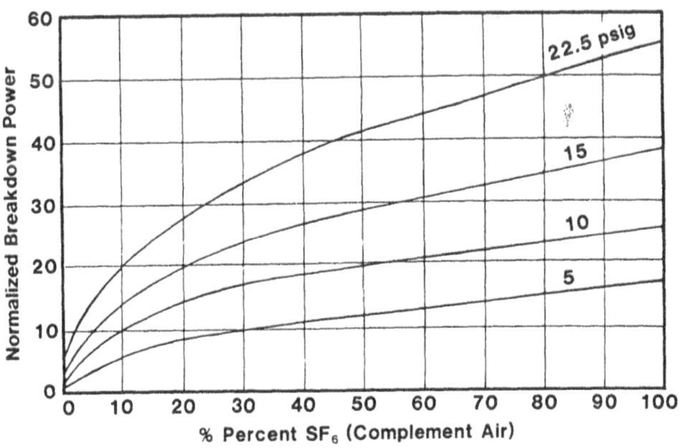

Figure 15.25 Breakdown Power for Mixtures of SF_6 and Air Normalized to Air at a Pressure of 5 PSIG

(From V.C. Vannicola and L.L. Stevens, Air Force Report No. RADC-TDR-64-67, March 1964.)

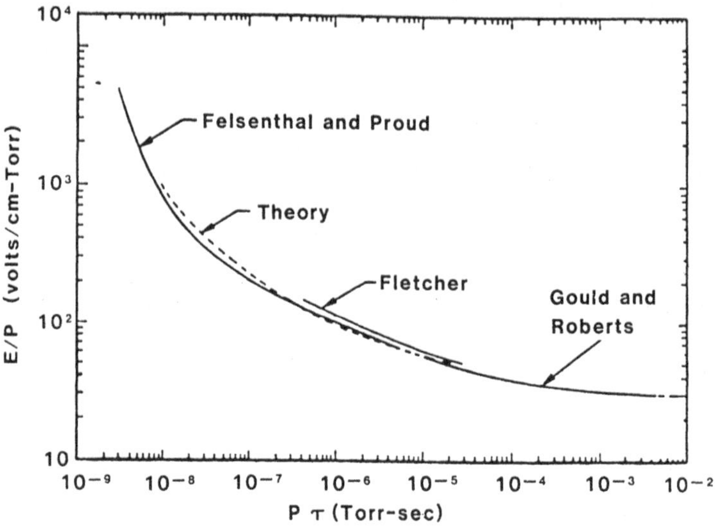

Figure 15.26 Formative Time in Air

(From P. Felsenthal and J.M. Proud, *Phys. Rev.*, 1965.)

The formative time for air can be determined from Fig. 15.26. In this figure, P is gas pressure, E is electric field, and τ is the formative time. For a 1 μs pulse, and atmospheric pressure, the $P\tau$ product is slightly less than 10^{-3} Torr-sec, which is near the right side of Fig. 15.26. Thus, at pulse lengths substantially shorter than 1 μs, the formative time is long enough so that the breakdown field increases.

An example of the change in breakdown power resulting from pulsewidth considerations is given in Fig. 15.27. Notice that the reduction in power capability when the pulse length is increased from 1 to 3 μs is about 30 percent.

The voltage standing wave ratio (VSWR) in the RF output system of a tube may cause breakdown to occur. The reason for this is, of course, that the reflected wave adds to the forward wave to produce regions of increased field level. A VSWR of 1.5:1 reduces the breakdown power by 35 percent, and 2:1 reduces breakdown power by 50 percent. If a system must be designed for operation with a significant VSWR, then the normal waveguide or coaxial line rating must be reduced to prevent breakdown as indicated by Fig. 15.28.

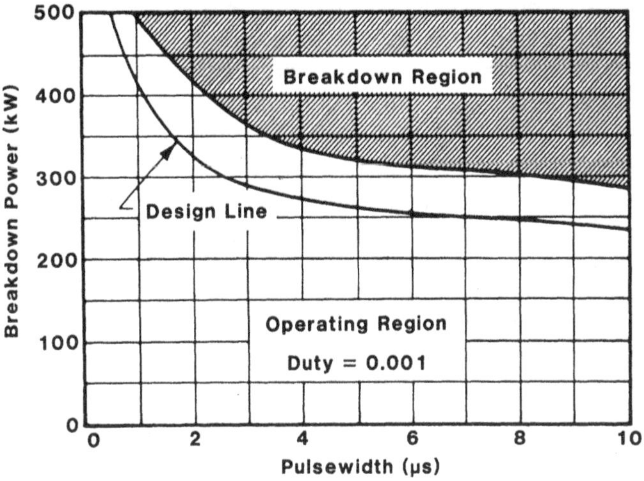

Figure 15.27 Effect of Pulsewidth in Breakdown Power in Waveguide.

(From Microwave Tube Manual, by Varian Associates. Air Force Publication T.O. 00-25-251, October 1979.)

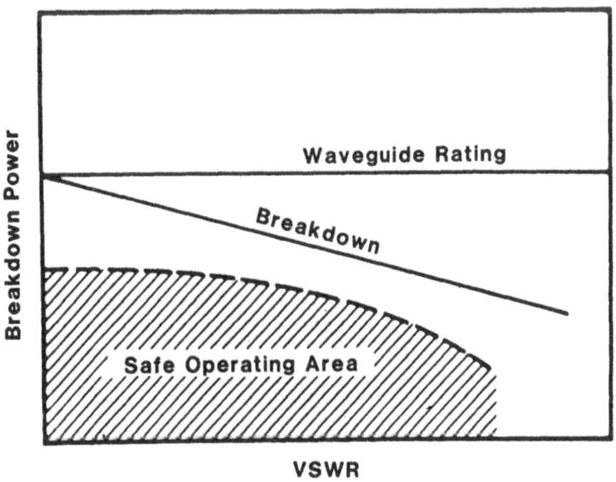

Figure 15.28 Effect of VSWR on RF Power System Rating

(From Microwave Tube Manual, by Varian Associates. Air Force Publication T.O. 00-25-251, October 1979.)

If breakdown does occur in the RF output system connected to a microwave tube, it is probable that the resulting arc will reflect most of the incident power back toward the tube. As a result, the VSWR will be very high and may produce breakdown at other points along the RF system as is indicated in Fig. 15.29. In this case, an arc at plane I produces a standing wave and causes the electric field to exceed the breakdown level of a critical element such as the output window of a tube located at plane IV.

An isolator is often used to protect a microwave tube from excessive reflected RF power. The isolator protects the tube output window and may impede or even stop a traveling arc. The effect of a ferrite isolator on VSWR is shown in Fig. 15.30. While the isolator is effective in protecting the tube, it will not prevent arcing in the remainder of the RF system.

In some microwave systems, an attempt is made to reduce arcing problems and arc damage by using a detector of some type. Reflected power can be sensed with a directional coupler and measured with a threshold detector that generates a signal to turn off the microwave tube if excessive reflections occur.

A waveguide arc may be sensed with an optical device for detecting visible arcs. The sensing portion of the device is usually mounted in a 90-degree waveguide bend located between the tube output flange and the waveguide transmission line. The output signal from the sensing device is normally used to turn off the high power tube.

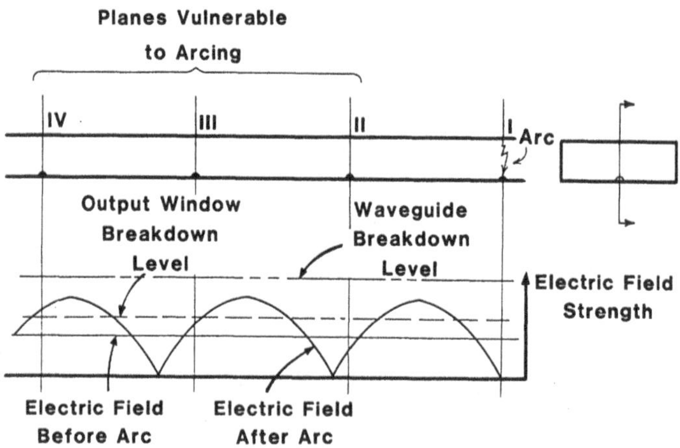

Figure 15.29 Arc Propagation in a Waveguide

(From R.H. Stone and R.M. White, Final Report TIS R61 ELM-205 General Electric Co., January 1961.)

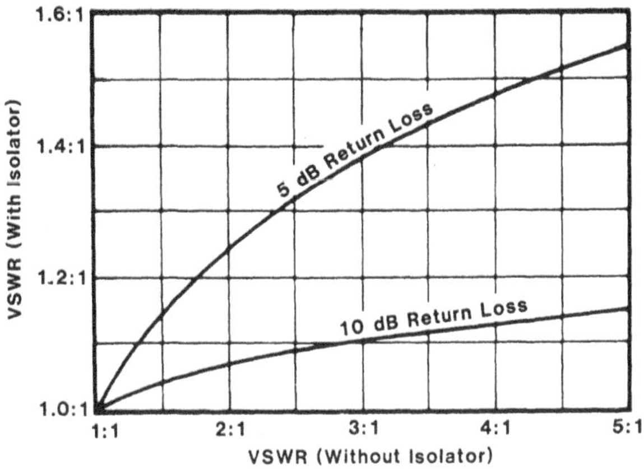

Figure 15.30 Effect of Ferrite Isolator on VSWR

(From Microwave Tube Manual, by Varian Associates. Air Force Publication T.O. 00-25-251, October 1979.)

In another technique sometimes used to protect the output window of a tube, a metal probe is inserted through, but insulated from, the waveguide in the region near the output window. When an arc occurs and moves toward the window, the electrical circuit between the probe and the waveguide is closed with the result that a relay can be activated to turn off the microwave tube.

15.4 DC BREAKDOWN IN VACUUM [15.5]

Under ideal conditions, the breakdown field level for vacuum exceeds that of all other media including liquids and solids, and may be 10 MV/cm or higher. The upper limit results from field emission (see Section 5.1.2) from the negative electrode. In practice, the field level at which breakdown occurs is found to be two to three orders of magnitude below that occurring under ideal conditions. For example, for average gap lengths ($\sim$ 1 cm) breakdown normally occurs at a field level ranging from a few tens to a few hundreds of kV/cm.

For many vacuum-tube applications, "Kilpatrick's criterion" [15.6] is used as the guideline to the maximum electric field that can be used in vacuum if breakdown is to be avoided. According to Kilpatrick, breakdown in vacuum varies as electrode spacing raised to the 0.75 power. Kilpatrick refers to field levels below this criterion as the "no-spark region" and to higher field levels as the "possible-spark region."

Figure 15.31 contains Kilpatrick's criterion and some results from the tube industry [15.7, 15.8]. The shaded areas indicate roughly the values of plate hold-off voltages for gridded tubes with thoriated tungsten (Th W) and oxide-coated cathodes, and cylindrical configurations.

The data points in Fig. 15.31 represent the experimental breakdown characteristics of Pierce guns. The data points represented by circles are for dc operation. The triangles and X's represent long-pulse and short-pulse operation, respectively. Recent results presented by Staprans [15.9] are in basic agreement with those given in Fig. 15.31. Note that, for pulse operation, field levels somewhat above the Kilpatrick curve are sometimes used.

The Kilpatrick criterion is based on results for parallel plane electrodes while the tube results in Fig. 15.31 are for various electrode configurations. As a result, there are varying degrees of field enhancement that contribute to the scatter in the data. In addition, apparently when oxide coated cathodes are used, decomposition products form surface layers of reduced work function, greater roughness, different hardness, *et cetera*, that result in a reduction of the breakdown voltage.

The mechanisms responsible for breakdown in vacuum at field levels that are orders of magnitude below the field emission limit have been under investigation for decades. Still, there is often controversy over which mechanism dominates in any circumstance. Those mechanisms thought most likely to occur in microwave tubes are discussed in this section..

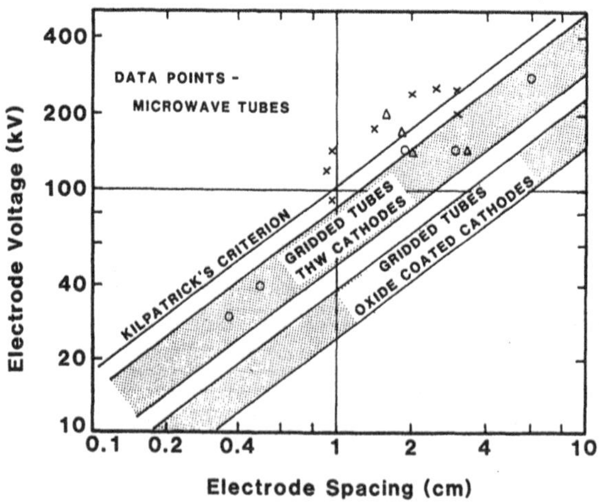

Figure 15.31 Kilpatrick's Criterion and Experimental Breakdown Characteristics of Tubes

Figure 15.32 Whiskers on Stainless Steel and Aluminum Produced by Electric Field

(From R.P. Little and W.T. Whitney, *J.A.P.*, August 1963.)

Conductive whiskers form in high-field regions on electrodes in vacuum. These whiskers can result from the migration of debris such as the decomposition products of a cathode or loose particles left from the manufacturing process. Whiskers can also grow from metallic surfaces subjected to high electric fields. The whisker shown in Fig. 15.32 was produced by subjecting an optically flat stainless steel surface to a field of somewhat less than 200 kV/cm [15.10].

The field level of about 200 kV/cm for Fig. 15.32 is above the nominal field levels indicated by Fig. 15.31. It must be kept in mind, however, that the field-enhancement factors resulting from electrode curvature, machining irregularities, *et cetera*, will produce fields on electrode surfaces well above those indicated in Fig. 15.31.

The field-enhancement factor at the tip of a microprotrusion or a whisker, like that shown in Fig. 15.32, can be extremely high. Values as high as 100 have been estimated. The resulting electric field at the tip of a microprotrusion can produce field emission of electrons. The current flow through the tip coupled with ion bombardment (ions produced when the field-emitted electrons strike gas atoms) produces heating of the tip, gas evolution, and additional electron emission. Further heating occurs which may produce an explosion of the tip. A microdischarge occurs which may or may not grow into a full scale breakdown between electrodes.

An example of the time profiles of current and voltage for a microdischarge that does not result in breakdown is shown in Fig. 15.33. At this point, it is appropriate to point out that microdischarges of the type indicated by Fig. 15.33 are common occurrences during the voltage conditioning of

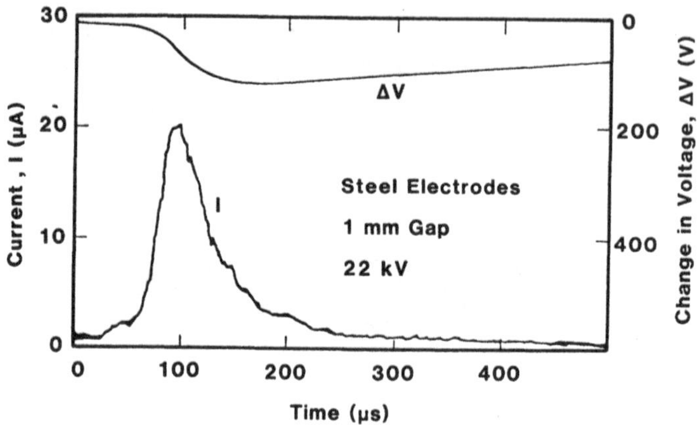

Figure 15.33 Microdischarge for 22 kV Applied to a 1 mm Gap between Steel Electrodes

(From High Voltage Technology Seminar, Ion Physics Corporation, 1969.)

electrodes in high voltage devices. During this process, high voltage is gradually applied to the electrodes as shown in Fig. 15.34. Microdischarges are produced that "clean up" the electrodes. That is, whiskers and other microprotrusions that are present on the electrode surfaces for any of many reasons are removed to permit high voltage operation. In a high voltage device, this process must be carried out when voltage is first applied and then must normally be repeated in an accelerated manner each time the device is turned on after being off for even relatively short periods (hours in some cases).

It should also be noted that, even for a newly conditioned gap with voltage applied, the voltage hold-off capability degrades with time as is shown by Fig. 15.35.

Several mechanisms have been proposed by which microprotrusions may produce full-scale breakdown in vacuum. Two of these are illustrated in Fig. 15.36. For small gaps (<1 mm) the microprotrusion may explode and produce a plasma cloud of sufficient density between the electrodes to cause a vacuum arc (also called a metal-vapor plasma arc) to form. For the relatively modest currents encountered in microwave tubes, even under fault conditions (kiloamperes at most) the arc that forms is on the negative-electrode surface and the resulting voltage drop between electrodes is from roughly 10 to 35 V, depending on electrode material. Vacuum arcs are discussed later in this section.

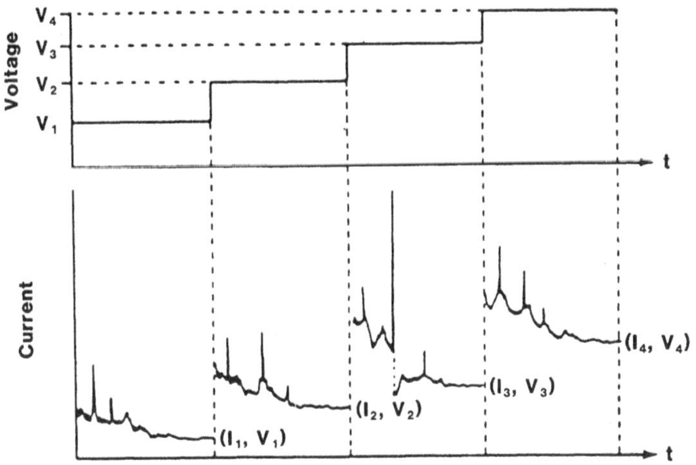

Figure 15.34 Typical Voltage and Current During Conditioning

(From *High Voltage Vacuum Insulation: The Physical Basis,* by R.V. Latham, Copyright 1981 by Academic Press, Reproduced with permission.)

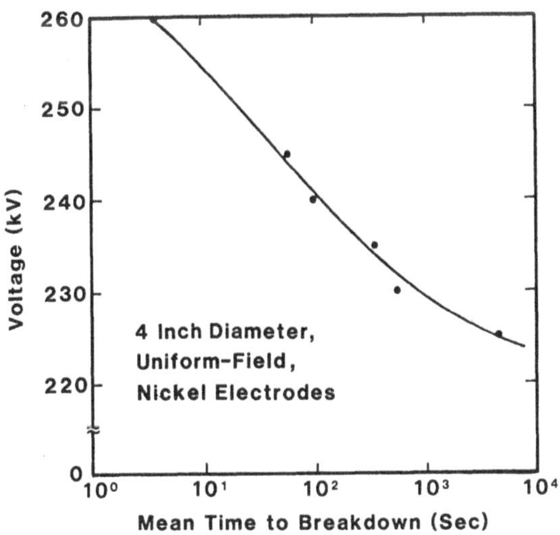

Figure 15.35 Breakdown Characteristics for a Conditioned Vacuum Gap

(From High Voltage Technology Seminar, Ion Physics Corporation, 1969.)

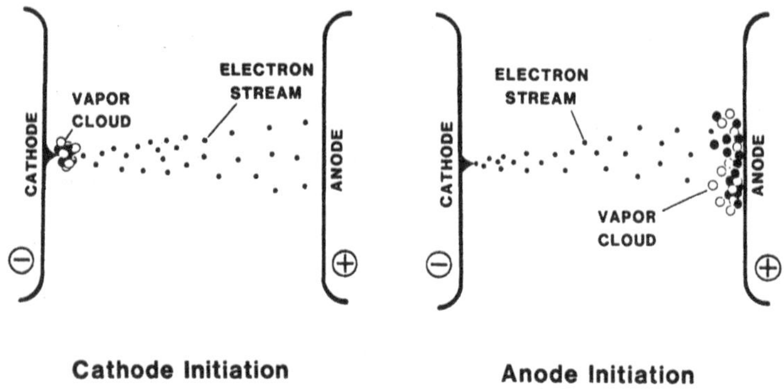

Figure 15.36 Breakdown Initiation Resulting from Field Emission from a Microprotrusion

(From *High Voltage Technology,* Chapter 4 by R. Hawley, edited by L.L. Ashton, Oxford University Press, London, 1968. Reproduced with permission.)

For a large gap (> 1 cm), field emission from the negative electrode may produce the release and ionization of vapor from the positive electrode. If the density of the resulting plasma cloud is sufficient, then a vacuum arc identical to the one described in the previous paragraph can form.

As Fig. 15.37 shows, the discharge current required to initiate an arc is only a fraction of a milliampere for electrode spacings approaching 1 cm. Figure 15.37 also shows that the value of β plays a role in breakdown as is expected.

It has been clearly established that loose particles in the interelectrode region can result in electrical breakdown of a vacuum gap. Several mechanisms by which the breakdown process can occur have been suggested. Some of these are listed below.

1. When voltage is applied, a surface charge density is induced in the electrodes and in particles on the surfaces of the electrodes. The particles may be drawn across the gap by the electric field as shown in Fig. 15.38. If sufficient vapor is generated when the particles strike the opposing electrodes, breakdown may occur.

2. For large particles, breakdown may occur between a particle in transit and the electrode surface that it approaches. This may lead to breakdown of the gap.

3. When a particle strikes a surface, a crater is produced. The splash rim of material surrounding the crater may have sharp protrusions that field emit and result in breakdown.

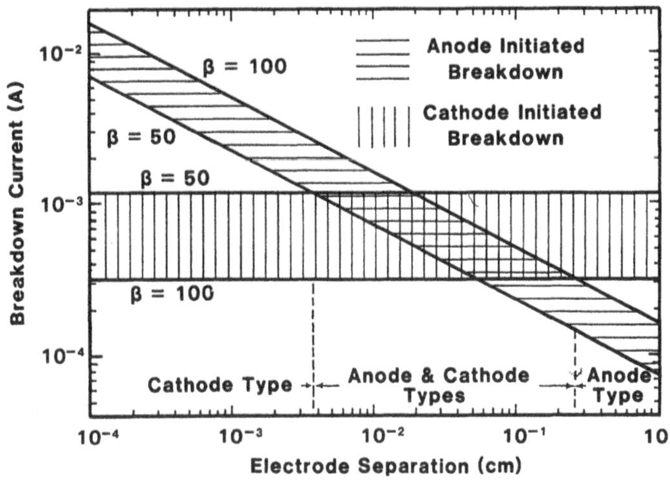

Figure 15.37 Electrode Dominance in Vacuum Arc Initiation

(From T. Utsumi, Technical Report RADC-TR-66-712, January 1967.)

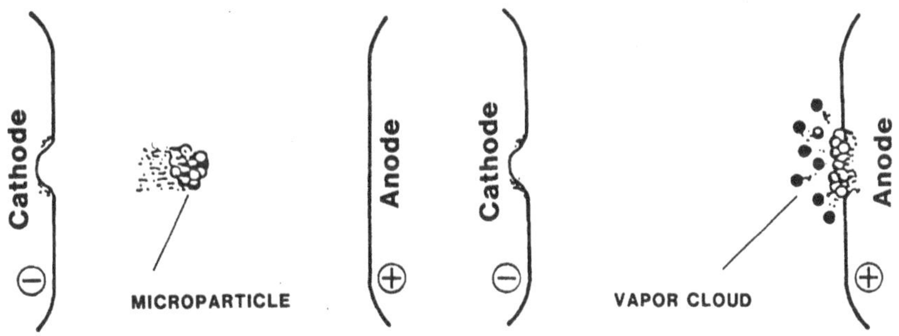

Figure 15.38 Breakdown Initiation by a Microparticle

(From *High Voltage Technology,* Chapter 4 by R. Hawley, edited by L.L. Ashton, Oxford University Press, London, 1968. Reproduced with permission.)

A very practical indication of the role of particles in breakdown may be had from a technique sometimes used by the tube industry, apparently, to eliminate breakdown caused by particles. An electron gun to be highly stressed is placed in a vacuum chamber, high voltage is applied, and it is vibrated until arcing stops. The process may be repeated with varying

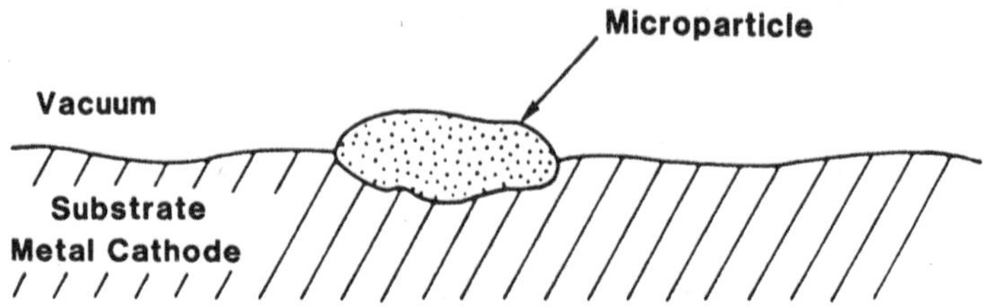

Figure 15.39 Partially Embedded Dielectric Microparticle

(From *High Voltage Vacuum Insulation: The Physical Basis,* by R.V. Latham, Copyright 1981 by Academic Press, Reproduced with permission.)

orientations of the gun. Particles are observed to fall from the gun during the process and it is presumed that they are the cause of arcing.

In addition to the effect of loose particles, it has been demonstrated that insulating particles like that shown in Fig. 15.39 that are partly imbedded in the electrode surfaces are sites for the initiation of breakdown. The bigger the particles are, the lower is the breakdown voltage.

The sources of particles that can lead to breakdown are many and varied. Some examples are:

- debris in gun from manufacturing process
- careless assembly procedures that cause chipping of insulators or other damage
- dust from inadequate cleanliness
- cathode evaporants that form loose films on various gun surfaces
- use of improper getters
- particles from heater potting material
- debris that enters gun through anode aperture
- debris from previous arc
- embedded particles (from grinding operation, for example) that are exposed by chemical etching

Regardless of the initiation mechanism, dc breakdown in vacuum leads to an arc on the negative electrode (at the current levels encountered under fault conditions in a microwave tube). This arc is normally known as a vacuum arc, but it is sometimes referred to as a metal-vapor plasma arc.

In a vacuum arc, the material of the negative electrode is vaporized and ionized by arc spots to form a conducting medium between the electrodes. Vacuum arcs can be established on any reasonably good conductor. The voltage between electrodes during a vacuum arc is dependent on the material of the negative electrode, typically between 10 and 30 V, and varies little with electrode spacing for typical (cms) spacings. As shown in Fig. 15.40, the arcing voltage is a slowly varying function of current.

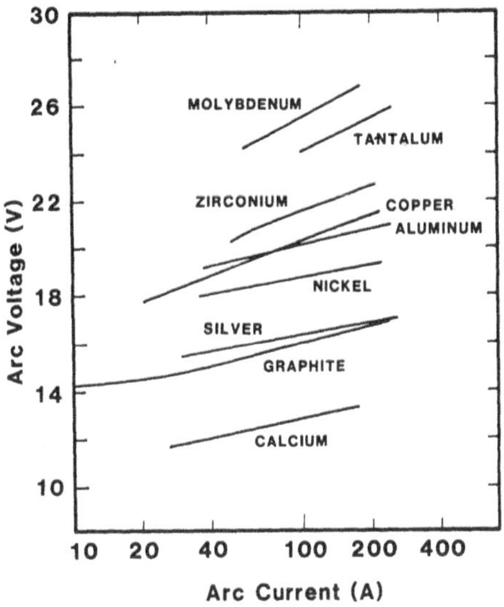

Figure 15.40 dc Volt-Ampere Characteristics of Vacuum Arcs

(From W.D. Davis and H.C. Miller, *J. Appl. Phys.*, 1969.)

Unlike an arc in gas (such as air at atmospheric pressure), which consists of one large spot anchored in position, many very small arc spots exist in a vacuum arc, as indicated in Fig. 15.41. These spots travel at high speed (meters per second to tens of meters per second) with apparently random motion across the electrode surface. On a macroscopic scale, a surface subjected to prolonged vacuum arcing has the texture of a piece of frosted glass. On a microscopic scale, individual arc tracks somewhat resemble the tracks of insects crawling in random directions across a freshly painted surface.

On a surface consisting of differing metals, a vacuum arc will operate on the low arcing voltage metals. For example, if a calcium film partially covers a copper surface, a vacuum arc will remove the calcium film before attacking the copper. This results from the relatively low arcing voltage of calcium (about 12 V) relative to that of copper (about 20 V). Similarly, the evaporation products from thermionic cathodes will be preferentially removed from nearby grid or control electrode surfaces by a vacuum arc. This is because barium, strontium, calcium, *et cetera* all have similar arcing voltages which are about one-half (or less) of those of the normal materials of grids and electrodes. As a side note, the arcing may produce additional deposition of the cathode material on critical surfaces such as insulators.

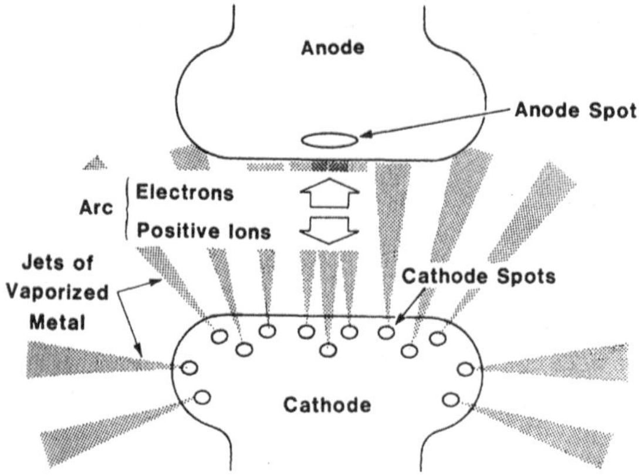

Figure 15.41 Details of a Vacuum Arc

(From *Vacuum Arcs, Theory and Application,* edited by J.M. Lafferty, Copyright 1980 by John Wiley and Son., Reproduced with permission.)

The rate at which material is removed from a surface by a vacuum arc varies somewhat from metal to metal but is generally in the neighborhood of 0.1 milligram per Coulomb of charge passed through the arc. Thus, if the time history of the current in a tube during an arc is known, the amount of erosion can be estimated with a reasonable degree of accuracy.

Similarly, if energy input to a vacuum arc in an electron gun is known, the amount of erosion can be estimated. This is because the arcing voltage is restricted to a reasonably narrow range of values. For example, if a 20 Joule arc occurs, the charge passed through the arc is estimated to be 1 Coulomb using an average arcing voltage of 20 V. For 1 Coulomb of charge, 0.1 milligram of electrode material is removed. For a density of 10 gms/cm^3 (most electron-gun metals are not significantly different from this value), the volume of electrode material removed is 10×10^{-6} cm^3. For 1 cm^2 of electrode material, the thickness removed is only 10×10^{-6} cm or 0.1 μm. Thus, the amount of material removed by a 1 Coulomb or a 20 Joule arc is extremely small.

Unfortunately, the small amount of material removed by the 20 Joule arc is not the only factor that must be considered. If the surfaces on which some of the arc material deposits are critical ones, such as insulators, leakage currents or insulator breakdown may result. Furthermore, gas molecules on the arced surfaces or nearby surfaces may be driven off by the arc, producing a gas burst. As an example, the release of a monolayer of gas from 1 cm^2 of area into a volume of 1 liter raises the pressure from 10^{-6} to 10^{-4} Torr.

This gas must be removed (by adsorption, an appendage pump or a getter) before tube operation can be resumed.

The maximum permissible energy input to a tube during an internal arc in the gun is sometimes specified by the manufacturer. This energy is based largely on the manufacturer's experience with the particular tube type. Maximum permissible energy levels from Joules to tens of Joules are not uncommon. Knowing the maximum permissible energy level, it is then possible for the tube user to design a protection circuit to interrupt the flow of energy before it becomes excessive.

Before leaving the subject of vacuum arcs, a comment should be made concerning the interruption characteristics of these arcs. As the current through a vacuum arc is reduced, the average arc lifetime becomes shorter and shorter. For example, Fig. 15.42 shows that when the current through a tungsten (W) arc approaches 5 A, the average lifetime approaches 100 μs. Thus, as current is driven toward zero by a protection circuit, the arc will extinguish well before the current reaches zero.

When a vacuum arc extinguishes, it does so very rapidly. In fact, the current is said to "chop" and the chopping time is usually less than 10^{-7} seconds. If there is significant inductance in the protection circuitry, the voltage generated by an arc that chops can be sufficient to cause insulation failure.

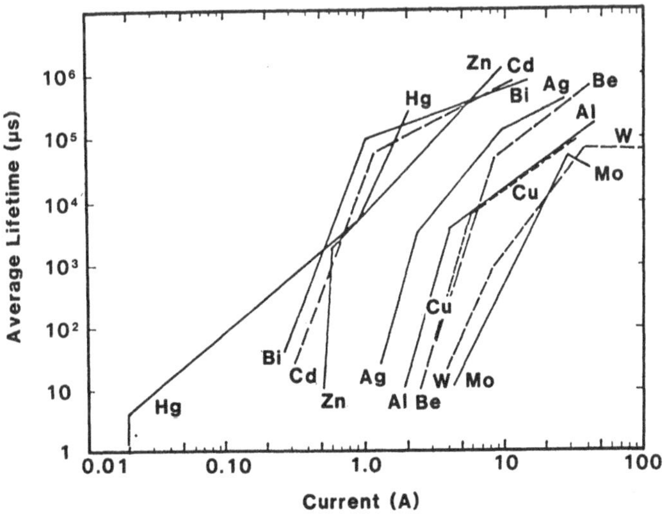

Figure 15.42 Average Arc Lifetime for Pure Electrode Materials

(From J.D. Cobine and G.A. Farrall, *J. Appl. Phys.*, Dec. 1960.)

Once a vacuum arc has extinguished, the voltage recovery rate across the electrodes between which the arc had occurred can be extremely rapid. The recovery rate depends on the duration and intensity of the arc, however, a value of 10 kV/ μs is not uncommon. Thus, it is often only necessary to cause a very brief interruption of current through a vacuum arc before voltage is reapplied to the electrode. In fact, this principle has been used in the development of vacuum arc interrupters [15.11].

15.5 RF BREAKDOWN IN VACUUM

Kilpatrick's criterion, discussed at the beginning of Section 15.4, is often used as a guideline to the maximum RF voltage that can be applied between electrodes. Very often, however, another kind of discharge develops at voltage levels well below the Kilpatrick level. This discharge is called a multipactor discharge and usually occurs when electrons move back and forth across a gap in synchronism with an RF field. If the secondary emission ratio of the gap surfaces is greater than unity, then the number of electrons involved in the process builds up with time.

It is also possible for a multipactor discharge to develop if the electron transit time across a gap corresponds to 3/2, 5/2, *et cetera* cycles of the applied RF voltage. In each case, electrons can receive a net acceleration in crossing the gap and can produce an avalanche buildup of current. The voltage, frequency, f, and gap separation, d, which result in a multipactor discharge are related as shown in Fig. 15.43.

During multipacting, large amounts of energy can be discharged in a small volume with the result that electrodes or RF windows can be damaged or destroyed. Even if excessive damage does not occur, multipacting can load a microwave tube so that the output power is reduced and modulated with a noise-like signal.

Techniques that are used for suppressing multipacting fall into two general categories:

1. alteration of electron dynamics;
2. reduction of the secondary emission ratio.

The most common ways to alter the electron dynamics are to adjust the magnetic focusing field (within the limits of acceptable tube operation), or to apply RF power rapidly enough so that the voltage rises above the multipacting zone before current has had time to build up.

The secondary emission ratio may be reduced by serrating the cavity drift tube tips on which the multipacting electrons land. Some of the electrons enter the notched regions of the serrations and the secondaries that are produced are trapped within the notches. Titanium-based films have been

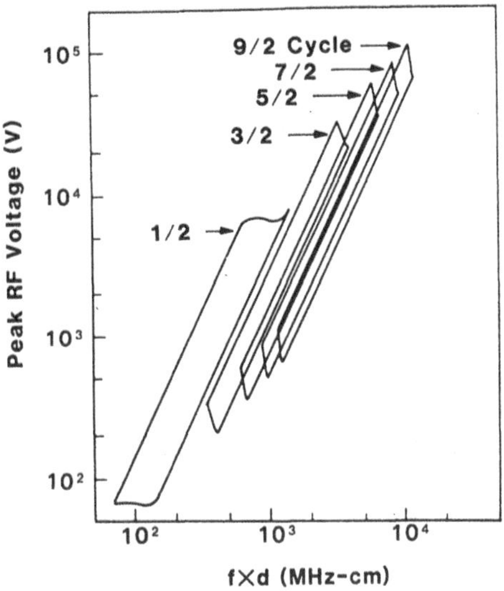

Figure 15.43 Theoretical Multipacting Zones

(From A.J. Hatch and H.B. Williams, *Phys. Rev.*, 1958.)

demonstrated to be very successful in suppressing multipacting in the high power klystrons in the Stanford linear accelerator [15.12].

If multipacting can be avoided, then the next problems to be faced are the vacuum breakdown mechanisms described in Section 15.4. The investigation of techniques that would permit the use of fields well above the Kilpatrick limit is an active area of research at the present time in the linear accelerator field [15.13].

15.6 INSULATOR FLASHOVER

One of the weakest points in a vacuum device from an electrical breakdown point of view is the interface between a metal, an insulator, and vacuum. This interface is often called the "triple junction" and is illustrated in Fig. 15.44. Because of electron launching mechanisms at the triple junction and surface effects on the insulator, the breakdown strength of the insulator (actually, its surface) is well below that of a vacuum gap or a solid insulator of equivalent length. Shannon *et al.* have suggested that through proper design, insulators in vacuum without conditioning should achieve 50 kV/cm, and conditioned insulators should be operable at 80 kV/cm [15.14].

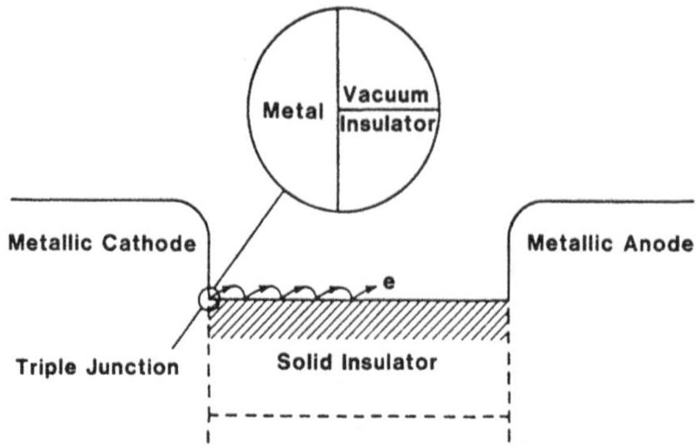

Figure 15.44 Triple Junction and Hopping Mode of Electron Travel

(From *High Voltage Vacuum Insulation: The Physical Basis,* by R.V. Latham, Copyright 1981 by Academic Press, Reproduced with permission.)

The initiation of surface flashover occurs at the triple junction. In this region, there are at least two situations that must be considered. In one, the insulator is brazed to the metal surface. In the process of metallizing the surface of the insulator and brazing it to metal, it is virtually impossible to avoid the formation of a fillet of braze material on the surface of the insulator. This fillet has a razor-sharp edge with a high field-enhancement factor. In the presence of the high field that results when voltage is applied, microprotuberances can grow, causing additional substantial increases in the enhancement factor. The result is thought to be the field emission of electrons and the generation of microparticles *et cetera* at the triple junction at relatively low applied field levels.

The second situation that must be considered is when the insulator is mechanically held in place, but not brazed. In this case, there are voids between the insulator and the metal along the triple junction. The field enhancement factor in these voids may be as high as the relative dielectric constant of the insulator. For aluminum oxide, which is commonly used in these situations, the relative dielectric constant is about 9. This field-enhancement factor, coupled with those of microprotuberances, is thought to lead to field emission and microparticle generation.

The second step in insulator flashover is thought to be the positive charging of the insulator in the vicinity of the triple junction. This occurs, as indicated in Fig. 15.45, because field-emitted primary electrons from the triple junction strike the insulator surface causing secondary emission to occur. The secondary emission coefficient for most insulating materials is

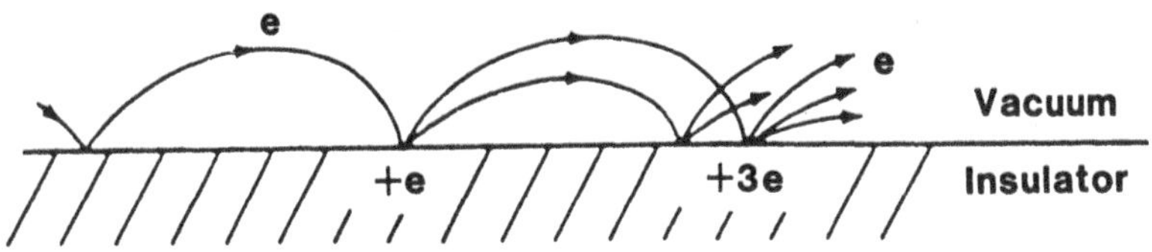

Figure 15.45 Insulator Charging by Hopping Electrons

(From *High Voltage Vacuum Insulation: The Physical Basis*, by R.V. Latham, Copyright 1981 by Academic Press, Reproduced with permission.)

relatively high (values from 5 to 10 are not uncommon, especially if the primary electrons strike the surface at a grazing angle). The secondary emission process is regenerative in that the secondary electrons produced are accelerated by the field at the surface to produce additional secondaries.

The result of the secondary emission process is that the insulator surface becomes positively charged in the triple-junction region, and this increases the field and the rate of electron emission at the triple junction.

The next phase in the flashover process is thought to be the desorption of gas molecules from the insulator surface and subsequent ionization by the hopping electrons. These processes are indicated in Fig. 15.46. The positive ions are accelerated toward the cathode and further enhance the field and increase electron emission.

Other processes possibly involving photon emission, are thought to occur and the end result is rapid breakdown across the surface of the insulator. The techniques that have been investigated for suppressing insulator flashover center on the prevention of the initial charging of the insulator in the region of the triple junction. Some of the techniques investigated or used are listed below:

1. The field is reduced at the triple junction by shaping the metal electrodes in this region. On a larger scale, it is common to place the output window for a tube in an enlarged circular section of the waveguide as is shown in Fig. 15.47. Not only does this reduce the field at the triple junction, it also facilitates the fabrication of the metal-to-ceramic braze joint.
2. The shape of the insulator, and in particular the angle its surface makes with the metal electrodes, is used to control the trajectories of secondary electrons. The variation of flashover field with insulator-to-metal surface angle is shown in Fig. 15.48.

3. A coating of material having a low secondary electron yield (ideally <1) is applied to the surface of the insulator. An important consequence of using these coatings is that they are normally slightly conductive so that residual charging effects are minimized.

In concluding this section, an observation by G. Farrall of General Electric Co. is pertinent. He notes that most of the work on insulators in vacuum has been conducted in unbaked systems and in systems sealed with "O" rings. As a result, the phenomena associated with breakdown are bound to be dominated by desorption of gas and a large variety of unknown impurity effects. Finally, Farrall states, "Perhaps the time has come to adopt a more scrupulous and possibly more fundamentally rewarding attitude toward the studies of insulators in vacuum."

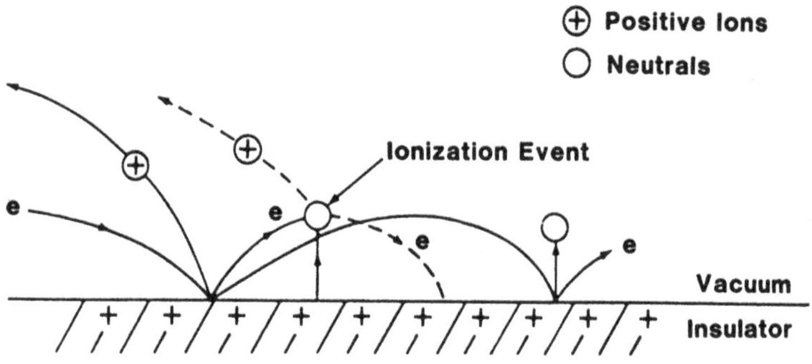

Figure 15.46 Desorption and Ionization of Gas Molecules by Hopping Electrons

(From *High Voltage Vacuum Insulation: The Physical Basis,* by R.V. Latham, Copyright 1981 by Academic Press, Reproduced with permission.)

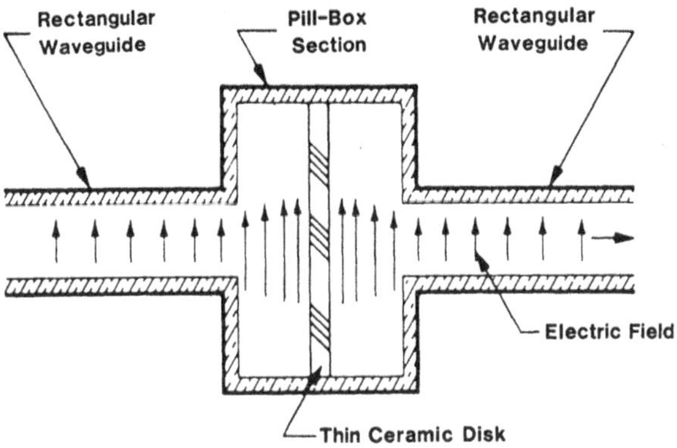

Figure 15.47 Output Window in Enlarged Section of Waveguide

(From Microwave Tube Manual, by Varian Associates. Air Force Publication T.O. 00-25-251, October 1979.)

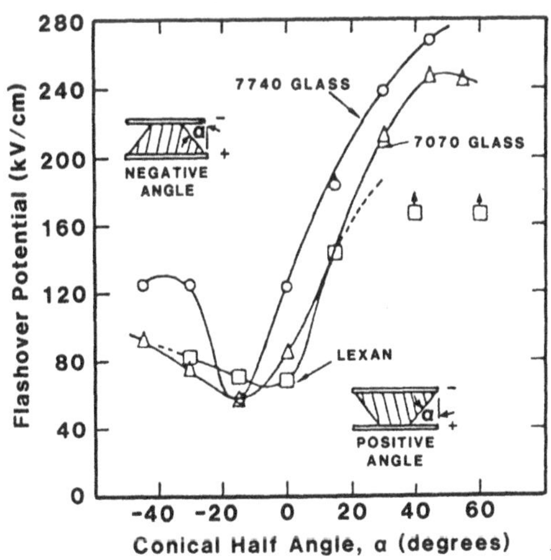

Figure 15.48 Effect of Insulator Geometry on Flashover Voltage

(From A. Watson and J. Shannon, *Proc. Second International Symposium on Insulation of High Voltages in Vacuum,* Cambridge, MA, Sept. 1966. Courtesy of A. Watson.)

REFERENCES

[15.1] James Dillon Cobine, *Gaseous Conductors, Theory and Engineering Applications*, New York, Dover Publications, 1958.

[15.2] M.N. Hirsh and H.J. Oskam, *Gaseous Electronics*, Volume 1, *Electrical Discharges*, New York, Academic Press, 1978.

[15.3] P.D. Ankrum, *Principles and Applications of Electron Devices*, Scranton, PA, International Textbook Company, 1959.

[15.4] Richard Woo, "RF voltage breakdown in coaxial transmission lines," *Proc. IEEE*, vol. 57, pp. 254–256, February 1969.

[15.5] Recommended for further reading:

J.M. Lafferty, *Vacuum Arcs, Theory and Applications*, New York, John Wiley and Sons, 1980.

R.V. Latham, *High Voltage Vacuum Insulation: The Physical Basis*, New York, Academic Press, 1981.

[15.6] W.D. Kilpatrick, "Criterion of vacuum sparking designed to include both RF and DC," *Review of Scientific Instruments*, vol. 28, no. 10, October 1957, pp. 824–826.

[15.7] Machlett Pulse Tubes for High Voltage, High Power, Video and RF Pulsing, The Machlett Laboratories, Inc., June 1964.

[15.8] A. Staprans, "Voltage breakdown limitations of electron guns for high power microwave tubes," *Proc. Second International Symposium on Insulation of Voltages in Vacuum*, M.I.T., Cambridge, MA, 1966.

[15.9] A. Staprans, "Electron gun breakdown," 1985 High Voltage Workshop, Monterey, CA, February 1985.

[15.10] R.P. Little and W.T. Whitney, "Electron emission preceding electrical breakdown in vacuum," *J. Appl. Phys.*, vol. 34, no. 8, August 1963, pp. 2430–2432.

[15.11] A.S. Gilmour, Jr., and D.L. Lockwood, "The interruption of vacuum arcs at high DC voltages," *IEEE Trans. Electron Devices*, vol. ED-22, no. 4, April 1975, pp. 173–180.

[15.12] R.W. Bierce, J. Jasberg, and J.V. Lebacqz, in *The Stanford Two-Mile Accelerator*, edited by R.B. Neal (Benjamin, New York, 1968), Ch. 10.

[15.13] William Peter, "Vacuum breakdown and surface coating of RF cavities," *J. Appl. Phys.*, 56 (5), September 1984.

[15.14] J.P. Shannon, S.F. Philip, and J.G. Trump, "Insulation of high voltages across solid insulators in vacuum," *Proc. First International Symposium on Insulation of High Voltages in Vacuum*," M.I.T., Cambridge, MA, October 1964, p. 281.

Chapter 16
Concluding Remarks

In many systems, the microwave tube is the heart of the system and, at the same time, it has the reputation of being the weakest link in the system. At first glance, it would appear that this should not be the case because the only wear-out mechanism is cathode depletion, but cathode failure is usually not the main cause for failure.

The list of other reasons for failure is long and includes items such as cracked output windows, internal arcing, leaks at various locations, and internal structural damage (melted drift tube tips in klystrons, *et cetera*). While some of these failures can be attributed to problems in design and manufacture, many result from abuse in the field. In addition, in many cases, it is estimated that from 25 to 40 percent of "defective" tubes were not defective after all. Something else in the system failed, but the tube was replaced first, probably because of its reputation as being a failure-prone item.

Then, in addition, there are the tubes that have external damage, clearly due to mishandling or abuse. RF connectors and coolant fittings are damaged, coolant lines are clogged, leads are broken, *et cetera*. If the truth were known, it could well be that over 50 percent of all failures are the fault of the users, rather than the tube manufacturers. This is not to say that the manufacturers should not continue their efforts to build more reliable tubes, but it is essential to realize that the "bullet-proof" tube is a virtual impossibility and that intense efforts must be made to better inform the users about the tube characteristics.

While Fig. 16.1 may be an exaggeration of the user situation, it is symbolic of the lack of understanding on the part of many users of the extreme complexity and fragile nature of some of the elements of many microwave tubes.

Figure 16.1 User Abuse is often a Major Factor in the High Failure Rate of Microwave Tubes

(From "On Target," January 1983.)

Index

www.ingramcontent.com/pod-product-compliance
Lightning Source LLC
LaVergne TN
LVHW102007210926
848483LV00009B/119
* 9 7 8 0 8 9 0 0 6 1 8 1 7 *